Steelwork Corrosion Control

Steelwork Corrosion Control

Second edition

D. A. Bayliss and D. H. Deacon

London and New York

First published 2002 by Spon Press
11 New Fetter Lane, London EC4P 4EE

Simultaneously published in the USA and Canada
by Spon Press
29 West 35th Street, New York, NY 10001

Spon Press is an imprint of the Taylor & Francis Group

© 2002 D. A. Bayliss and D. H. Deacon

Typeset in Times by Wearset Ltd, Boldon, Tyne and Wear
Printed and bound in Great Britain by St Edmundsbury Press,
Bury St Edmunds, Suffolk

British Library Cataloguing in Publication Data
A catalogue record for this book is available from the British
Library

Library of Congress Cataloging in Publication Data
A catalog record for this book has been requested

ISBN 0-415-26101-5

Contents

Acknowledgements

The original version and first edition of this book, published in 1985 and 1991 respectively, were largely the inspiration and work of the late Ken Chandler. Ken's objective was never to provide a comprehensive text book on coating technology but, instead, an easy to read reference for engineers, architects and others, for whom the protection of steelwork is an important, although often a comparatively minor, part of their total professional activities.

Nowadays, not only are new materials and methods being developed constantly but the increased emphasis and legislation on health, safety and environmental issues have made even more radical changes necessary in paint materials, surface preparation and paint application. It has become even more difficult for the non-specialist to keep abreast of the situation.

The sudden death of Ken Chandler in 1995 was not only a personal loss of a friend and colleague, but deprived the Industry of somebody of great integrity and very long and valuable experience. When requested to produce this new edition I was able to persuade David Deacon, somebody with similar long experience, to become co-author, this despite the many other calls on his time. Fortunately we were both able to gain the services of yet another colleague, namely Garth Cox, whose experience as a senior paint chemist for both major paint manufacturers and raw material suppliers, has been of invaluable help.

I also take this opportunity to acknowledge the work of colleagues in this field. They are too numerous to mention, but many of the views expressed in this book have arisen from discussions with them and the study of their contributions to journals and conferences over many years.

Derek Bayliss
Woodbridge, Suffolk

October 2001

Chapter 1

Introduction

Many Resident Engineers for major building projects nowadays echo the sentiment that 'Painting amounts to 10% of the job but provides 90% of the problems.' Yet at the same time there are, unquestionably, large areas of coated steelwork withstanding the most adverse conditions for surprisingly long periods. A typical example is the coating of offshore platforms in the North Sea. Even in less exotic circumstances, for example bridge structures inland, engineers have successfully extended repainting cycles by as much as three times compared with practices of only thirty years ago. The majority of coatings used nowadays have considerably improved properties over the materials used then. However, this is not the sole reason for the success. The following factors have become even more important than previously:

   (i)   Coating specifications should say what they mean and mean what they say. See Chapter 8.
   (ii) Coating manufacturers' recommendations regarding application and suitability of their products for the relevant conditions should be followed as closely as possible. See Chapter 14.
   (iii) Materials supplied should be of a consistent standard of quality. See Chapter 16.
   (iv) Surface preparation should be no more or no less than is required to achieve the specified durability. See Chapter 3.
   (v)   Because of the variable performance that can be obtained owing to sometimes even minor differences in the preparation and application process, the whole operation, ideally, should be monitored by a competent and qualified coating inspector. See Chapter 9.

It can be shown, even to the satisfaction of the accountants, that in the majority of cases there are substantial economic benefits to be gained in the long term by adopting sound coating practice. In the short term, however, the costs are likely to be higher and this must be appreciated by all concerned (see Section 14.5.2).

As things stand today, there is little wonder that even relatively young engineers seem to yearn for the days when painting steel structures meant little more than a perfunctory wire brushing followed by red lead in oil primer, followed by two coats of gloss paint. Indeed, only about thirty years ago this was the standard paint system for many major structures, such as the structural steelwork of new power stations in the UK. The advantages of such a system were cheapness and foolproofness. Durability was generally only about 4 years for exterior exposure but the process could be repeated without much expense or trouble. However, there is little question that in modern times the slow drying, toxicity (see Chapter 4) and repeated cost of maintenance (see Section 14.5) would be unacceptable. In contrast, today there are several office buildings in London where the exposed steelwork has been painted with a zinc silicate primer and two-pack urethane top-coats. That is a coating system with probably the greatest durability and least tolerance during application (see Chapter 4). Also there are structures with very limited and expensive access that are being painted with three coats of moisture-cured urethane during the course of one day, even under damp and other adverse conditions (see Section 4.9.3.5).

In fact, this revolution in the paint industry could really be said to have started back in 1938. Dr Pierre Castan of Switzerland, a chemist working for a firm making dentures, invented epoxy resins. A Swiss patent for the invention was filed in 1940 and for a curing agent in 1943. However, the firm was unable to exploit the discovery sufficiently in the dental field and sold the rights to what is now CIBA-Geigy. They started to market epoxy resins, under the tradename Araldite, in 1946. Subsequently Shell (USA) also purchased the patent and marketed under the tradenames Epon and Epikote. However, it was at least a decade later that such materials began to be accepted for such specialist operations as tank lining.

Other synthetic resins also became available, including chlorinated rubbers and vinyls, etc., but it was urethanes and epoxies in their infinite variety that led to many complications. Then, in the 1990s, it was the impact of environmental legislation and health and safety controls with the requirement for low VOC-compliant coatings (see Chapter 4) which resulted in major changes in coating technology. Many favourite coatings such as chlorinated rubber and coal tar epoxy, with a proven history of long-term durability for wet or immersed surfaces, had to be phased out because of high solvent content and toxicity problems. The result has been substitution with newer, less familiar materials or methods of application. It is no wonder that engineers and others often find the subject confusing.

The development of quick-drying, high-build coatings has also brought another complication, namely the need for high standards of surface preparation. Even for the old-established oleo-resinous systems, it has been known for many years that application on to cleaned surfaces, such as blast-cleaned (see Section 3.2.3) or pickled (see Section 3.2.6) gives consistently greater durability than with hand cleaning.

The production of the first edition of the Swedish Standard photographs of rust and preparation grades in 1946 was a far-sighted work of exceptional quality for its day. This standard was soon used, or at least paid lip-service, by most of the industrial nations.

In 1978 there was the first meeting of the International Standards Organisation, Sub-committee ISO/TC 35/SC12. The objective of this committee was to produce standards for the preparation of steel substrates before the application of paints and related products. At that date the very few Surface Preparation Standards that existed dealt solely with the visual cleanliness of the surface after cleaning and in particular the absence of millscale. For example, no regard was paid to invisible contaminants such as soluble iron corrosion products hidden at the bottom of corrosion pits on rusted steel, particularly under maintenance conditions. Also the Standards for abrasives specified only their size requirements.

At the 17th meeting of the International Standards Committee (ISO) TC 36/SC12 ISO committee, held in Sydney, Australia in March 2000, forty-four Standards on 'Preparation of steel substrates before application of paints and related products', covering visual cleanliness, tests for surface cleanliness, measurement of surface profile, surface preparation methods and specifications for metallic and non-metallic abrasives, had been published. All but one (a Japanese apparatus for measuring chloride by ion detection tube) had become identical British Standards and it was anticipated that all would become CEN Standards. Other important Standards then still in draft form included: guidance levels of water-soluble contamination, preparation grades for welds, cut edges and other steel surface defects, preparation grades after high pressure water jetting and measurement of surface profile by replica tape. Unfortunately it takes several years before Committee drafts eventually emerge as published ISO Standards.

All this may well make the specifier's task more effective, but certainly not easier. One suspects that there will be a real danger of demanding the lowest limits of contamination in all circumstances and regardless of the actual requirements. Clearly there is a need for specifiers, who may be practising engineers, architects, designers and others, for whom corrosion control and steelwork protection are a comparatively minor part of their overall, professional responsibility, to understand the implications of coating technology.

Those fully involved in the field, such as steel fabricators, paint applicators, galvanisers and paint manufacturers, have much expertise and advice to offer. Full benefit should be taken of such information and, again, a background knowledge of the subject will be of considerable assistance.

The principles of good corrosion control do not change but the development of new techniques and materials is a continuing process. It is hoped that specifiers and others who study the relevant chapters in this book will be in a sound position to judge the merits of the alternatives that will be offered to them by a range of different suppliers.

Steel protection and corrosion control are essential elements in most modern structures and the authors have aimed to present the basic facts in a concise manner and, where relevant, to provide references for those who wish to study the matter in more detail.

1.1   Health and safety considerations

It could be said that corrosion prevention of steel structures is an above-averagely dangerous occupation in the construction industry. Most of the coating materials used are flammable, toxic and explosive. Methods of surface preparation involve propelling hard particles at high velocity into the atmosphere. Additionally there are the normal perils of falling from heights or being hit by falling objects. Everybody concerned therefore must be responsible for their own safety, report unsafe practices to the appropriate authority and follow all the required national and industrial safety rules and requirements.

The current trend in practically all countries is to increase the scope and tighten the limits of any legislation on matters of health and safety. In any commercial organisation involved in the construction industry, there should be a person, or persons, solely concerned with health and safety matters. It is from this source that advice should be obtained for specific working practices.

In previous editions of this book, the subject of safety was dealt with in the final chapter. However, with the increased awareness of their importance and to avoid them being overlooked, specific health and safety matters are highlighted at the end of each appropriate chapter.

Chapter 2

The corrosion of steel

Coatings are used to prevent or control corrosion, so an appreciation of the basic principles of corrosion is advantageous to those concerned with coatings technology. When coatings break down, then the steel will corrode, and the nature and anticipated extent of corrosion may well determine the types of coating to be used. Furthermore, the type and degree of corrosion under a paint film will influence its protective value to a marked degree and will, to a considerable extent, affect maintenance decisions. In the case of bare metal coatings, their performance will depend entirely upon the amount of corrosion that occurs in the specific environment of exposure. Again, a basic understanding of corrosion principles is necessary in order to appreciate the way in which cathodic protection operates and the situations in which it can be used.

The aim here is to provide a general account in relation to the selection and performance of coatings and the operation of other control processes that may be used for structures and buildings.

2.1   Corrosion: the basic process

The corrosion of steel arises from its unstable thermodynamic nature. Steel is manufactured from iron, which is made in a blast furnace by reducing ores such as haematite (Fe_2O_3) with carbon in the form of coke. This can be illustrated in simple chemical terms as follows:

$$2Fe_2O_3 + 3C \rightarrow 4Fe + 3CO_2$$
(iron ore)   (coke)   (iron)   (↑gas)

This reaction occurs at a very high temperature but the final products, iron and eventually steel, are unstable, a great deal of energy having been supplied in the process. Consequently, when steel is exposed to moisture and oxygen it tends to revert to its original form. Again, in simple chemical terms,

$$Fe + O_2 + H_2O \rightarrow Fe_2O_3.H_2O$$
(iron)   (rust)

Rust is a hydrated oxide, similar in composition to haematite. This explains why steel tends to rust in most situations and the process can be considered to be a natural reversion to the original ore from which it was formed. It does not, however, explain why steel corrodes more rapidly than most other constructional alloys. All of these, with the exception sometimes of copper, are found in nature in the form of minerals or ores, i.e. they are combined as oxides, sulphides, etc. Energy is expended in producing them either by heating, as with steel, or by some other method. As the natural mineral is more stable, all constructional metals have a tendency to revert back to their original form. However, this tendency, which can be calculated from the thermodynamics of corrosion processes, is concerned with the equilibrium state of a chemical system and the energy changes that occur. Although thermodynamics provides information on the tendency of a reaction to occur, it provides no data on the rate of reaction or, in chemical terminology, the reaction kinetics.

It may be known that steel, if exposed to moisture and oxygen, will rust, but in practice the important point is usually how fast it will rust. A piece of steel left in a damp garage during the winter months may exhibit some surface rust, whereas the same piece of steel left out in the garden may have rusted to a much greater extent. Again, a piece of galvanised, i.e. zinc-coated, steel left in the garden may show some surface deterioration which can easily be rubbed off leaving the zinc barely corroded. These simple examples illustrate the following points:

(i)   The same alloy will corrode at different rates in different situations.
(ii) Different metals and alloys corrode at different rates under the same conditions of exposure.

The second of these points arises not, as might be supposed, because, for some reason, alloys have different intrinsic corrosion characteristics: in practice, some of the most reactive metals actually corrode at a low rate. It is because the corrosion reaction with air (oxygen) often results in the immediate formation of an oxide film on the surface, which protects the metal. A typical example is aluminium, which forms a thin surface film (Al_2O_3) on exposure to the atmosphere and so tends to insulate the aluminium metal or alloy from the environment. In some situations these films are either not formed or are not particularly effective in stopping reactions between the environment and the metal. An oxide film, basically Fe_2O_3, is formed on steel but in most situations it is not particularly protective, so the environment can react with the metal, leading to rusting. However, the surface film can be improved by adding certain elements to steel in sufficient amounts. For example, the presence of 12% chromium results in the formation of a more protective film, Cr_2O_3, which acts as a very good barrier, reducing the corrosion rate by a considerable amount.

Such a ferrous alloy, containing 12Cr, is 'stainless steel', although generally there is a much higher percentage of alloying elements in the more corrosion-resistant stainless steels, typically 18% chromium, 10% nickel and 3% molybdenum.

It will be gathered from the above discussion that, generally, the corrosion rate of a metal or alloy will be determined by the formation of surface films and their ability to protect the metal from the environment. This is not a full explanation of the situation but is sufficient to show the importance of the environment in determining the rate of corrosion. Corrosion can be defined in various ways, but an acceptable definition is 'a chemical or electrochemical reaction between a metal or alloy and its environment'. The chemical or electrochemical requirement differentiates corrosion from other forms of deterioration of metals, e.g. wear and abrasion, which involve mechanical effects.

It follows that the corrosion characteristics are not an intrinsic property of an alloy as are, for example, strength or hardness at ordinary temperatures. Although some alloys are considered to be more corrosion resistant than others, it should not be assumed that in all circumstances this will be the case. In some chemical solutions, e.g. certain concentrations of sulphuric acid, a protective surface film is produced on ordinary steel and this reduces corrosion to a level below that sustained by stainless steels.

It can be seen that the rate of corrosion depends upon the environment to which the alloy is exposed, and this will be considered in some detail for ordinary carbon steels (Section 2.4). The electrochemical nature of corrosion can also be explained with reference to steel.

2.2   The electrochemical nature of corrosion

As discussed above, steel produces a rather poor protective film on its surface so that, in the presence of moisture and oxygen, corrosion occurs. This corrosion is electrochemical, i.e. it is basically the same process as that occurring in a simple electrolytic cell (see Figure 2.1). The essential features of such a cell are two electrodes, an anode and a cathode, joined by an external conductor, e.g. copper wire, and immersed in an electrolyte. An electrolyte is a solution capable of carrying current, e.g. rain or tap water. The processes involved are complex and will not be discussed in any detail. However, if pieces of copper and zinc are joined together and immersed in an electrolyte with an ammeter in the external circuit, a current will be detected. The copper becomes the cathode of the cell and the zinc the anode. The potentials of the two metals are different and this provides the driving force for the cell. Current similarly flows if steel is joined to zinc; again, zinc acts as the anode and steel as the cathode. This experiment can be used to illustrate the principle of two important

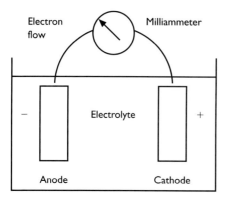

Figure 2.1   A simple electrolytic cell.

methods of corrosion control: cathodic protection (see Chapter 12) and protection by zinc coatings (Chapter 7). However, if copper is substituted for the zinc and connected to steel, the copper is the cathode and steel the anode of the cell, and so the steel corrodes.

Corrosion occurs at the anode of the cell; little or no corrosion occurs at the cathode. The simple experiments discussed above illustrate another important phenomenon – bimetallic or galvanic corrosion. If different metals or alloys are joined in the presence of an electrolyte, one will corrode at an increased rate whereas the other will corrode at a lower rate or will not corrode at all. This arises because of the potential difference set up when different metals are joined. If two pieces of steel are joined in a cell, there may be sufficient variations in the surface condition to produce a small potential difference. However, if one of the two pieces of steel is oxygenated, i.e. if air is blown around it or the electrolyte is heated locally near one of the pieces of steel, a current will flow. Summarising, variations in either the metal or the environment may well produce the conditions required to set up a cell with corrosion occurring at the anode.

In practice there are small variations over the steel surface. If a piece of steel is polished and etched, then examined under a microscope, the structure will usually be seen to consist of grains (Figure 2.2). These produce small potential differences on the surface. If an electrolyte – this may be rain or dew – is present on the steel surface, then small cells can be set up with corrosion occurring at the anodic areas. The corrosion reactions can be illustrated using chemical terminology as follows:

anodic reaction

$$\underset{\text{(iron metal)}}{Fe} \rightarrow \underset{\text{(ions)}}{Fe^{2+}} + \underset{\text{(electrons)}}{2e^-}$$

Figure 2.2   Diagrammatic representation of a steel surface, showing anodes and cathodes.

This is a simple way of describing the process where iron is removed as charged particles called ions (Fe^{2+}) and electrons (e^-) carry current to balance the electric charge.

Clearly a balancing reaction must occur at the cathode and under ordinary natural exposure conditions this can be represented as follows:

cathodic reaction

$$\tfrac{1}{2}O_2 \underset{\text{(oxygen)}}{} + \underset{\text{(moisture)}}{H_2O} + 2e^- \rightarrow \underset{\text{(hydroxyl)}}{2OH^-}$$

In short, hydroxyl ions are produced at the cathode. These two reactions can be combined in a chemical equation:

$$Fe + \tfrac{1}{2}O_2 + H_2O \rightarrow 2OH^- + Fe^{2+}$$

The ferrous and hydroxyl ions react together to form ferrous hydroxide:

$$2OH^- + Fe^{2+} \rightarrow Fe(OH)_2$$

This is a simple form of rust which is unstable and is eventually oxidised (i.e. reacts with oxygen) to form the familiar reddish brown rust, chemically denoted as FeOOH, or more commonly $Fe_2O_3.H_2O$. This is the form of rust usually produced in air, natural water and soils. However, under acidic conditions hydrogen is produced at the cathode and the corrosion product may be Fe_3O_4 (magnetite).

2.3   Corrosion terminology

Terms frequently used in relation to corrosion are discussed briefly below. Full explanations are available in standard text books.

2.3.1   Potential

There is a theoretical e.m.f. series of metals (not alloys) called 'standard equilibrium potentials'. These are important in purely electrochemical terms for the understanding of processes but are of little importance so far as practical corrosion problems are concerned. More useful are potentials experimentally measured using a suitable reference electrode and published in tables such as the 'Galvanic Series in Sea Water'. These have some practical value because the extent of differences in potential between different alloys provides an indication of the effect of coupling them (see Section 2.2).

Potentials are also important in determining the operating effectiveness of cathodic protection systems (see Chapter 12).

2.3.2 Polarisation

The potential difference between the two electrodes in a cell provides the 'driving force' for the current, which determines the extent of corrosion at the anode. However, when the cell is operating, i.e. when current is flowing, the e.m.f. of the cell is different from that theoretically predicted by taking the difference in potentials of the two metallic electrodes. *Polarisation* occurs at both the anode and the cathode.

Polarisation, sometimes termed overpotential or overvoltage, can be defined as the difference of the potential of an electrode from its equilibrium or steady-state potential. This can be considered in terms of the energy required to cause a reaction to proceed. An analogy would be the initial energy required to push a car on a level path. Once the car is moving, less energy is required, but if a slope is reached the energy required on the level is not sufficient to push it up the slope, so it tends to slow down and eventually stops.

Once the cell is operating, changes occur in the cell; ions tend to collect near the anode and reactants tend to surround the cathode. The net result is reduction in the potential difference between the electrodes.

2.3.3 Passivity

Under certain conditions a corrosion product forms on the surface of a metal, providing a barrier to the environment, i.e. it acts in a similar way to a coating. To achieve passivity the corrosion product must adhere to the surface and be stable both chemically and physically so that it does not disintegrate. Such products are sometimes produced near anodic sites and so tend to passivate these areas. Iron becomes passive when immersed in concentrated nitric acid because a thin film of ferric oxide is formed, which, provided it is not disrupted, isolates the iron from the corrosive environment.

A good example of a passive film is that produced on stainless steels where, because of the chromium content of the alloy (over 12%), a very resistant film, basically Cr_2O_3, is formed. This not only stops stainless steels from corroding to any extent in air, but also has the capability of rapidly reforming if it is damaged. It is interesting to note that the corrosion performance of stainless steel is determined by its ability to maintain a passive surface film and, if it is broken down and not repaired, the potential of the alloy changes dramatically and is moved towards the anodic end of the galvanic series.

2.4   Corrosion in air

Clearly, there is a plentiful supply of oxygen in air, so the presence or otherwise of moisture determines whether corrosion will occur. Steel is often visibly moist after rain or when there has been fog or dew. However, the water vapour in the air can also cause steel to rust even though no visible moisture is present.

The amount of water vapour in the air is indicated by the relative humidity, and Vernon carried out some experiments which showed the effect of relative humidity on rusting.[1] He showed that in pure air there was little corrosion below 100% relative humidity (r.h.), but in the presence of small concentrations of impurities, such as sulphur dioxide, serious rusting could occur above a certain critical humidity, which was about 70%. Below this level, rusting is slight provided moisture from other sources is not present (see Figure 2.3). These experiments showed the importance of two factors that determine the corrosion rate in air, i.e.

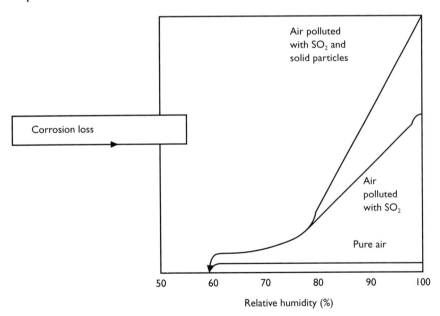

Figure 2.3   Simplified diagram showing the effect of relative humidity and pollution on the corrosion of carbon steel.

Source: Vernon.[1]

(i) relative humidity, and
(ii) pollutants and contaminants

The effect of moisture is related to the length of time it is in contact with the steels, so the influence of relative humidity is generally more important than that of precipitation processes such as rain, because the relative humidity may remain above 70% for long periods, particularly in the United Kingdom and other northern European countries. However, in the absence of pollution such as sulphur dioxide (SO_2), corrosion is only slight, but a reasonably linear correlation has been shown to exist between the corrosion rate and the amount of SO_2 in the air.[2] Although SO_2 can dissolve in moisture to form acids, the effect is not to produce a direct attack on the steel, but rather is the formation of salts such as ferrous sulphate ($FeSO_4$). These compounds, sometimes called corrosion salts, are able, by complex reactions, to produce further rusting. Additionally, they are hygroscopic and so can trap further moisture on the steel surface. Such salts are of more than academic interest because their presence in rust is one of the main causes of coating breakdown when paints are applied to rusted surfaces (see Chapter 3). When chlorides are present on the steel surface, typically near the coast where sea salts (sodium chloride) are prevalent, corrosion may occur at relative humidities as low as 40%.[3] Generally, the amount of chloride in the air drops off rapidly as the distance from the coast increases. The effects of this drop in corrosion are illustrated in Tables 2.1 and 2.2.[4]

Results from overseas sites showed clearly that in warm, dry, unpolluted inland sites, such as Khartoum in the Sudan and Delhi in India, the corrosion rate was negligible compared with that occurring in industrial areas. Probably the most interesting result arising from these tests was that for the surf beach at Lagos, where the corrosion rate was over 0.6 mm per year – nearly five times that at Sheffield. Conditions where sea spray continuously reaches the steel surface always lead to severe corrosion. The splash zone on offshore structures is always a critical area for corrosion; see Figure 2.4.

Table 2.1 Effect of sea salts on the corrosion of steel[a]

Distance from coast (yards)	Salt content of air[b]	Corrosion rate (mm per year)
50	100	0.95
200	27	0.38
400	7	0.06
1300	2	0.04

a Based on tests carried out in Nigeria.
b Expressed as a percentage of content at 50 yards.

Table 2.2   Corrosivity of environments

Class	Annual metal loss	Exterior
Very low–low	<10–200 g/m²	Rural areas, low pollution, dry
Medium	200–400 g/m²	Urban and industrial atmospheres Moderate SO_2 pollution Moderate coastal Cl^-
High	400–650 g/m²	Industrial and coastal
Very high industrial	650–1500 g/m²	Industry with high humidity and aggressive atmosphere
Very high marine	650–1500 g/m²	Marine coastal, offshore, high salinity

The size, shape and orientation of the steel all influence the corrosion rate to varying extents because they affect the local environment at the steel surface. The orientation of the steel has most influence because it has a marked effect on the 'time-of-wetness' of the surface. In the Northern Hemisphere, north-facing steelwork remains moist for longer periods than south-facing steelwork and so tends to corrode more. Again, on horizontally exposed steelwork the upper surface may corrode less rapidly than the groundward side because corrosive particles are washed off by rain

Figure 2.4   Corrosion in splash zone.

and the sun dries the surface more quickly. Tests on specimens exposed at an angle of 30° to the horizontal confirmed this;[5] 60% of the corrosion occurred on the underside. Tests on steel plates of different thickness showed a slightly greater corrosion rate on thicker plates, e.g. 95 μm per year on 55-mm plate compared with 75 μm per year on 5-mm plate.

2.4.1   Steel composition

Minor variations in the compositions of commercial carbon steels generally have little effect on the corrosion rate, the one exception concerning copper. Additions up to 0.2% provide a marked reduction in the corrosion rate in air, but further additions have little or no subsequent effect. The addition of copper has little or no effect on steels immersed in water or buried in soils. Copper is used as one of the alloying elements in low-alloy steels called 'weathering steels', which are sometimes used without coatings for structures. This is discussed more fully in Chapter 12.

2.4.2   Rust

Rust is, of course, the corrosion product of the processes considered above. Although it is generally considered to have the composition Fe_2O_3. H_2O, other minor constituents will also be present in the rust and will have a marked effect on the course of corrosion and the performance of coatings applied over the rust. Rust also causes problems because it has a much greater volume than the steel (or iron) from which it is produced. This can result in the buckling of thin steel sections or sheet if rusting occurs at crevices or overlaps. Under paint films, rust formation can result in blistering and cracking of the coating.

2.5   Corrosion in water

The basic corrosion reaction is the same for steel immersed in water as for steel exposed to air. However, there are differences in the processes that occur. In water the availability of oxygen is an important factor, whereas in air corrosion does not occur in the absence of moisture. Hence, in water corrosion is generally inappreciable in the absence of oxygen. Under immersed conditions there are more factors to be taken into account than with atmospheric corrosion. The environment itself is more complex and the rust does not necessarily form on the steel surface because the products of the corrosion reaction, e.g. Fe and OH ions, may diffuse from the steel itself and react in the solution.

In view of the complexity of the corrosion process in water, only a few basic points will be considered. A short list of books is provided at the end of the chapter for those wishing to study the matter in more detail.

2.5.1   Composition of water

Water is presented chemically as H_2O but, of course, there are many other salts, solids and gases present in the various waters of practical concern. Even fairly pure tap water has a complex composition. Water from rivers, sea, estuaries and wells covers a range of compositions and properties.

The pH of water usually falls within a neutral range (pH 4.5–8.5), but some types are acidic and these can be particularly corrosive to steel. Generally, however, the main factors in determining the type and extent of corrosion are the dissolved solids (which influence the conductivity, hardness and pH of the water), dissolved gases (particularly oxygen and carbon dioxide) and organic matter. The conductivity is important and the presence of salts, such as sodium chloride (NaCl), tends to make seawater more corrosive than fresh water.

Corrosion can be prevented by making water alkaline, but in some situations the alkalinity is such that only a partial passive film is formed and this can result in pitting corrosion.

Hardness is a particularly important property of waters. This determines their ability to deposit protective scales on the steel surface and is influenced by the amount of carbon dioxide and the presence of salts such as calcium carbonate and bicarbonate. The scale formed in what are termed 'hard water' reduces the rate of corrosion, and 'soft waters' can be treated with lime to make them less corrosive. Although the formation of protective scales reduces corrosion of steel, it may have other less advantageous effects. For example, it may reduce the efficiency of heat exchangers and may eventually lead to the blockage of pipes.

In seawater the formation of protective calcareous scales has an important influence on corrosion. Their formation on the immersed parts of offshore platforms is one of the reasons why many such structures can be cathodically protected without the requirement for applied coatings.

The presence of organic matter, particularly in seawater and estuarine waters, can have both direct and indirect effects on corrosion. Living organisms result in what is termed 'fouling', i.e. marine growths on steel or on the protective coatings applied to steel. This fouling is a particular problem when occurring on ships' hulls because its effect is to increase drag and so increase fuel consumption if speed is to be maintained. Special anti-fouling coatings have to be applied to ships (see Section 15.3.4). A particular problem may occur if certain bacteria are present, particularly in mud and around harbours. These can cause bacterial corrosion (see Section 2.7).

2.5.2   Operating conditions

The corrosion of steel under static conditions in water may be quite different from that experienced in practice. Many factors will influence the type and rate of corrosion, in particular the temperature and velocity of the water. The velocity or rate of flow will be particularly affected by design features such as sharp bends in pipes, and may lead to a number of special types of corrosion, such as erosion–corrosion, impingement and cavitation. These will not be considered here, but may particularly affect the operation of a process plant.

Apart from special effects of velocity, the rate of flow is always likely to influence corrosion. It may be sufficient to remove protective coatings, both scale-formed and applied, particularly if abrasive particles are entrained in the water. It will also have an effect on the supply of oxygen, which may directly influence corrosion. Although, in fresh waters, a high flow rate may provide sufficient oxygen at the surface to cause passivity, generally the corrosion rate increases with velocity. In one series of tests the corrosion rate under static conditions was 0.125 mm per year compared with 0.83 mm per year at a velocity of 4.6 m/s.

2.5.3   Steel composition

Generally, small variations in the composition have no influence on the corrosion rates of steels immersed in water. Small amounts of copper, which has an effect on corrosion under atmospheric conditions, do not improve corrosion to any significant extent under most immersed conditions.

2.5.4   Corrosion rates of steel in water

Although corrosion is generally reasonably uniform on steel immersed in water, there is more tendency for it to pit because of the effects of design, scale formation and variations in rates of flow. In particular, the presence of millscale (see Chapter 3) may lead to serious pitting. This may arise particularly in seawater but also in other waters, where the steel is virtually coated overall with millscale but with a few small areas of bare steel. At such areas the galvanic effect of large areas of cathodic material (millscale) in contact with small anodic areas (steel) can lead to severe pitting. Many tests have been carried out on steel specimens immersed in waters of different types in order to determine corrosion rates. Generally, under fully immersed conditions in seawater, rates from 65 μm per year to 100 μm per year have been measured. The rates at half-tide immersion are much higher than these. In fresh water, lower corrosion rates around 45 μm per year have been obtained, although in river water rates similar to the lower end of the range in seawater are not uncommon.

2.6   Corrosion in soil

The corrosion process in soil is even more complex than that in water although, again, the basic electrochemical process is the same and depends on the presence of electrolyte solutions, i.e. moisture in the soil. Soils vary in their corrosivity. Generally, high-resistance soils, i.e. those of low conductivity, are the least corrosive. These include dry, sandy and rocky soils. Low-resistance soils such as clays, alluvial soils and all saline soils are more corrosive. The depth of the water table has an important influence on corrosion and the rate will depend on whether steel is permanently below or above it. Probably, alternate wet and dry conditions are the most corrosive. In many soils the effect of the water table leads to variations in corrosion with depth of burial of the steel.

Steel buried in soil tends to be in the form of pipes or piles. Both may be in contact with different layers and types of soil, but pipelines in particular will be influenced by variations in the soil over the long pipe distances. The effect of differences in soil resistivity, water content and oxygen availability all lead to the formation of differences in potential over the pipeline and the formation of electrolytic cells. Generally, this is not a serious problem provided sound protective coatings in conjunction with cathodic protection are used. Usually a soil survey is carried out before determining the necessary protective measures. With piles the problem is more acute because although protective coatings are used they are often damaged during driving operations. In practice, however, this does not appear to be particularly serious. A number of piles have been withdrawn and examined; corrosion has been lower than might have been anticipated.

Stray current corrosion is a form of attack that can occur when a steel structure or pipeline provides a better conducting path than the soil for earth-return currents from electrical installations or from cathodic protective systems in the neighbourhood. If such currents remain unchecked then there is accelerated corrosion of steel in the vicinity of the stray currents.

Some indications of the corrosion rates of steel in soils have been published.[6,7] The highest rates obtained were 68 μm per year in American tests and 50 μm per year in British tests. Pitting was 4–6 times greater than these general rates.

2.7   Bacterial corrosion

Although the presence of oxygen and moisture or water is generally necessary for corrosion, there is an important exception which is worth considering because it occurs in a number of situations under immersed and buried conditions.

The nature and extent of the corrosion will be determined by the form of microbiological activity. The form most commonly encountered is that arising from the presence of sulphate-reducing bacteria (*Desulfovibrio desulfuricans*). They derive their name because they reduce inorganic sulphates to sulphides and are able to cause corrosion under anaerobic conditions, i.e. in the absence of oxygen. Generally, under immersed conditions in water or burial in soil, oxygen is essential for corrosion. However, in the presence of these bacteria, corrosion can occur without oxygen because the process is different.

A number of investigations into the mechanism have been carried out and the most likely explanation of the process is as follows:

anodic reaction

$$4Fe \rightarrow 4Fe^{2+} + 8e^-$$

cathodic reaction

$$8e^- + 4H_2O + SO_4^{2-} \xrightarrow{\text{bacteria}} S^{2-} + 8OH^-$$

Combining these equations, the overall reaction is represented as follows:

$$4Fe + 4H_2O + SO_4^{2-} \rightarrow 3Fe(OH)_2 + FeS + 2OH^-$$

This represents the corrosion products obtained when bacterial corrosion occurs and is a more likely reaction than the direct one, i.e.

$$Fe + H_2S \rightarrow FeS + H_2$$

The exact mechanism is not of practical importance but the reaction products indicated above do provide a means of detecting the presence of sulphate-reducing bacteria, which is usually associated with a distinct 'sulphide' smell and black corrosion products on the steel. Although sulphate-reducing bacteria do not necessarily attack coatings, they are capable of attacking coated steel if the protective film is porous or damaged. Sulphate-reducing bacteria are found in clays, muds, silts and seawater. Coating manufacturers should be consulted to ensure that specific coatings are suitable for conditions where bacteria are present. Cathodic protection can effectively prevent attack by sulphate-reducing bacteria. Other forms of bacteria can attack coatings, but these are not of the types that corrode steel.

2.8   Health and safety considerations

The loss of strength in a steel structure due to corrosion wastage may be obvious, but it should be remembered that it is possible for severe corrosion to be masked by some thick flexible coatings, particularly plastic coatings with poor adhesion or thick layers of materials such as bitumen. Corrosion occurring in a sealed space can use up all the available oxygen. Precautions must be taken before entering these areas.

References

1. Vernon, W. H. J., *Trans Faraday Soc.*, **31**(1) (1935) 668.
2. Chandler, K. A. and Kilcullen, M. B., *Br Corres. J.*, No. 3 (March 1968) 80–4.
3. Chandler, K. A., *Br. Corres. J.*, No. 1 (July 1966) 264–6.
4. Chandler, K. A. and Hudson, J. C., *Corrosion*, Vol. 1. Newnes and Butterworths, 1976, p. 317.
5. Larrabee, C. P., *Trans. Electrochem. Soc.*, **85** (1944) 297.
6. Romanoff, M., *J. Res. Nat. Bur. Stand.*, **660** (1962) 223–4.
7. Hudson, J. C. and Acock, J. P., *Iron and Steel Institute Special Report No. 45*, London, 1951.

Further reading

Fonatana, M. G. and Greene, N. O. (1967). *Corrosion Engineering*. McGraw-Hill, New York.
Scully, J. C. (1975). *The Fundamentals of Corrosion*. Pergamon Press, Oxford.

Chapter 3

Surface preparation

The long-term performance of a coating is significantly influenced by its ability to adhere properly to the material to which it is applied. This is not simply because the coating might flake away or detach from the surface but because poor adhesion will allow moisture or corrosion products to undercut the coating film from areas of damage.

The adhesion of some coating materials, such as hot-dip galvanising, is due to the formation of a chemical bond with the surface. For example, in hot-dip galvanising the zinc combines with the steel to form iron/zinc alloys. This is undoubtedly the most effective adhesive bond. However, for the most part organic coatings adhere to the surface by polar adhesion which is helped or reinforced by mechanical adhesion.

Polar adhesion occurs when the resin molecules act like weak magnets and their north and south poles attract opposite groups on the substrate. A few organic coatings have no polar attraction at all, for example some formulations of vinyl coatings can be stripped from a steel substrate in sheets and are therefore used as temporary protectives. But in all cases the attraction is only effective to a molecular distance from the steel, and films of dirt, oil, water, etc. can effectively nullify all adhesion.

Mechanical adhesion is assisted by roughening the surface and thereby increasing the surface area, for example by two or three times when abrasive blasting, upon which the coating can bond. Some coatings, for example the unsaturated polyesters used in glass-fibre laminates, develop excessive shrinkage on curing and require a high surface profile – the term used to denote the height from peak to trough of a blast-cleaned surface. The majority of organic coatings can obtain adequate adhesion on surface profiles in excess of $25\,\mu$m. Another important factor of mechanical adhesion is the firmness or stability of the substrate. For example, unstable substrates would include millscale, rust scales or old paint that is liable to flake and detach from the substrate, and friable, powdery layers of dirt, rust, etc. Modern, fast-drying, high-build, high-cohesive-strength coatings, such as epoxies applied by spray, put considerably greater stress on the adhesive bond during their drying and curing process than do 'old-fashioned', brush-applied, slow-drying, highly penetrating materials such as red lead in oil primers.

Because of the cost and difficulty of carrying out surface preparation to the required high standards, in recent years there has been a move towards the use of so-called 'surface-tolerant' coatings. These perform in two ways, firstly by containing hydrophilic solvents or surface-active agents which combine with the moisture on a surface and disperse it through the paint film. These can be effective providing that the amount of moisture present does not exceed the available solvent or agent or that they are not trapped in the film by premature overcoating or skin curing. The second method is to use two-pack epoxies, often formulated with aluminium pigmentation, to give good 'wetting' and penetration. These materials often cure more slowly than other primers, in order to facilitate penetration and to provide a dry film, free from internal stresses.

In general, all paint systems, including the 'old-fashioned' systems, will give improved performance on surfaces prepared to a high standard of cleanliness. The use of 'surface-tolerant' coatings should be the unavoidable exception rather than the rule.

3.1   Steel surface contaminants and conditions

The effects of steel surface contaminants and conditions on coating performance are described below. Methods of detection or measurement are described in Chapter 9.

3.1.1   Oil and grease

Residues of oil, grease, cutting oils, silicones, etc. left on a steel surface, for example after fabricating operations, will weaken the adhesive bond of subsequent coatings (see Figure 3.1). Such residues must be removed before any further surface preparation operation, such as mechanical cleaning or blast-cleaning, since these are likely to spread the contamination over a wider area. Where abrasive is re-used, as in centrifugal blast-cleaning, this can also spread the contamination onto erstwhile clean surfaces.

3.1.2   Millscale

Steel sections and plates are produced by rolling the steel at temperatures in the region of 1200°C. The temperature at the rolling is likely to be well over 1000°C, so the steel reacts with oxygen in the air to form oxide scales. In general terms, the reaction can be expressed as follows:

$$x\text{Fe} + \tfrac{1}{2}y\text{O}_2 \rightarrow \text{Fe}_x\text{O}_y$$

Figure 3.1   Flaking of paint from surface contaminated with grease.

In practice, the scale is composed of a number of layers, the thickness and composition of which will be determined by factors such as the type and size of the steel, the temperature of rolling and the cooling rate. At the temperatures generally used for rolling, three layers are present. The proportion of each layer will vary, but the general proportions are as follows:[1,2]

FeO (wüstite)	40–95%
Fe_3O_4 (magnetite)	5–60%
Fe_2O_3 (haematite)	0–10%

FeO (wüstite) is unstable below 575°C, so scales produced at temperatures lower than this do not contain this layer. Alloy steels form a scale with somewhat different properties, although basically of iron oxides; often they are thinner and more adherent than those formed on unalloyed steel. In practice, the composition and formation of millscale is not of particular importance other than in determining its ease of removal or its effect on coatings if allowed to remain on the steel to be painted. The latter point is particularly important and will be considered below.

Millscale is a reasonably inert material and in principle, if it adheres well to the steel surface, might prove to be a highly protective coating. However, it is brittle and, during handling of steelwork, parts of the scale tend to flake off comparatively easily. Experiments have been carried out in the rolling mill in an endeavour to produce scales with improved properties, so as to provide a sound base for paint coatings. However, little success has been achieved in this area of research.

The presence of millscale on steel has two important effects on coating performance.

(i)   Although millscale is an oxide, when it is in contact with steel a galvanic cell is set up (see Chapter 2). The millscale is cathodic to the steel and a potential difference of as much as 0.4 V may be set up in seawater. Consequently, at breaks in the millscale quite deep pitting of the steel may occur, particularly under immersed conditions and particularly when the area of the cathode (the millscale) is large relative to the area of the anode (the bare steel). Even if the scaled steel is painted, some moisture will reach the steel surface because all paint films are to some extent permeable. Furthermore, coatings can become damaged, thus allowing moisture to remain in contact with the scale. This galvanic effect can be serious under immersed conditions but it is not the only problem caused by the presence of millscale, as discussed below.

(ii)   During handling, scale on the steel surface is inevitably damaged and if exposed to a corrosive environment, e.g. that of a stockyard, then the steel will corrode. The general course of corrosion follows a pattern:

(a) The steel section will be carrying some intact millscale and some areas of bare steel where the scale has cracked or flaked off.
(b) Although the millscale will not rust, the steel will react with the atmosphere and the rusting will tend to undermine the scale, leading to blistering or flaking. The amount of flaking will depend upon the environment of exposure and the time the steel is left to rust. In tests

carried out in Sheffield in the UK, on $\frac{3}{8}$-in (9.5 mm) plates, 66% of the scale was removed in 2 months but only an additional 10% was removed over the next 8 months.

(c) If the steel is cleaned manually, e.g. by scrapers and wire-brushes, loose scale and rust will be removed but intact scale is merely burnished.

(d) If the steel is then painted, the performance will depend on a number of factors, in particular the chemical and physical state of the surface, the environment of exposure and the paint systems used.

Where the scale has been virtually all removed, then the surface will be covered with rust and the effects will be as discussed below (Section 3.1.3.1).

On the other hand, if the scale is practically intact with little or no rusting, then the paint performance may be reasonably good. This situation rarely occurs in practice, but in tests carried out in Sheffield on three groups of specimens coated with similar paint systems, the results shown in Table 3.1 were obtained.

However, the most common situation with scaled steel is where it has been weathered and some intact scale remains with some rust. This covers a range of conditions and possible effects. Generally, there is a considerable reduction in the life of the system but a potentially disastrous situation can arise. After normal cleaning, with removal of loose scale, a considerable percentage of apparently intact scale may be left on the surface. However, this scale may have been undermined to a great extent by rust and, after painting, the scale carrying the paint may flake off within a few weeks. In tests carried out in a coastal atmosphere on steel specimens weathered for a period of 3 months, leaving some apparently intact millscale, wire-brushed and then painted, virtually all the paint flaked off within a few months. On similar specimens where all the scale had been removed by weathering, the paint coating lost adhesion due to rust after about 2 years.[3]

Clearly, painting over steel carrying millscale is likely to cause problems and it is generally accepted that visible and identifiable scale should be removed. Before blast-cleaning facilities were as readily available as they are today, this was carried out by leaving steel sections exposed to the

Table 3.1 Life of paint system on different steel surfaces[4]

Condition	Average life of paint system (years)
As rolled – intact millscale	8.5
Weathered – rust and scale	2.6
De-scaled – no rust or scale	10.0

atmosphere so that all the scale eventually flakes off. However, this is achieved only by allowing the scale to be undermined by rusting, so leaving a layer of rust on the steel and some pitting. This is not a sound substrate on which to apply coatings, as discussed below.

3.1.3 Surface cleanliness

3.1.3.1 Rust

Rust is the corrosion product formed when steel reacts with oxygen and water. This has been discussed in Chapter 2. The corrosion reaction is generally denoted as follows:

$$4Fe + 2H_2O + 3O_2 = \underset{\text{(rust)}}{2Fe_2O_3 . H_2O}$$

Although rust is primarily hydrated ferric oxide, it also contains other compounds. Rusts have a wide range of composition, depending on the conditions under which they are formed. Typical compositions cannot, therefore, be given but analyses of a range of rusts have indicated that air-formed rusts generally contain about 5% of compounds other than $Fe_2O_3.H_2O$. These derive in part from the steel, which contains elements other than iron, e.g. copper, silicon and manganese, and in part from atmospheric contaminants and pollutants, mainly sulphates and chlorides, although other pollutants such as ammonium salts are also generally present in rust.

3.1.3.2 Water-soluble contaminants

The main constituents of rust, i.e. iron oxides, are not themselves the main problem in determining the performance of paint applied over rust. In fact, iron oxides are commonly used as pigments in paints. The constituents formed by reactions between steel and pollutants such as sulphur dioxide, sometimes called 'iron salts' or 'corrosion salts', cause most problems.

Sulphur dioxide in the air reacts with moisture to form acids. Considerable publicity is frequently given to what has been described as 'acid rain' arising from such reactions. Weak sulphuric acid solutions react with the steel to form ferrous sulphate, the presence of which can be readily detected in rust. These salts tend to form in shallow pits at the steel surface and the corrosion process is such that the sulphates tend to move inwards to the anodic areas, which are likely to be in crevices, the bottom of pits, etc. The salts are also not 'rust' coloured, being white or light coloured. They are very difficult, if not impossible, to remove with tools such as scrapers and wire brushes and are often difficult to remove even with blast-cleaning. The presence of salts such as ferrous sulphate leads to rather complex reactions involving the regeneration of the sulphuric acid from which they were formed. This in turn causes further corrosion and the production of more rust (see Figure 3.2). As rust has a considerably greater volume than the steel from which it is produced, this leads to disruption of the paint film applied over it by cracking, blistering and eventually flaking (Figure 3.3).

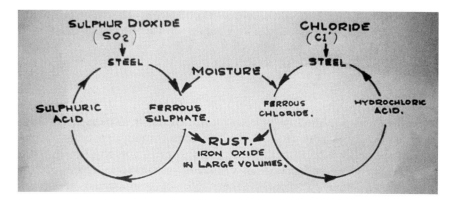

Figure 3.2   The cyclic process of rusting.

Figure 3.3   Blistering of paint resulting from the presence of soluble iron salts under the coating.

Ferrous sulphate is the salt most commonly found in rusts formed in industrial-type atmospheres. Near the coast, chlorides are likely to be a greater problem. The reactions arising from the two types of salt, sulphate and chloride, are not necessarily the same. Chlorides are hygroscopic, i.e. they absorb moisture. It has been shown in laboratory tests[5] that whereas rusting may occur at relative humidities below 70% with sulphates, the presence of chlorides in rust can result in corrosion of the steel at relative humidities as low as 40%. Chlorides may, therefore, be a greater immediate problem than sulphates, but all salts present under a paint film will lead to a reduction in the coating's life. ISO and BS Standards call these ferrous salts 'soluble iron corrosion products'. In American literature they have been called 'non-visible contamination' which is particularly appropriate.

The rusting of steel is complex and in many ways unpredictable. An investigation of the process was carried out by the former British Iron and Steel Research Association (BISRA) and some of the results have been published.[6] These show that the amount of rust formed on a steel surface is not necessarily related to the length of time the steel has been exposed and, perhaps even more important, the amount of sulphate in the rust also does not relate to the length of exposure. In rusts sampled in January, about $8\,g/m^2$ of sulphate were measured in rusts formed over a period of 2 months, i.e. from steel exposed initially in November of the previous year. This rose to a figure of about $12\,g/m^2$ for rusts formed over a period of a year. However, for rusts sampled in the summer months much lower sulphate contents were obtained. In July, rusts formed over a period of a year contained about $6\,g/m^2$, and over 2 months contained $2.5\,g/m^2$. It follows, therefore, that irrespective of the period of rust formation, the amount of sulphate is higher in winter. Consequently, painting over rusted steel is a somewhat haphazard operation because, without carrying out chemical tests on the rust, it is virtually impossible to know the extent of iron salt formation. Painting in the summer at inland sites in the UK is likely to provide better performance from the paint coating than in winter.

Similarly, chloride contamination in coastal areas can depend upon prevailing wind direction and even the steepness or shallowness of the coastline. Against this must be set the fact that the chloride has a higher solubility than the sulphate and therefore more is washed from the surface by rainfall. For both contaminants, the situation where rusted surfaces are subject to atmospheric pollution and are not washed by rain is the most aggressive.

The reduction in durability of coatings due to soluble iron corrosion products trapped beneath coatings is most obvious for surfaces exposed to severe marine environments or frequent condensation, and also for the linings of storage tanks containing aqueous liquids. This effect is less obvious on coatings subject to normal weathering but this can also depend upon the type of coating. Some coatings are thick and have a high cohesive strength and others are thick and are very flexible; in both these cases substrate corrosion can be masked until it has reached an advanced state. For conventional paint systems, such as alkyds, the effect can generally be seen within a few years by the appearance of corrosion blisters in the paintwork. The corrosion blisters form from the underlying corrosion pits.

The fact that the presence of ferrous salts in pits, even after blast-cleaning to a high visual standard, can seriously affect coating durability has been known since the work of Chandler in 1966.[7] It is still not possible to quantify permissible levels of soluble salts for different coatings in different environments.

Most of the reliable information on the effects of water-soluble salts on performance of coatings is from the marine industry and relates to coatings subjected to immersion. Here the soluble contaminants are largely chlorides from seawater and therefore the detection methods used for monitoring are either specifically for chlorides or, more commonly, by measuring conductivity. The higher the quantity of dissolved salts in water, the lower the resistance. Typically levels of 5 μg/cm^2 of chloride or less are considered acceptable maximum levels by Jeffrey.[8]

Information on acceptable levels of soluble salts is scarce for less demanding environments. It would be advisable to assume that, for any paint system where the longest durability is required, the initial state of the steel is Rust Grade D[9] or worse, and the coating is subjected to some degree of wetness, that a maximum level of 15 mg/m^2 of soluble iron corrosion products as measured by ISO/TR 8502-1 would be desirable.

However, a requirement for excessively low limits for non-aggressive environments could be very costly and probably not justified. A Working Party of the ISO Committee dealing with surface preparation of steel substrates is charged with providing guidance levels, but the general opinion is that it may be some time before this is possible.

There is also now ample evidence from the Highways Agency[10] experience that conventional oleo-resinous paint systems, normally with a life expectancy of 5–7 years, have lasted at least 2 or 3 times longer. One of the essential ingredients for such success is the monitoring of blast-cleaned surfaces to ensure they are free from contamination.

All of the current methods of determining soluble contaminants are discussed in Chapter 9, but as a guide it can be assumed that the deeper the corrosion pitting before surface preparation, the greater the problem. This is another sound economic reason why no area, however small, of a painted structure should be allowed to deteriorate into severe corrosion before maintenance is carried out.

3.1.3.3 Non-water-soluble contaminants

In the oil industry the handling of sour crudes is a major source of hydrogen sulphide. Sulphides also occur in certain process plants such as sewage treatment and metal refining. Although iron sulphide is insoluble, it is cathodic to steel and therefore will initiate corrosion.

Contamination of surfaces by fatty acids can occur in several types of process plants, for example those dealing with food, paper and grain. Fatty acids can react with a steel surface to form insoluble soaps which are difficult to remove completely but if overcoated can cause loss of adhesion.

Silicones are used in many industrial applications. Silicones possess a special affinity for steel and are difficult to remove. Left on the surface they can readily cause severe loss of adhesion of subsequent coatings.

Oils, greases and waxes are frequently used as temporary protective films on steel surfaces. Wax is also a constituent of many crude oils and may be left on the surface after the oil has been removed. Generally, all oils, greases or waxes must be removed before further coating and this is sometimes difficult to accomplish because of the tenacity with which such substances cling to steel surfaces. Even slight traces of wax, not visible to the eye, can cause overall loss of adhesion.

3.1.4 Roughness

The degree of roughness of a surface to be coated is important since too smooth a surface will impair adhesion and too rough a surface can leave prominent areas of the steel inadequately coated.

When steel is blast-cleaned the surface is inevitably roughened and this characteristic has been described as the 'anchor pattern', 'surface roughness' and 'surface profile'. The latter term is the one now used in International and British Standards. Over past years the term 'anchor pattern' has been officially discouraged because it was felt that it led to operators trying to obtain as rough a surface as possible. The disadvantages of very rough surfaces are that they use more paint and there is always the likelihood with these films that peaks of the metal profile are not being adequately protected. However, there is some evidence that the 'anchor' formed when a paint film penetrates into the three-dimensional irregularities that form a blast-cleaned surface is very beneficial for the adhesion of modern high-build systems. It is necessary to be able to specify the surface profile because some coatings, for example very high-build paint coatings and metal coatings, need a surface with a high profile, and therefore a large surface area per unit area, in order to ensure that the adhesive bond is greater than the cohesive bond. Indiscriminate use of large abrasives, however, particularly for relatively thin coats of priming paint, means that the peaks of the blast-cleaned surface are not adequately covered and rust spot if not quickly overcoated.

Methods for the determination of surface profile are described in Chapter 9.

3.1.5   Surface defects and welds

Defects on the surface of hot-rolled steel, such as laminations and shelling, can often remain undetected until the surface has been cleaned, particularly by blast-cleaning. Apart from the need to check the extent and depth of the defects to ensure they do not impair the strength of the steel, such sharp slivers of steel and the accompanying crevices cannot be coated satisfactorily. Generally the most effective method of removal is by mechanical grinding or discing. Some authorities specify the reduction in thickness which is allowable before the item is rejected. Any area that has been ground smooth may be too smooth for the adhesion of high-performance coatings and may require to be roughened, for example, by abrasive blasting. Careless grinding can also leave a sharp edge which is more difficult to coat adequately than the original defect.

Any burrs around cut or drilled edges, for example around bolt holes, should be removed before coating.

Weld areas require particular attention during surface preparation. Frequently in service the paint coating on a weld area will fail while the remainder of the paintwork is still in excellent condition. Much of this breakdown must be attributed to the roughness of the weld protruding into the paint film thickness but contamination left by the weld process also plays its part.

Manual metal arc welding tends to leave particularly rough surfaces depending upon the difficulty of the operation and the skill of the welder. Some of the coated electrodes give acidic weld flux (slag) deposits, but mainly the contamination is alkaline. This latter contamination can cause relatively rapid failure of alkyd or oleo-resinous paints. Undercut edges of welds need to be ground flush where necessary but deep-seated undercutting or other similar serious defects require grinding out and re-welding.

Coatings will not penetrate properly into any surface blowholes on welds. Coated over, there will be entrapped air which can then cause pinholes in the coating. Such blowholes should be filled either by welding or with a filler such as a solventless epoxy.

Gas welding or cutting with an oxyacetylene, or other type of flame, leaves an oxide film or thick scale on the weld or near the cut edge and this should be removed before painting.

TIG and MIG welding with effective gas shielding give only a very thin adherent layer of oxide and, depending upon the coating to be used and the service required, may not need to be removed.

Fully automatic submerged arc welding should produce a weld with a smooth dome-shaped weld surface which is normally satisfactory for painting with the minimum amount of weld dressing.

Weld spatter is a problem because it is seldom removed by abrasive blast-cleaning. Weld spatter is not only large in relation to most paint film thickness but like weld slag is generally cathodic to steel and therefore a source of initial corrosion under wet conditions. Anti-spatter coatings, which can cover a wide range of types, can cause loss of adhesion of subsequent coats and should be removed. Certain inorganic zinc pre-fabrication primers prevent weld spatter and may not, necessarily, need to be removed.

Non-continuous welds, i.e. skip or spot welding, may be used where continuous welding is not required for strength purposes. Such welding is impossible to protect from corrosion and should not be used in aggressive environments.

With all welding it is important to remember that, while the requirements of welding may be met satisfactorily, the surface produced may be unsuitable for long-term durability of corrosion-protective coatings.

Regarding surface preparation of weld areas, ideally all welds should be first dressed as described above, all loose slag and all weld spatter removed by chipping and then the weld should be abrasive blast-cleaned. For manual welds, even after suitable dressing, it is still advisable to apply an extra coat of primer to the weld area, i.e. stripe coating.

3.2 Surface preparation methods

3.2.1 Degreasing

The four main methods of degreasing steel surfaces prior to painting are liquid solvent cleaning, solvent vapour cleaning, alkaline cleaning and detergents. In all cases it is advisable to remove excessive deposits of oil and grease by scraping before any other operation.

3.2.1.1 Liquid solvent cleaning

Liquid solvent cleaning is used for degreasing on site, using such solvents as white spirit or solvent naphtha. Petrol must not be used because of the fire and explosion hazard.

At works the cleaning may be carried out by immersion or spraying the solvent over the surface. On site it is normally by scrubbing with rags. Too frequently the technique is to saturate a rag with solvent, wipe the contaminated area and its surrounds and allow the solvent to evaporate. This procedure does little more than spread the contamination in a thinner layer over a wider area. The following procedure should be followed:

1 Wipe or scrub the surface with rags or brushes wetted with solvent.
2 Wipe the surface with a clean, lint-free rag.
3 Repeat steps (1) and (2) until all visible traces of contaminant are removed.
4 Carry out final wiping with a clean rag and clean solvent.
5 If considered advisable, test surface to ensure it is oil free.

This method is becoming less used these days because of the increasing number of regulations restricting the use of organic solvents.

Some proprietary solvent cleaners contain emulsifying agents and this has the advantage that instead of wiping with rags the work can be hosed off with clean water. Extreme care must be taken if solvent cleaning is to be considered viable either in works or at site.

3.2.1.2 Solvent vapour cleaning

This is a procedure for works use only. The items to be degreased are suspended in a specially designed vapour degreasing plant which has an atmosphere of solvent vapour in equilibrium with boiling solvent in a heating tank. When condensation of solvent virtually ceases on the steel surface the work is removed. If a non-greasy residue remains after processing it may be removed by wiping with a clean, lint-free cloth. If greasy residues remain the process should be repeated.

Initially, trichloroethylene was almost universally used for this type of plant: nowadays there are a wide range of halogenated solvents with different degrees of efficiency, toxicity and potential damage to the ozone layer.

3.2.1.3 Alkaline cleaning

Alkaline cleaners for steel are mainly based on sodium hydroxide as 20–60% by mass of the active constituents, which also generally include sodium carbonate, sodium tripolyphosphate and surface-active agents. They are specially formulated to avoid any appreciable attack on the steel and are generally used at elevated temperatures, i.e. 80–100°C. Before the application of coatings, any cleaned surfaces should be washed with clean water, or steam, until free of alkalinity.

3.2.1.4 Detergents

There are a number of proprietary cleaners, with a wide range of formulae, mainly based on detergents and emulsifiers. They work by wetting the surface and emulsifying the oil and greases so that they can be washed from the surface. The method is generally very effective and relatively free from toxic hazard. Its disadvantage, in some circumstances, is the need to use water washing.

3.2.2 Hand- and power-tool cleaning

Basically any hand-held tool falls into this category; included are scrapers, wire brushes, chipping hammers, needle guns and abraders. Some of these can be used manually or as power tools. Generally speaking, power tools give an improved degree of cleanliness, and certainly a higher rate of working, than hand-operated tools. However, care must be taken with power tools to ensure that the steel surface is not damaged, producing sharp ridges or gouges, for example.

Most hand- or power-tool cleaning methods cannot give a standard or visual cleanliness comparable with abrasive blast-cleaning and in particular such methods are generally incapable of removing rust and soluble iron corrosion products from pits.

Hand cleaning is ineffectual in removing intact millscale from as-rolled steel, i.e. unweathered steel from the mill. The practice of weathering as-rolled steel to make it easier to remove the millscale is undesirable since it encourages pitting corrosion, which is even more difficult to clean adequately. Hand-tool cleaning in particular requires the subsequent use of primers with good surface wetting ability. Hand- or power-tool methods of cleaning may be suitable for steel that is to be fully encased, for example in concrete, or for use in dry, warm interiors of buildings.

For steel that is to be exposed in other than the mildest of environments, abrasive blast-cleaning should be the first choice of surface preparation method. However, abrasive blasting has its limitations. These include the generation of dust and abrasive particles that can damage motors, pumps and other machinery that may be in the vicinity. Furthermore, there is generally a large quantity of spent abrasive to be removed. Abrasive blast-cleaning also has more safety hazards than hand- or power-tool cleaning. For these reasons power-tool cleaning sometimes has to be used for important structures such as off-shore oil platforms.

Power tools can be divided into two main categories: impact and abrading. An old-fashioned but still used impact tool is the needle gun. A typical model consists of a bundle of 65 2-mm diameter flat-ended, hardened steel needles held in individual slots in a tube suitable for holding in the hand. These tools require approximately $130\,kg/m^3$ of compressed air per minute at $0.6\,N/mm^2$ pressure. The needles are propelled and retracted by a spring-loaded piston about 2400 times per minute. The advantages of the method include the facts that (i) there is no loose abrasive; (ii) it produces a surface profile; and (iii) it has some ability to penetrate awkward corners, shapes, pits, etc. Its disadvantages include the facts that (i) blunt needles tend to impact contamination into the surfaces;

(ii) the method generates a considerable amount of noise; and (iii) it is tiring to use.

Modern pneumatic abrading tools are available that use tungsten carbide discs mounted on a flap assembly or a rotating nylon web impregnated with silicon carbide abrasive. A further refinement is to incorporate a vacuum shroud to capture as much debris and dust as possible (see Figure 3.4).

The abrading tools are suitable for use on previously rusted surfaces where the corrosion has produced a roughened surface. For smooth surfaces such tools may not produce a satisfactory surface profile. In all cases care must be taken not to produce a burnished or polished surface of the rust.

NACE International of America claim that visual standards of cleanliness equal to Sa2 of ISO 8501-1: 1988 can be achieved when using either impact or abrading tools that incorporate the new abrasive materials, but not if the rusted surface is deeply pitted.

Figure 3.4 Vacuum shrouded coatings removal, mechanical cleaning system.

Source: Trelawny Surface Prepapation Systems.

3.2.3 Abrasive blast-cleaning

This method of surface preparation is essentially one of mechanical removal of scale and rust by continuous impact of abrasive particles onto the surface. There are two methods of achieving this: (i) by using equipment to carry the abrasive in a stream of compressed air through suitable hoses and nozzles, and (ii) by using impellers to throw the abrasive onto the surface by centrifugal force. The latter method requires comparatively large static equipment and is used mainly in works. The compressed air method is also used in works but is essentially portable and so can be utilised on site.

The effectiveness of the method depends upon the energy produced at the steel surface; this is related to the mass of abrasive particles and their velocity, i.e.

$$\frac{MV^2}{2}$$

where M is the mass of the particle and V is its velocity.

Because, for a particular machine, velocity is effectively constant, the impact energy is determined by the mass of the abrasive, i.e. its volume and density. However, although the energy of impact is critical, other factors also influence the effectiveness and type of surface finish obtained with blast-cleaning. Both the shape and type of abrasive are important, and these are discussed in Section 3.2.3.5. The superficial area cleaned, i.e. the coverage by the abrasives, also determines the overall effectiveness of the operation.

There are no standard names for these methods but the one using compressed air is generally called either air blast or open nozzle blast (Figure 3.5), and the method using blast wheels is called centrifugal or airless blast.

3.2.3.1 Air blast-cleaning

The sizes of blast-cleaning machines vary, but essentially the equipment consists of the following parts:

(i) a pressure tank or 'pot' which contains the abrasive and the necessary valves, screens, etc., for correct operation;
(ii) air hoses;
(iii) hoses for carrying the abrasive onto the steel surface through a nozzle;
(iv) moisture and oil separators;
(v) control valves;
(vi) a compressor of suitable size.

All the equipment must be capable of withstanding the pressures involved, i.e. a pressure at the blast nozzle of 620–689 kPa (90–100 psi) which will be

Figure 3.5   Air blast-cleaning on site.

correspondingly greater at the compressor. Operators carrying out the blast-cleaning operation must be suitably clothed and use a proper helmet into which air is fed. The air supply to the blast-cleaning machine provides the pressure necessary to carry the abrasive onto the steel surface.

The efficiency of the blast-cleaning operation is governed largely by the pressure at the blast nozzle. Increase in the pressure at the nozzle results in a greater force per unit area that can be applied to the surface to be cleaned. The pressure also governs the size of the surface area that can be covered. Ideally the pressure should be as close to 689 kPa, i.e. 100 psi, as possible. Each blast nozzle may require an airflow of 0.05–0.2 m^3/s depending upon size. The compressor must be of adequate size to supply a sufficient volume of air for any given nozzle diameter to ensure that the correct pressure is maintained. A compressor should always be selected to supply more air than is required by theoretical calculations in order to have a reserve to allow for worn nozzles, extra length of hose, restrictions in the air line, and so on.

There are two main types of blast hose. Four-ply hose is for heavy-duty use and when there is a danger that the operator will allow the hose to rest in a right-angled bend. If the latter occurs, the abrasive impinges directly upon the hose wall and can cause it to perforate. Two-ply hose is lighter and easier to handle but is obviously not as strong.

The diameter of the blast hose should be as large as possible in order to reduce loss of air pressure by friction losses. It is recommended that the hose should not be smaller than 320-mm i.d. but operators often prefer to have 3 or 4 m of 190 mm i.d. hose at the operating end, for ease of handling. This will cause a pressure drop and should only be used where it is necessary. Generally speaking, the internal diameter of the blast hose should be at least three times the orifice size of the blast nozzle.

To avoid contamination of the work surface, the compressed air should be free of condensed water or oil. Adequate separators and traps should be provided in the air lines.

Blasting pressures at the nozzle are normally limited to 689 kPa. Higher pressures could be an advantage and work carried out by Seavey[11] on blasting pressures up to 1034 kPa (150 psi) demonstrated that as the pressure increased both productivity and efficiency continued to increase. The main problems at the higher pressures were: (i) a shortened life of compressors which were only rated for the lower discharge pressures; (ii) increased thrust at the nozzle so that extra operator strength and technique was required; and (iii) some abrasives were too friable and shattered to dust on impact. In these tests copper slag was unsatisfactory and silica sand was satisfactory but, particularly for the latter, there are a wide range of materials with different properties that could give different results.

In works situations the steel may be cleaned in a cabinet, so containing the spread of abrasive. However, on-site the abrasive is not usually contained within the immediate confines of the machine. Screens, e.g. tarpaulins, are frequently used to prevent wide dispersal of the abrasive. In some situations it may be essential to employ such screening to comply with health and safety regulations. The abrasives are not re-circulated during site cleaning and so expendable types are used.

3.2.3.2 Vacuum blast-cleaning

A special form of this equipment has been designed to collect the abrasive by vacuum recovery after cleaning (Figure 3.6). A rubber shield around the nozzle allows abrasive to be sucked back and screened so that it can be re-used. This has the advantage of containing the abrasive and restricting dust, but is a good deal slower than the open method. Although various types of nozzles are available for complex shapes, the vacuum recovery method is most suitably employed on fairly large flat areas or on very narrow contained areas such as welds. The pressure type of equipment is most commonly employed for blast-cleaning heavy steelwork but another form of equipment using suction is also available. The abrasive is contained in an unpressurised tank and is sucked through a nozzle and blown by a jet of air onto the steel surface. It is slower than the pressure method and is generally used for component cleaning in cabinets and for touch-up work. The vacuum recovery method can also be adapted for this type of cleaning.

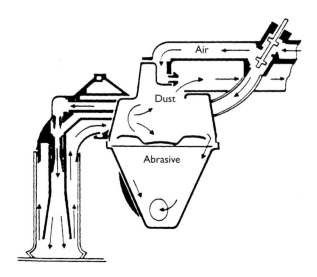

Figure 3.6   Diagram of vacuum recovery system.

Source: Reproduced by permission of Hodge Clemco

3.2.3.3   Centrifugal blast-cleaning

Although portable equipment has been developed for some purposes, most cleaning of this type is carried out in fairly large static equipment in fabricators' works (Figure 3.7). The essential features of the method are:

(i)   Wheels with blades, fitted radially, onto which the abrasive is fed from the centre of the wheel. As the wheel revolves, the abrasive is thrown onto the steel surface (see Figure 3.7). The force of impact is determined by the size of the wheel and its radial velocity. Usually a number of wheels are employed, placed at different angles relative to the steel being cleaned. Commonly 4- or 8-wheel units are used and a range of different sized wheels operating at different velocities is available.

(ii)   An enclosure or cabinet to contain the wheels.

(iii)   A recycling system for the abrasives with separator screens to remove fine particles and dust and allow reclamation of suitable sized abrasives which are then fed back into the hopper.

(iv)   A system of dust collection.

(v)   Suitable methods of feeding the steelwork into the equipment.

The method is particularly suitable for cleaning steel plates because of their simple geometry. On more complex shapes it may be necessary to hand-blast re-entrant angled steelwork. Centrifugal cleaning is faster than air blast-cleaning, particularly for large sections, although the main control of

Figure 3.7 Centrifugal blast-cleaning at works.

the efficiency of the cleaning process is the speed at which the article is transported through the equipment. Often the cleaning installation is arranged as part of a line in which the steelwork emerging from the cleaning process is immediately protected with a coat of quick-drying priming paint.

Large plants containing many blast wheels have been constructed to clean specific fabricated items, for example, railway wagons. In the USA the larger plants are used to clean fabricated sub-sections of ships and up to 40 centrifugal wheels may be used. Normally, however, for structural steel the method is used prior to fabrication. The advantages of the method compared with the air blast type, include:

(i)   Greater economy.
(ii) Relatively automatic and in-line operation.
(iii) Containment of dust from blast operation.
(iv) More efficient metallic abrasives can be used because abrasive can be recycled easily.
(v)   The plant does not require compressed air.

The disadvantages include:

(i)   High initial cost of the equipment.
(ii) Higher maintenance cost due to the wear and complexity of the equipment.

(iii)   Difficult or impossible to clean complicated steel shapes and even relatively simple shapes such as girders, may require supplementary air blast-cleaning.

(iv)   The use of steel shot, in order to reduce wear on the plant, can peen the steel surface and also drive contaminants into the surface.

(v)   The separators used on these machines remove large particles of fines, they do not 'clean' the abrasive. If the abrasive becomes contaminated with oil or soluble iron corrosion products, for example from cleaning old, rusty surfaces, such contamination will be carried on to future work.

The method is also used for repetitive cleaning of small items and there is a wide range of plants for use at works, such as tumbling mill machines where the wheel units are mounted on the roof of a cabinet in which parts are tumbled in a revolving mill below. There are several designs of table machines where items are swung or turned underneath fixed blast wheels (see Mallory[12]).

In 1996 a new type of blast wheel called a Rutten was devised and is now in use in particularly large blast rooms in the Netherlands and Belgium. Rutten blast wheels are designed with curved blades in extremely hard metal which provides higher than normal projection speeds. For large installations using conventional blast wheels, the distance between the wheels and the workpiece can cause such a loss of impact speed that it falls below the optimum for effective cleaning of the steel. The use of Rutten wheels can overcome this problem.

3.2.3.4   Wet blasting and water jetting

Two factors have led to the introduction of water into the blast-cleaning operation; firstly the increasing environmental requirements to reduce the dust hazard, particularly when removing old lead-based paints, and secondly the realisation that coatings can only achieve their optimum life when applied to surfaces substantially free of water-soluble contaminants. Therefore, wet blasting methods are essentially for maintenance painting.

The methods of surface preparation using water can be divided into two main groups: water jetting, i.e. high-pressure water without the addition of abrasive (see Figure 3.8) and low-pressure wet abrasive blasting, i.e. a combination of water and abrasive.

3.2.3.4.1   LOW-PRESSURE WET/ABRASIVE BLAST-CLEANING

This uses standard air blast equipment with a modification that introduces water into the hose just before the nozzle. The system allows either water or abrasive to be fully excluded during the cleaning process. Air/water

Figure 3.8   Ultra-high pressure water blasting on London's Thames Barrier.

pressure at the nozzle is relatively low, up to a maximum of $7\,kgf/cm^2$ ($100\,lbf/in^2$) and is adjustable below this level to enable selective coating removal.

This method effectively removes water-soluble salts from all but narrow-necked, deep pits and also keeps the dust down. Its disadvantages are that because of the small water/abrasive ratio of the system, fine particles of wet abrasive can remain on the cleaned surface and have to be removed by washing down with air and water. The wet abrasive slurry from the operation is also much more difficult to clear up from adjacent surfaces than dry abrasive. Also the operatives, which unlike water jetters are more used to a dry operation, can resent the wetness and, due to the water spray, the increased difficulty of seeing the surface to be cleaned. Without constant monitoring it is impossible to tell if the operator has merely wetted an otherwise dry blasted surface. Wet blasting obviously leaves a wet surface which, as it dries, will flash rust. However, if the rerusting is very rapid, this will indicate that soluble salts remain on the surface. Providing that it is not a powdery surface, light flash rusting, often called 'gingering', is generally considered harmless, but paint manufacturers can give guidance on this regarding their products.

Rerusting can be avoided by the addition of inhibitors, but this is generally not recommended. Evaporation from the surface could result in concentrations of inhibitor, itself water soluble, being left under a subsequent coating and affecting its adhesion. Despite the various problems, wet abrasive blasting remains the preferred method of site-based surface preparation for Highway Agency bridge structures in the UK and has proved a valuable asset. Their procedure is to remove flash rusting by flash dry blast just prior to painting.

3.2.3.4.2   WET BLASTING WITH SOLUBLE ABRASIVE

Water blasting and water jetting systems that employ water-soluble abrasive, such as bicarbonate of soda, have recently been introduced. It is claimed to be an effective method of removing old paint without damaging the substrate, that the abrasive media is safer than other types and that, in most cases, special ventilation, dust collection or abrasive reclamation are unnecessary. It is also claimed that paint debris can be filtered from the used water and then discharged into the normal drainage system without problems. Its disadvantage is that it will not remove firmly adherent metal coatings or millscale.

3.2.3.4.3   HIGH-PRESSURE WATER JETTING

High-pressure water jetting (up to 1300 bar) is effective in removing surface contamination such as algae, marine growth and loose and flaking top coats. It does not abrade the steel surface and therefore is not effective in removing water-soluble corrosion salts from pits. It is still a dangerous operation and the higher the pressure, the greater the fatigue for the operator. Typically the system will use between 8 and 25 litres per minute of water.

3.2.3.4.4   ULTRA-HIGH PRESSURE WATER JETTING

Uses pressure of 2000 bar and above. It is a relatively new operation for surface preparation and is fast gaining in popularity prior to maintenance painting. It does not abrade the steel surface, but is very effective in providing a 'white metal' finish, plus the removal of water-soluble corrosion products. In the early days of water jetting the operation was limited to a pin jet or a fan jet. A pin jet was a straight jet which concentrated the output of the pump onto a single point. Because its cleaning path was only about 3 mm wide, it was found to be totally impracticable for cleaning large surfaces. A fan jet spread the jet at angles between 15° and 90°. This lost so much power that it was inefficient for the removal of either paint or corrosion. The latest equipment is in the form of rotary cleaning heads, comprising many pin jets mounted on a head, which can rotate over the surface. This means the power of a pin jet can be utilised in a manner so as to give the coverage of a fan jet. Small heads, having a cleaning width of approximately 50 to 100 mm can be mounted on the end of a hand-held lance. Devices with larger cleaning heads have been constructed, but these are generally used mounted on wheels to be pushed manually, or attached to remote controlled vehicles or robots which crawl over the surface. Goldie[13] has described and compared six such systems available in Europe. In addition, it is possible to shroud the cleaning head with a vacuum to recover the water and debris as was carried out in 1999 on the UK's Thames Barrier (see Figure 3.9). A further advantage of this method is that it raises the surface temperature of the steel to about 20°C so that surfaces dry rapidly with minimum 'gingering' or 'bloom' and are warm for subsequent painting.

The disadvantages include high maintenance of the equipment which are classed as pressure vessels for insurance purposes and high-pressure leaks can cause excessive downtime. UHP leaks can be both unexpected and dangerous. Noise levels are high. In some delicate areas it is not always possible to use water.

There are no National or International Standards for the acceptable appearance of the cleaned surface. With regard to this, there are some published by marine paint manufacturers that are acceptable and will probably form the basis of any future International Standard. These are: Hempels Photo reference from Hempel Paints, Denmark, International Hydroblasting Standards from International Paints/Akzo Nobel UK, STG Guide from Schiffbauttechnische Gesellschaft, Germany and SSPC-SP12/Nace 5 from SSPC and NACE International, USA.

Figure 3.9   Vacuum UHP water blasting on London's Thames Barrier.

These Standards define acceptable levels of surface preparation by water jetting and provide illustrations of how substrates cleaned to those levels of preparation should look. Some of the Standards also give acceptable levels of flash rusting. A Standard from Jotun Paints, Norway, gives references only for flash rusting.

A variation to the wet abrasive blasting process, but not widely used in the UK, is where the water and abrasive are mixed together in the blast pot. This is called slurry blasting. These units are designed for high-production work and typically have several nozzles and hoses connected to a single control. They are frequently operated at lower pressures than in conventional dry blasting.

A further variation, now generally going out of favour, is to add the water to the stream of abrasive after it leaves the nozzle. This is accomplished by a simple water ring adapter fitted over the standard blast nozzle. This method reduces the dust hazard but has little effect on the cleaning efficiency, since the water does not mix with the abrasive.

3.2.3.5   Abrasives for cleaning steel

The cleaning of steel by abrasives is a straightforward concept depending upon sufficient impact energy at the surface to remove scale, rust and other deposits. However, in practice the types of surface to be cleaned vary and a number of different processes are used for the blast-cleaning. Consequently, a range of different abrasive particles has been developed.

Basically, abrasives are used not only to clean the surface but also to roughen it so that coating adhesion will be satisfactory. This roughened surface sometimes termed 'etching' produces a 'profile' or 'anchor pattern'. The profile is discussed in more detail later, but the size of the abrasive used will clearly have a marked influence on it. Figure 3.10 is a metallurgical cross-section of grit-blasted steel. This shows clearly why it is difficult to determine a 'blast profile'.

Steel surfaces before cleaning are not perfectly smooth; even where they are covered with virtually intact millscale, which can be removed comparatively easily with reasonably coarse abrasives, there will be small depressions or pits in the steel. These require small abrasive particles for thorough cleaning. Where steel has rusted, there will be considerable shallow pitting containing various iron salts. Again, finer abrasives will be required for cleaning such areas. Apart from their effect on cleaning steel surfaces, the abrasives will, by their nature, cause abrasion and wear on the blast-cleaning equipment, so they must be chosen in relation to the equipment used. All abrasives have a limited life because during blast-cleaning operations they fracture or disintegrate in the process. Where they are recovered and recirculated, this must be taken into account by the addition of new abrasive particles to ensure that a suitable mix is main-

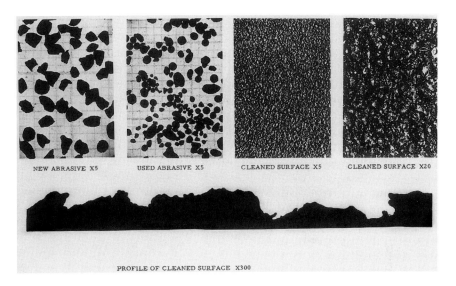

NEW ABRASIVE X5   USED ABRASIVE X5   CLEANED SURFACE X5   CLEANED SURFACE X20

PROFILE OF CLEANED SURFACE X300

Figure 3.10   A metallurgical cross-section of steel after grit blasting ($\times$300).

tained. The mix used for cleaning is important to ensure that a suitable size distribution of particles is maintained during the process. Many factors will influence the size distribution of particles, such as the original size and shape, hardness of the abrasive, the velocity of the blast (i.e. the energy of impact at the surface), the type of material used for the abrasives and the type of surface being cleaned. The last factor is important because steels vary in their hardness and the composition and rolling procedures used to produce a particular section will influence the thickness and adhesion of the millscale. Some indication of the variations in profile obtained with shot and grit is shown in Figure 3.11.

In addition to particle size, other important factors in the choice of abrasive are hardness, specific gravity and whether the abrasive is metallic or non-metallic. Relative hardness and specific gravity figures for commonly used abrasives are shown in Table 3.2.[14]

Increasing hardness has several consequences: harder abrasive tends to clean faster, owing to the sharpness and energy of impact; it can reduce the usable life of the abrasive since increasing hardness can result in increased brittleness and the particles tend to fracture, rather than wear away; the harder the abrasive, the greater the wear on the blasting equipment.

Table 3.2 also indicates the specific gravity of some abrasives. Cleaning efficiency is determined by the energy imparted to the surface, i.e. $E = \frac{1}{2}mv^2$ where E is energy, v is the velocity of the abrasive particle and m is the mass

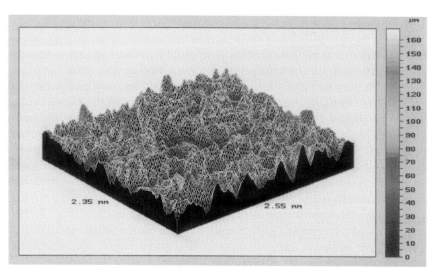

Figure 3.11 Axonometric diagrams of shot (S330) and grit (G17) blast-cleaned steel surfaces.

Source: Corus Research, Development and Technology.

Table 3.2   Properties of abrasives

Abrasive	Hardness	Specific gravity
Metallic	Rockwell C	
Chilled iron	60+	7.0 kg/dm^3
Steel shot/grit	42–66	7.0 kg/dm^3
Non-metallic		
Iron furnace slag (calcium silicate)	6–7	2.5 kg/dm^3
Copper refinery slag (iron silicate)	7–8	3.7 kg/dm^3
Coal furnace slag (aluminium silicate)	6–7	2.5 kg/dm^3
Fused aluminium oxide	9–9.2	4.0 kg/dm^3
Olivine sand	6.5–7.7	3.3 kg/dm^3
Garnet	7–8	4.0 kg/dm^3
Calcium carbonate	4–5	2.7 kg/dm^3
Walnut shells	3–4	0.5 kg/dm^3

of the particle which is proportional to its specific gravity. The pressure at the blast nozzle determines the velocity of the abrasive for open nozzle or airblast cleaning. This is generally maintained as near to 100 p.s.i. (7 bar) as possible to achieve the maximum cleaning rate. There is little advantage to exceed this pressure, since most abrasives will then fracture excessively and energy will be lost. Metallic abrasives are normally more expensive than the commonly used non-metallic types. However, because metallic abrasives are capable of withstanding repeated impact, they are most often used for cleaning processes where recovery and reuse is possible. The most widely used metallic abrasives are iron or steel cast materials. Cast iron which is supplied as grit, i.e. angular particles, is brittle and fractures on repeated impact, but cleans the surface rapidly and provides a sharp angular surface profile. Steel abrasive is available in either high or low carbon content, cleaning rate is comparable, but the low carbon type, normally supplied as shot, i.e. rounded particles, gives up to 20% greater durability because it is less likely to fracture. It also causes less wear on the equipment. However, high carbon steel abrasive is still the most commonly used reusable type. It is supplied in both shot and grit form. Grit is used mainly for open nozzle type blasting, because of its superior cleaning rate and shot for wheel blast equipment, so as to reduce the wear on the equipment itself.

The ISO Standard Specifications for metallic abrasives includes: chilled iron grit, high carbon cast steel in shot and grit, and low carbon cast steel shot (ISO 11124/1 to ISO 11124/4). These Standards specify definition, identification, size, hardness, chemistry and soundness, i.e. freedom from defects such as cracks. There are also ISO Standards of test methods of determining: particle size, distribution, hardness, apparent density, percentage defective particles and microstructure, foreign matter and moisture (ISO 11125/2 to ISO 11125/7).

Of the non-metallic abrasives, silica sand is relatively cheap and very effective. It is still used widely throughout the world. However, because of the danger of silicosis from inhaling finely divided particles, wet or dry, there is an increasing number of prohibitions and regulations regarding its use. It is virtually banned in Europe.

Standards for visual identification of surface cleanliness have suffered problems due to sand's popularity as an abrasive in the 1950s. In all cases during the blasting process, abrasive particles become embedded in the steel surface. The Steel Structures Painting Council (SSPC) has reported[15] that it appears that this is more likely with non-metallic abrasives than with metallic, typically to the extent of over 10 particles per square centimetre. Providing such particles do not contain corrosion-promoting substances, such as chlorides, there is no evidence that they affect subsequent coating performance. However, they do affect the colour of the cleaned surface. Sand abrasives give an excellent bright and white finish compared with, for example, copper refinery slag, which is black and gives an overall darker finish. This has been recognised in ISO 8501-1 'Specification for Rust Grades and Preparation Grades of Uncoated Steel Substrates after Overall Removal of Previous Coating'. The photographs of cleaned surfaces in the Standard are the originals from the former Swedish Standard SIS 055900 which were prepared by sand blasting. To counteract this the ISO Standard provides a Supplement, which shows some samples of how other types of abrasive affect the whiteness of the finish. It indicates that several commonly used abrasives give a darker hue to the cleaned surface than the difference in shade between the grades Sa3 and Sa2½. The non-metallic abrasive that has largely replaced sand is copper refinery slag (iron silicate) which is a by-product of copper smelting. It has a high specific gravity and is available in most countries.

Coal furnace slag (aluminium silicate) is basically similar although, because it derives from burning coal in power stations, there is some variation in these slags because of the differences in the quality of the coal burnt. The major production area in Europe is Germany. Like copper refinery slags, these abrasives are mainly used for blast-cleaning steel. However, because of their lighter colour they are commonly specified for cleaning non-ferrous metals. Iron furnace slag (calcium silicate) is a by-product of iron smelting. It is a versatile abrasive, which can be used on virtually any substrate.

Fused aluminium oxide is a synthetic material similar to carborundum. It is available in a variety of grades and its main use is a recyclable abrasive in blast cabinets. However, it is generally more expensive than metallic recyclable abrasives.

Staurolite is a dark-coloured mineral that is a silicate of aluminium and iron. It has some free silica but much less than silica sand. It is marketed mainly in the USA where it occurs naturally, but it is relatively highly priced.

Olivine sand, which is pale green, is a mineral from Norway and is a silicate of iron and magnesium. It is marketed as a silica-free sand. This abrasive is often used for cleaning buildings and sometimes for non-ferrous substrates. It is very hard and tends to fracture on impact, creating a lot of light-coloured dust. It is also relatively expensive.

Garnet is a hard silicate mineral, quarried in Australia, India, USA and South Africa. The almandite form is the most useful as an abrasive, but it must be suitably processed to remove any soluble salt content. Garnet is more expensive than commonly used non-metallic abrasives but its ultra high cleaning efficiency means less abrasive needs to be used. Because it produces less dust and is of a paler colour, it is a useful material for environmentally sensitive areas.

Agricultural shell products are also sometimes used as abrasives for specialist applications. Walnut shells, olive stones, peach stones and corn cobs are examples. They are relatively soft and are generally used for polishing or for deflashing. However, Bennett[16] claims that in the petroleum industry, walnut shells provide a safe, non-sparking abrasive for use in certain hazardous areas, providing the parts to be cleaned are earthed and there is adequate cross ventilation. It is considered that they provide an excellent alternative to bronze or copper hand or power tools used in tedious hand cleaning.

There are ISO Standard Specifications for copper refinery slag, coal furnace slag, iron furnace slag, fused aluminium, olivine sand, staurolite and almandite garnet (ISO 11126/2 to ISO 11126/10). Of the test methods for non-metallic abrasives, there is determination of: particle size distribution, apparent density, hardness by glass slide test, moisture, water-soluble contaminants by conductivity measurement and water-soluble chloride (ISO 11127/2 11127/7). Because of the way particles settle in storage, it is important to ensure that samples taken for test are not unrepresentative. The ISO Standards with suitable procedures for sampling are ISO 11125/1 for metallic and ISO 11127/1 for non-metallic abrasives.

3.2.4   Innovative methods

For many years an alternative to abrasive blasting has been sought, if not to replace it entirely, at least to provide an alternative and effective method of achieving a high standard of cleanliness in sensitive situations. Ideally the method should not add to the waste disposal problem.

3.2.4.1   Cryogenic blast-cleaning

Cryogenic blast-cleaning is still under development and involves blasting with either ice crystals or carbon dioxide pellets. It is claimed that ice crystals can remove old paint without damaging the substrate. It is a safe system to use and apart from paint debris, the waste produce is only water. At the moment in its development it is best suited for cleaning soft metals or composite materials.

For certain applications, carbon dioxide has advantages over standard abrasives. Because it is non-conductive it is possible to clean electrical equipment even while it is in operation. Carbon dioxide blasting has been used in the nuclear industry and other applications where cleanliness is crucial, e.g. for turbine blades in a power station. Because the equipment generates little airborne carbon dioxide, it is also claimed that the blast-cleaning operatives need only standard eye and ear protection. Providing that they are not working in a confined space, e.g. a tank or a ship's hold, where the carbon dioxide might displace breathable air, blasting suits and respirators are usually not necessary. The disadvantage of cryogenic blast-cleaning at present is that it is slower than most other methods and needs further development before it can be considered for cleaning structural steelwork.

3.2.4.2   Laser cleaning

Laser cleaning is another relatively new technology still under development but it is considered that its main use would be to strip old paint from steel surfaces. Two promising types of laser are: CO_2 laser and Xenon flashlamp.

CO_2 lasers generate more power than xenon flashlamp lasers and, at present, are more effective for coating removal. The laser's energy agitates the molecular structure of the coating and explodes the molecules apart and, at the same time, instantaneous combustion incinerates the paint binder, leaving a dry ash residue that can be vacuumed from the surface.

Xenon lasers work on the same principle as the flashlamp with a camera, but producing a light several thousand times more intense. The laser intensity is controlled by adjusting the discharge voltage which passes through the xenon gases in the lamp and can be optimised for the specific type of coating to be removed. The coating is burnt in a matter of microseconds and then the dry ash residue can be removed by vacuum.

At present the slow speed of cleaning and high capital costs puts laser technology at a disadvantage. However, lasers do have the advantage over conventional cleaning methods in that the waste products are disposed of easily. With several manufacturers now involved and possible military applications, primarily in the United States, it is thought only to be a matter of time before laser cleaning becomes commercially viable.

3.2.4.3   Sponge media blast-cleaning

Sponge media blast-cleaning, a process emanating from the United States, has been designed with health and safety and environmental restrictions in mind. A synthetic open-cell polymer sponge impregnated with an abrasive is impacted on the steel surface by compressed air. Unlike conventional hard abrasives, which ricochet from the surface, the impact energy is absorbed by the deforming sponge to shear the paint film, rust or millscale from the surface. As the sponge leaves the surface, debris and dust are captured in the voids of the sponge. The sponge is then collected, cleaned and reused. Different grades of sponge are available, impregnated with various abrasives such as iron furnace slag, garnet, aluminium oxide, steel grit or plastic chip, depending on the surface to be cleaned and the required surface profile. Sponge without abrasive is also available for cleaning surfaces of oil, grease and other contaminants. The sponge can then be recycled through a washing and rinsing chamber to separate out the contaminants.

3.2.5   Flame cleaning

Flame cleaning is an old-fashioned method of trying to achieve the same objective as laser cleaning. In this method an oxyacetylene or oxypropane flame is passed across the steel. The heating causes millscale and other scales to flake off as a result of the differential expansion between the scale and the metal. In addition, any rust present is dehydrated. Immediately after the passage of the flame, any loose millscale and rust that remains is removed by wire-brushing. This generally leaves a powdery layer which must also be removed by dusting down.

The level of cleanliness obtained is generally considered to lie between that obtained by abrasive blast-cleaning and that resulting from manual cleaning, and the method is used for maintenance work rather than for cleaning new steelwork. The advantages of the process are:

(i)   The relative mobility of the equipment enables it to be used at any stage of fabrication or erection.
(ii)   It can be used under relatively wet and damp conditions and helps to dry the surface.
(iii)   If priming paint is applied while the surface is still warm, this ensures that there is no condensation and it also speeds up slightly the drying of the primer.

The disadvantages are:

(i)   If the flame is traversed too slowly, unbonded scale or other foreign matter is fused to the surface of the steel, or alternatively thin sections of the steel are warped.
(ii) It is a fire hazard.
(iii) If not carefully controlled it can affect the metallurgical properties of the steel and should never be used near high strength friction grip bolts.
(iv) It is an expensive method taking into account its comparative inefficiency in cleaning steel.

Over the years this method has been in and out of fashion. It has been dropped from the SSPC and NACE Specifications because it is considered a method no longer in use in North America. It has been incorporated in ISO 8501-1: 1988 because it is apparently used in Germany as a surface preparation method for the maintenance of road bridges.

3.2.6   Pickling

Before the advent of blast-cleaning as an effective method for the thorough cleaning of structural steelwork, the removal of rust and scale by immersion in dilute acids was sometimes employed. This method of cleaning steel is called *pickling*. Nowadays it has only a limited use on structural steel to be painted, although it is employed on steel to be hot-dip galvanised. It is also the method most commonly used to clean steel sheet and strip, usually as part of a continuous treatment process.

The composition of scale has been discussed in Section 3.1.2. Both the magnetite (Fe_3O_4) and haematite (Fe_2O_3) layers are relatively insoluble in acids. The layer of scale nearest to the steel surface, wüstite (FeO), is generally partially decomposed and can be fairly readily removed provided acid can diffuse through the cracks in the brittle layers of magnetite and haematite. The acid attack is partly on the wüstite layer and partly on the steel substrate; this undermines the scale, allowing it to become detached from the steel. Hydrogen is produced during these reactions and this leads to the removal or 'blowing off' of the scale during the process. Rust is also removed during the pickling process but, depending on its form and thickness, may require longer pickling times than do many scales.

Scale is not removed in even layers from the steel, so where it is quickly lifted away the acid can attack the steel. Sometimes scales are readily removed, whereas at other times the steel may need to be immersed in the acid for some time before it is fully descaled. This could lead to an unacceptable level of attack on the steel itself. To prevent this happening *inhibitors* are used with the acids. These may be organic or inorganic in nature.

3.2.6.1   Inhibitors

Inhibitors are used to stop attack on the steel and to save unnecessary use of acid. The attack on the steel is likely to be localised, producing surface pitting, which may affect coating processes. Furthermore, such attack will be influenced by the steel composition and the type of scale produced during processing, so the addition of inhibitors adds a degree of control to the pickling operation. Inhibitors also reduce the amount of carbonaceous matter left on the steel surface.

A wide range of organic inhibitors is marketed, many under proprietary names. The requirement is to reduce attack on the steel, but at the same time they must not slow down the pickling process to any marked extent. If electroplating is to follow inhibited pickling, it may be necessary to clean the surface, e.g. by an acid dip to remove traces of the inhibitor remaining on the steel surface. Inorganic inhibitors are not widely used for commercial pickling although they are used for laboratory de-rusting.

3.2.6.2   Hydrogen embrittlement

A detailed discussion of hydrogen embrittlement is outside the scope of this book, but a few points are worth noting. During the pickling process hydrogen is evolved and plays an important role in removing the scale. Most of it is evolved as a gas but some may diffuse into the steel in the atomic form. This can lead to hydrogen embrittlement of the steel which, as the name implies, will affect the mechanical properties of the alloy. With some steels this may result in actual cracking of the material. Additionally, the atomic hydrogen that has diffused into the steel may combine in voids to produce gaseous hydrogen, which sets up a pressure that may be sufficient to cause blistering of the steel. Generally, the effects can be ameliorated by suitable heat treatment, but advice should be sought if high-strength or high-carbon steels are to be pickled.

Although inhibitors reduce the amount of hydrogen evolved, their presence in the acid solution does not necessarily reduce the likelihood of hydrogen embrittlement of the steels. In fact some may increase the absorption of hydrogen.

3.2.6.3   Pickling procedures

Although the basic pickling process is simple, in practice considerable control is required to ensure efficient and economic cleaning of steelwork. Generally, sulphuric acid is used for pickling structural steelwork, although both hydrochloric (muriatic) acid and phosphoric acid are employed to some extent. Hydrochloric acid may be preferred for the continuous pickling of steel strip because, although it is usually more expensive than sulphuric acid, the waste liquor can be removed more cheaply. Phosphoric acid is too expensive for use as the main scale-removing acid but may be used as a final treatment to produce a thin iron phosphate coating on the steel. This acts as a good basis for paint, provided the phosphate layer is not too thick. Thick phosphate layers have poor cohesive strength and can delaminate if a paint coating is applied over them, so causing loss of adhesion.

During pickling in a bath containing sulphuric acid, iron is dissolved, mainly from the steel itself, and reacts with the acid to form iron salts, e.g. ferrous sulphate ($FeSO_4$). These salts influence the pickling procedures and tend to slow them down. Consequently, the acid must be discarded before it becomes saturated with the salts. Steel must be washed after pickling to ensure that such salts are removed. Hydrochloric acid similarly produces chloride salts, but these are more easily removed from the steel surface.

The general procedure for acid pickling is as follows:

(i) Removal of all grease, oil, etc., by suitable solvent cleaning. Heavy deposits should be scraped off before the application of the solvent.

(ii) Removal of heavy deposits of rust, scale and paint by suitable manual methods, e.g. scraping.

(iii) Immersion of steel in a bath containing the acid, with inhibitors, at a suitable temperature. For sulphuric acid a 5% (volume) solution is often used at a temperature of 75–80°C.

(iv) After scale removal, the steel is thoroughly rinsed or immersed in clean water.

(v) Where appropriate, immersion in an aqueous solution containing inhibitors in suitable concentrations, e.g. 0.75% sodium dichromate.

Generally, the pickling process is improved with agitation of the acid bath. The Steel Structures Painting Council of America (SSPC) has published a specification for pickling, 'Surface Preparation Specification No. 8' (*Steel Structures Painting Manual*, Vol. 2, Steel Structures Painting Council, Pittsburgh, USA).

3.2.7 Iron and zinc phosphating

Many smaller piece parts used for architectural steelwork – as well as aluminium windows and certain galvanised sections used for shop fronts – are prepared for subsequent organic powder coating by a chemical conversion phosphating process. The substrate is subjected to a degreasing process at elevated temperatures of around 60°C, using materials previously described in Sections 3.2.6.3 and 3.1.6.4. The process is carried out by immersion, rather than the conveyorised spray process used to prepare refrigerator bodies. Following a hot rinse in demineralised water, items are transferred to a tank containing a solution of iron or zinc phosphate also operating at between 50° and 60°C. A zinc phosphate process is often used for steelwork, whilst iron phosphate tends to be the preferred pretreatment chemical for non-ferrous parts; certain manufacturers have combined the degreasing and phosphating chemicals into one 'bath' in which case a 'fixing' rinse is required. There is then generally a final rinse, where the conductivity of the rinse water, constantly overflowing by addition of fresh demineralised water, is monitored to ensure cleanliness of the prepared item. To avoid 'water marks', hot air drying completes the process.

This method of providing a conversion coating to the steel or non-ferrous piece parts requires the application of an ultra-thin phosphate layer, yielding just 1.5 to $3.5 \, \text{g/m}^2$ phosphate coating and is used only for subsequent organic polyester, polyurethane or epoxide based powder coatings.

3.3   Health and safety matters

3.3.1   General

Only personnel who have been instructed and trained in its use should operate any of the surface preparation equipment, most of which is potentially dangerous. This particularly applies to High Pressure (HP) or Ultra High Pressure (UHP) water jetting. The operator should be warned of the danger of getting any part of the body in front of the jet and particularly of injury to the feet when standing on the surface being cleaned.

Protective clothing is essential for the operator and should include waterproof clothing, helmet or visor, heavy-duty gloves and safety boots. When HP water jetting during hours of darkness, the area of operation should be clearly illuminated. When the HP water jetting pump is unattended the compressor should be stopped, the fail-safe valve should be in the safe position and the jet nozzle removed from the lance. Should a person be struck by a HP water jet, urgent medical attention is essential and the medical personnel concerned should be informed of the nature of the equipment which caused the injury. This is vital in order that the appropriate medical treatment can be given speedily.

Some metallic abrasives may contain amounts of heavy metals such as lead, beryllium, cadmium and even arsenic. In most cases the amounts are less than the Threshold Limit Values (TLV) or Operational Exposure Limits (OEL) required for safe use. It is also claimed that the manufacturing process bonds such metals within a complex matrix so that they are not available as free metals. However, occasional chemical analysis would be wise to ensure that the levels are within the specified limits. In any case, approved respirators and proper ventilation should always be used. Another source of toxic dust during blasting operations is that derived from old paint being removed. There is a special emphasis on lead, since so many old paint systems contain a lead-based primer. Abrasive blasting of these painted surfaces produces fine lead-containing particles which can be inhaled or ingested if the workers fail to wash the dust off their hands or smoke dust-contaminated cigarettes. Lead is a particularly insidious poison. Exposure to small doses over a long period of time can cause severe and permanent damage to the central nervous system, kidneys, urinary and reproductive systems and the brain before any symptoms are felt. Compliance with the current Lead Regulations is essential.

During the removal of lead-based paints the operators should use respirators. The use of the simple type of face mask has limited effectiveness and, for the removal of large quantities of lead-based paints, the operators should wear air-fed respirators and be properly trained in their use and maintenance.[17] Other pigments may also be a hazard, for example chromate, which in the early days was a common replacement for lead. Certain types of hexavalent chromium compounds have been identified as potential carcinogens.

Before removal of large quantities of any type of paint from a structure it is advisable to obtain a full analysis of its composition and obtain advice from the appropriate health authority.

The use of blast-cleaning in atmospheres that may be flammable or explosive is also a matter of concern. However, Singleton[18] considers that, although grit blasting of rusted steel produces numerous sparks, they are dull and not capable of igniting flammable gas mixtures. In American practice, it seems that the use of high-pressure water jetting or mechanical cleaning using tools made of beryllium–copper alloys is preferred to dry abrasive blasting in potentially explosive atmospheres.[19]

Another health hazard with practically all surface preparation methods is the level of noise generated. This ranges from 102 to 106 dBA at 1 mm for mechanical chipping and 102 to 104 dBA for grit blasting.

3.3.2   Open nozzle blast-cleaning

Abrasive issues from an open blast nozzle at up to 450 mph (600 km/h) and can inflict serious injury on a human body. If the nozzle is accidentally dropped during operation, it does not lie still on the ground, but writhes like a snake, spraying all and sundry. Nozzles should therefore always be fitted with a safety cut-off valve, or 'deadman's handle' as it is generally known, to shut off the pressure instantly. Some operators resent the extra effort required to keep the handle closed and consequently fix it to remain open permanently. This dangerous practice should be banned because it relies on constant vigilance by the person at the blast pot, who is responsible for turning off the pressure. It is much safer for the operator at the nozzle to carry out that function.

As abrasive passes through the hose, it can generate static electricity. A shock from this source might cause the operator to fall or drop the hose. Blast hoses should therefore be adequately earthed. Generally, this is an integral part of the hose construction but, if not, it is necessary to install an external earth. Quick-release external couplings are used to connect lengths of hose. It is obviously undesirable that hoses should come apart during operation and couplings should be wired together as an additional safeguard. Also, care should be taken to ensure that the cut ends of the hose, fitted to the couplings, have been sealed. Unsealed ends allow the ingress of air, which can cause blisters in the hose wall, or water, which can rot the fabric. Also, the screws used for fitting the couplings should not penetrate right through the hose wall. Before use the blast hose should always be inspected for wear or damage. The blast operator should always wear a full, protective hood, with a filtered and regulated supply of air to provide a positive pressure and prevent entry of harmful dust and abrasive. If the air is coming from a diesel compressor, an air purifier and carbon monoxide monitor are normally required. Full protective clothing – gloves, safety boots, etc. – must be worn for this type of operation.

Abrasive blasting produces voluminous quantities of fine dust that can be an irritant to the skin but are particularly damaging to the eyes. Eye protection should be worn, not only by the blast operator but also by those nearby. Its use may be restricted in applications in closed environments, where dust suppression or extraction facilities are unable to meet permissible environmental contamination levels.

The use of silica sand as an abrasive for blast-cleaning is prohibited or restricted in many countries. The reason is to protect workers and the general public from the dangers of lung injury or silicosis as a result of inhaling fine silica or quartz dust. In the UK the use of sand or other substances containing free silica is prohibited for use as an abrasive blasting medium.

3.3.3   Pressure water jetting

This method uses water at very high pressures and requires the greatest care in use. At high pressures, the water lance is difficult and stressful to handle. Within at least a 5 m range, anybody hit by high-velocity water is likely to be seriously injured. Typical safety procedure recommendations include the need for the provision of firm footing for the operator and ensuring that the equipment has a safety cut-off valve. It is also advisable that if high-pressure water jetting pumps are left unattended they are switched off and the jet nozzle removed from the lance.

Protective clothing is essential for the operator and should include waterproof clothing, helmet or visor, heavy-duty gloves and safety boots.

3.3.4   Flame cleaning

Obviously, when flame cleaning is used as a method of surface preparation, precautions should be taken against fire or explosion by removing or shielding any nearby stocks of paint or solvent.

The operators should always wear safety goggles, to protect the eyes from injury by scale or dust. The goggles should also be suitable to protect against the light of the flame. If the method is used for the removal of old paint, apart from the risk of flame spread there is also a possibility of a toxic hazard from the fumes evolved by the combustion of the binder.

Fardell[20] has specified the toxic vapours evolved from polymers burning in a fire.

Particular care is required if flame cleaning is carried out in a confined space and adequate ventilation and operator protection should be provided.

References

1. Iron and Steel Institute, Second Report of the Corrosion Committee, Special Report No. 5, 1934, p. 245.
2. Smuts, J. and de Villeas, P. R., *J. Iron Steel Inst.*, **204** (1966) 787–94.
3. Chandler, K. A. and Reeve, J., *J. Oil and Colour Chemists' Assoc.*, **49** (1966) 464–76.
4. Fancutt, F., Hudson, J. C., Rudram, A. T. S. and Stanners, J. F., *Protective Painting of Structural Steel.* Chapman and Hall, London, 1968, p. 33.
5. Chandler, K. A., *Br. Corros. J.*, **1** (1966) 264–6.
6. Chandler, K. A. and Stanners, J. F., *Second International Congress on Metallic Corrosion*, NACE, Houston, 1966, pp. 325–33.
7. Chandler, K. A., *British Corrosion Journal* (July 1966) 264–6.
8. Jeffrey, J., *J. Protective Coatings and Linings* (Nov. 2000) 17.
9. Preparation of steel substrates before the application of paints and related products. BS 7079: Pt A1: 1989. ISO 8501-1: 1988.
10. Durability of Steel Bridges: A survey of the performance of protective coatings. Pub. 241 Steel Construction Institute.
11. Seavey, M., Abrasive blasting above 100 psi. *J. Protective Coatings and Linings* (July 1985) 26–37.
12. Mallory, A. W., Description of centrifugal wheel (airless) blast systems. *Materials Performance*, **21**(10) (1982) 15–24.
13. Goldie, B., *Protective Coatings Europe* (April 2000) 15.
14. Paddison, R. D., *Protective Coatings Europe* (Sept. 1999) 9–18.
15. Keane, J. D., Bruno, J. A. and Weaver, R. E. F., *Survey of Existing and Promising New Methods of Surface Preparation*. Steel Structures Painting Council, Pittsburgh.
16. Bennett, P. J., Non-metallic abrasives for surface preparation. *J. Protective Coatings and Linings* (April 86) 32–9.

17. Hower, H. E., The dilemma of removing lead-based paint. *J. Protective Coatings and Linings*, **5** (1988) 30–7.
18. Singleton, D. W., Blast cleaning in inflammable atmospheres. *J. Oil and Colour Chemists' Assoc.*, **59** (1976) 363–8.
19. Problem Solving Forum. How is surface preparation accomplished in non-sparking atmospheres? *J. Protective Coatings and Linings*, **4** (1987) 21–5.
20. Fardell, P. J., Toxic hazards from burning polymers. *J. Oil and Colour Chemists' Assoc.*, **6** (1989) 223–5.

Chapter 4

Paints and paint coatings

Paint is the material most commonly employed to protect steel, although in practice the term 'paint' covers many different materials with a range of properties. They are comparatively easy to apply with no limitation on the size of steelwork that can be treated. Furthermore, paint coatings can provide a decorative finish to steel structures.

There should be a clear differentiation between paints and paint coatings, the former being the liquid material and the latter the protective film on the steel surface. Many of the advantages of paint can be lost during conversion of the liquid paint to the dry protective coating, e.g. by employing poor application techniques.

Modern paints have been developed to provide improved properties; however, these improvements will be achieved only by careful attention to factors such as surface preparation, paint application and selection of paints suitable for a specific situation.

4.1   General requirements

The paint coating has to have certain properties such as reasonable hardening and the ability to be repainted, but most of all it must provide the protective properties required for the particular set of conditions to which it will be exposed. The paint material must also meet certain requirements, including ease of application, reasonably long storage life, fairly quick drying properties and, of course, a suitable price.

Paint basically consists of solid particles, called pigments, dispersed in a liquid. When mixed and applied to steel the liquid paint dries and binds the pigments into a coherent film; hence the liquid is called the binder. In practice, as will be discussed later, paints generally contain additional substances to improve application and other properties.

The binder provides the properties required to resist attack by the environment, so it is generally the most important constituent of the paint. Naturally occurring oils, such as linseed oil, were originally used as binders. However, they had a number of disadvantages, e.g. slow drying, so a range of synthetic binders has been developed over the last 50 years. Virtually all paints now used are totally or partially manufactured from synthetic materials.

When oil was used as the only binder, many painters mixed their own paints and there was no paint technology as we know it. The lack of knowledge of chemistry probably did not hinder the manufacture or use of paints to any extent. However, as organic chemistry and, in particular, polymer chemistry developed, the possibilities of producing a vast range of materials suitable for use in both the solid and liquid form, by using solvents, became apparent.

These binders with names such as alkyd, epoxide and urethane have a range of properties and can be used without a detailed knowledge of their chemical structure. Paint manufacturers supply data sheets which provide the basic information necessary to use the paints to the best advantage. Nevertheless, some broad understanding of the nature of paint is advantageous because, in the context of protection of steelwork from corrosion, paint has what can be termed an 'engineering dimension'. Engineers are called upon to specify protective systems for structures and, while they will usually seek advice, they accept the final responsibility for deciding on the coatings to be used. The costs of protecting large structures may, of course, run into hundreds of thousands of pounds and some background knowledge of paint coatings must be considered as essential for specifiers.

Unlike many engineering materials, which are purchased in the finished form and can be tested to ensure that the properties are as specified, paint coatings are applied by the user and this, at least to some extent, determines the final properties. In this sense there are similarities with cement and concrete.

The protective properties of coatings applied from a can of paint can vary, depending on application procedures and cleanliness of the steel surface. However, it is not always by any means certain that paints with the same general description will in practice produce similar coatings, even under the most carefully controlled conditions. There are few British specifications for paints; other countries have many more standards and specifications for paints, but within the overall requirements of such standards it is possible to make paints varying quite considerably in properties. In the final analysis, the performance in practice is the important property, but this is virtually impossible to predict except in the most general way. Fortunately, paint companies have a vested interest in producing sound products, but the broad generic terminology used for paints may well confuse many specifiers. There are, no doubt, good reasons for the overall lack of standards for paints in many countries. Nevertheless, the generic terminology could be improved with certain minimum requirements regarding composition and properties. Paint companies claim, with some justification, that this would lead to many of their competitors producing a product of lower quality which, nevertheless, would fall within the standard requirements. However, improvements could usefully be made because, in some cases, the generic terminology may give a false impression of the particular paint being considered.

4.2   The nature of paint

As noted earlier, paints are essentially a dispersion of solid particles (pigment) in a liquid. These, after application, dry to provide a protective film on the surface of the steel or, in a paint system, on the previously applied paint coating. To achieve these requirements, the paints must:

  (i)   be capable of application under a specified set of conditions;
 (ii)   dry within a specified time limit;
(iii)   be able to provide the dry film with suitable properties of hardness, gloss, etc.;
(iv)   provide the necessary decorative requirements; and
 (v)   react with or 'wet' the surface so that the dry film adheres to the substrate.

The dry film must also have a range of properties which include:

  (i)   durability in the particular service environment;
 (ii)   protection of the steel from corrosion for a suitable period;
(iii)   formation of a coherent film that remains adherent to the surface;
(iv)   physical properties capable of resisting impact and mechanical damage; and
 (v)   low permeability to moisture, oxygen and corrosive ions, e.g. those arising from sulphur dioxide (SO_2) and sea salts.

To achieve these requirements, suitable pigments and binders are selected. However, certain properties such as application and drying may require additional constituents, so the paint usually includes ingredients other than those that would ideally appear in the final protective dry film.

Some of the more common constituents in a paint are considered below.

Binder. This is the film former and contributes mainly to the durability of the paint coating providing the necessary mechanical and physical properties, cohesion and flexibility. It is also the factor that determines the adhesive qualities of the film. It may vary from 20 to 50% by weight of the paint, but will be a higher percentage of the dry film.

Pigment. This provides colour and opacity to the dry film. It contributes to the hardness and abrasion resistance of the film and reduces its permeability. Also, since pigments are generally inert to the effects of ultraviolet light, their presence in a paint film reduces the degradation of the primer by sunlight. For that reason, binders without pigment, i.e. varnishes, have a shorter life outdoors than the comparable paint. However, the proportion of pigment to binder is a critical factor in paint formulation.

Some pigments have a flake or lamellar shape and it is claimed that such pigments, e.g. micaceous iron oxide, may improve the durability of the film. Most pigments are inert but a few used in priming paints may have inhibitive properties. Others may react in acidic or alkaline environments; such reactive pigments can reduce the durability of the paint film.

The percentage of pigment may vary from about 15% to 60% in the paint. In zinc-rich paints the pigment may be over 90% by weight in the dry film.

Extenders. These are similar to pigments; they are also insoluble in the binder, but have little or no opacity or colouring function. They may be added as a form of cheap pigment, but generally are used to modify the paint properties and are usually present in comparatively small amounts. Extenders may be employed to add bulk to the paint, reduce settlement, enhance abrasion resistance and provide a degree of 'false body' or thixotropy to the paint.

Solvent. This is used to reduce viscosity to a level suitable for application and to assist in manufacture. It evaporates completely and plays no part in the dry film. Although essential to ensure application, it is preferably kept to as low a level as possible. Solvents should evaporate completely in a reasonable time and hence play no part in the dry film. Problems can arise if solvents become trapped in the protective film.

There may be 5–40% by weight in paint and the type used will depend upon the binder. Some paints are 'solventless' but these are not commonly used for structural steelwork. (Thinners may be added prior to application but are not part of the paint composition.)

Diluent. These are volatile organic liquids, which are not capable of dissolving the binder but which are added to improve application or to improve evaporation at a lower cost than by using true solvents. The rate of evaporation of diluents and solvents must be balanced to ensure that sufficient true solvent is always available during application. Generally, as much diluent as is practicable will be added because of the saving in cost. Legislation has enforced the reduction of diluents, as well as solvents, in paints used for protective purposes. The limits presently in force for VOCs include the use of any diluent, summed with the quantity of solvent, within the paint formulation at point of use.

Driers. Originally salts of lead, manganese, cobalt and zinc, with suitable organic acids, e.g. naphthenic acids, were used to accelerate the oxidation of the oil portion of oleo-resinous, alkyd and epoxy ester binders. Lead, used for through drying of the film, and cobalt, for surface drying have now tended to be replaced with a single drier based on zirconium. The use of zinc naphthenate has been augmented by zinc complexes, which provides faster drying and less potential wrinkling should too rapid overcoating be carried out. Cobalt naphthenate is still employed in the base components of unsaturated polyesters, to provide free radicals for polymerisation.

Anti-skinning agents. These prevent skin formation on the surface of paint in partially filled containers. They act to inhibit oxidation and drying and, generally being fairly volatile, evaporate when the paint is applied to the steel.

In some paints a thixotropic agent is added to control sagging of the paint and to allow application of coatings of high thickness (high-build paints). Binders, pigments and solvents will be considered in more detail in later sections. Despite concerns over toxicity, methyl ethyl ketoxime remains the most used anti-skinning agent. For some moisture-sensitive paints, a molecular sieve is added to the paint formulation as a water scavenger. Moisture-curing paints are often put into cans that have had 'dry air' passed through them or the cans of filled paint are given a shot of nitrogen gas, to ensure suitable shelf-life stability.

4.2.1   Paint systems

Most paints do not provide protection to steelwork with a single coat of paint, although it may be possible with some two-pack materials to build up very thick coatings in one application. Generally, however, a number of coats are applied. The paints used for the various coatings may be different and they will be considered below.

4.2.1.1   Priming coat

Strictly, the priming coat is the one that is applied to the steel substrate, but the term is generally used to indicate the types of paints that can be used for that purpose, and sometimes two coats of primer are applied. In some paint systems there is little difference in composition between the primer and other coats, but with most of the conventional one-pack paints an inhibitive pigment is used (see Section 4.7.1). The priming coat is the foundation for the whole system and must 'wet' the surface and have good adhesive properties to the substrate.

4.2.1.2   Undercoats

Although there may be no essential difference between the priming coat and undercoat in some paint systems, the undercoat performs a different role from the primer. The pigments in undercoats are inert and the coating is used to build up the overall thickness of the system. In oleo-resinous and alkyd systems its formulation is similar to that of the finishing coat, although it does not have the same gloss properties. It is advisable to use different tints if more than one undercoat is to be used, in order to ensure proper coverage.

4.2.1.3   Finishing coat

Sometimes called the 'topcoat' or 'weather coat', it serves to protect the system from environmental factors such as ultraviolet light from the sun and provides the main abrasion resistance. It also is, where appropriate, the decorative coat containing the pigments to impart the required colour. In decorative systems it often provides an appearance of high gloss. In some systems, to provide the required appearance it may be of a type of paint different from that of the rest of the system.

Problems can arise when different types of binder are used within a paint system. This is discussed in Chapter 13.

In the next section the mechanism by which paint protects steelwork is considered.

4.3   Protection by paint films

Paint films protect in three general ways:

(i)   by insulating the steel from the environment (barrier coatings);
(ii) by inhibiting the attack on the steel substrate (inhibitive primers);
(iii) by galvanic action (zinc-rich paints).

In all cases (i) is involved; (ii) and (iii) may be involved with certain paint systems.

The corrosion mechanism for steel has been considered in Chapter 2. The overall reaction can be represented as

$$4Fe + 3O_2 + 2H_2O \rightarrow 2Fe_2O_3.H_2O$$

This can be divided into two reactions: one producing electrons, the anodic reaction, and the other consuming electrons, the cathodic reaction:

$$Fe \rightarrow Fe^{2+} + 4e^- \quad \text{(anodic)}$$

$$O_2 + 2H_2O + 4e^- \rightarrow 4OH^- \quad \text{(cathodic)}$$

The corrosion process can be prevented or retarded by suppressing either the cathodic or anodic reaction or by inserting a high resistance in the path of the corrosion current flowing in the electrolytic cell. The cathodic reaction can be suppressed by preventing the passage of oxygen and moisture to the steel. It has been demonstrated by a number of workers that organic paint films allow diffusion of sufficient water and oxygen to allow the steel substrate to rust at the same level as it would if the steel had not been painted.[1,2,3] It can, therefore, be concluded that paint films are too permeable for suppression of the cathodic reaction. The anodic reaction can be suppressed by supplying electrons from an external source and so making the potential of the iron sufficiently negative to prevent corrosion. This is the basis of cathodic protection.

4.3.1   Zinc-rich pigments

Polymers do not contain free electrons, so cathodic protection will not operate unless there is sufficient metallic pigment in the paint film capable of supplying the necessary electrons. Zinc fulfils this role, being less noble than iron. However, it must be present in sufficient concentration in the film to allow direct contact between the iron (or steel) and the zinc particles. Although zinc-rich paints act in this way in both organic and inorganic binders, their effectiveness depends upon the conductivity of the electrolyte present at the surface. Consequently, this mechanism may operate when steel coated with zinc-rich paint is immersed in seawater, but to a lesser extent if exposed in a fairly mild atmospheric environment. Such coatings act in a manner similar to, but not as effectively as, hot-dip galvanised zinc coatings and will protect the steel, for a time, where the paint is scratched or damaged.

Although zinc-rich paints act to some extent to suppress the anodic reaction, this is only a limited part of their function, occurring in the early stages or if the paint film is damaged; the suppression of the anodic reaction by this method is very limited as a mechanism of protection. However, zinc-rich paints are infrequently used under water, as the ultimate formation of zinc salts will tend to promote osmotic blistering, a phenomenon explained in Section 4.3.3.

4.3.2   Inhibitive pigments

The anodic reaction can also be suppressed by the use of inhibitors. Essentially solutions of certain salts and compounds called inhibitors act to passivate the anodic areas on the steel surface, so reducing or preventing corrosion. Some pigments have these inhibitive properties and are used in paints, generally in the priming coat of systems used to protect steel in air. They are not usually employed for immersed paint systems. The best-known inhibitive pigment is red lead, which was at one time widely used in oil-type paints. It was very effective and red lead primers performed reasonably well on rusted surfaces. The method of inhibition was somewhat complex depending on the lead salts produced in linseed oil media. There are disadvantages with red lead in oil paints. They are slow drying, difficult to spray and toxic. Consequently, they are not now widely used. Red lead pigments can also be used in non-oil media, e.g. chlorinated rubber, but are unlikely to provide the degree of inhibition experienced in oil paints.

Other inhibitive pigments based on salts of hexavalent chromium have also been widely used, e.g. zinc chromate. However, these pigments also cause toxicity problems, particularly in manufacture. In view of the various toxicity problems that may arise with these pigments, they have, to a large extent, been replaced by zinc phosphate. This is also inhibitive but less effective than red lead or zinc chromate in many binders.

4.3.3 Barrier coatings

Many paint systems do not contain either an inhibitive primer or a zinc-rich primer and it is, therefore, probable that they act to suppress corrosion because of their electrical resistance, which impedes the movement of ions. The water molecule is small and can penetrate into organic coatings. The moisture may be absorbed in the intermolecular spaces or pass through the coating. The moisture in the film tends to stabilise, with the movement into the coating being balanced by evaporation. Consequently, the water content tends to be reasonably constant. This absorption of water is not necessarily a serious problem. The electrical resistance remains at a level where the driving corrosion current is small. This means that corrosion of the steel substrate is also low. However, this situation can change if ions such as chloride (Cl^-) or sulphate (SO_4^{2-}) penetrate the paint film, because they will lower the electrical resistance.

It has been shown that the rate of diffusion of ions is much lower than that for either oxygen or water. Furthermore, although it is probable that electrolytes such as sodium chloride (NaCl) can penetrate the film, this tends to occur only where the density of cross-linking is low. Nevertheless, over a period of time the entry of ions will reduce the resistance of the film and this will lead to corrosion and deterioration of the coating. It is also possible that there is a degree of ion exchange as aggressive ions permeate through a paint film, and some paint formulations have been developed to take this aspect into account.

Clearly, if the film is able to prevent ingress of corrosive species it will protect the steel from corrosion. However, apart from diffusion processes through the molecular structure of the paint, ingress of moisture and other corrosive elements may occur because of defects concerned with the physical and mechanical integrity of the film. These may be related to the formulation of the paint, the ageing process which is, of course, inevitable, or the problems of applying some films to provide a reasonably compact coherent film. The effects of temperature variations may, particularly on thicker films, produce stresses that lead to loss of adhesion or even cracking of the film. Some coatings tend to retain solvent which, as it slowly evaporates from the film, leaves small holes. These small cavities may not always seriously affect the protective value of the film, but in some cases it seems likely that they would increase the possibility of greater diffusion or take-up of corrosive species.

Paint films may be influenced by osmotic effects. Osmosis is the term used to describe the passage of water through a semi-permeable membrane when the solutions on either side of the membrane have different concentrations. Water moves from the less concentrated solution to the one of higher concentration. This may be a problem if there are salts on the steel surface under the paint film, particularly under immersed conditions or where a good deal of condensation occurs on the paint film. The salts increase the concentration at the steel surface, so increasing the movement of water through the paint film. This can result in blistering of the film. The salts may arise from atmospheric deposits, e.g. sodium chloride in marine environments or from sweat which may leave acidic salts on steel. Salts may also be present on the steel surface as a result of corrosion processes, particularly where steel is allowed to rust before cleaning.

Work carried out by a number of investigators has demonstrated that rusts produced when steel corrodes in the atmosphere contain salts such as ferrous sulphate ($FeSO_4$).[4-6] If these remain on the surface after cleaning off the rust, then they can affect the life of the coating subsequently applied. When manual methods such as wire-brushing and scraping are employed, a considerable amount of adherent rust remains on the steel surface and this includes a comparatively large amount of $FeSO_4$ (up to 5%). However, it might be assumed that after blast-cleaning all these salts would be removed. This, though, is not necessarily the case. If, before blast-cleaning, steel has been allowed to rust to an extent where it is pitted, then some salt formation in the small pits can be anticipated. It may prove to be very difficult to remove these salts by blast-cleaning and even if the steel surface meets the requirements of the various visual cleaning standards, e.g. ISO 8501-1 there may still be small quantities of salts present. These can sometimes be seen on the surface if viewed with a magnifying glass but cannot usually be detected by the naked eye.

If such salts are present then, because both oxygen and moisture are able to penetrate paint films, corrosion reactions can occur. Mayne[5] has suggested that salts such as $FeSO_4$ short-circuit the resistance of the paint film and then become oxidised and hydrolysed, with the production of voluminous rust which can disrupt the paint film. Other salts such as ferrous chloride ($FeCl_2$) can act in the same manner and are likely to be found in marine environments.

This mechanism of paint breakdown undoubtedly occurs when paints are applied to rusty surfaces. Generally, in these situations oleo-resinous and alkyd paints are used to a comparatively low film thickness (more resistant paints such as most epoxies would not usually be applied to rusted surfaces). The volume of the rust formed is sufficient to cause first blistering, then cracking of the paint film.

On blast-cleaned surfaces with such salt deposits the reaction may be stifled at the steel surface, depending on a number of factors such as the following:

(i)   concentration of salts present;
(ii) distribution of the salts on the surface, e.g. whether in small pits;
(iii) type of paint applied to the surface;
(iv) thickness of coating;
(v)   general standard of the coating, e.g. pore-free, no entrapped solvents, etc.;
(vi) adhesion of coating to the steel substrate.

A thick, well-applied, resistant coating with sound adhesion to the steel may still allow diffusion of oxygen and water in amounts capable of producing rust. However, the mechanical nature of the film will play an important role in determining the reaction at the surface. Rust has a greater volume than the steel from which it is produced and in any confined situation, such as a crevice or under a paint film, the course of the corrosion reaction will be determined by physical and mechanical restraints. The rust can only form if certain conditions are fulfilled. Either the rust as it forms exerts sufficient force to deform the paint or, once the rust has filled the voids present, it will act to stifle further reactions. On comparatively thin, reasonably elastic films, the force will be sufficient to deform and eventually disrupt the coating. On thicker well-adhering coatings, however, the force produced by rust formation may be insufficient to disrupt the protective film. The main problem with coating disruption from corrosion processes occurring on the steel surface is likely to arise when coatings are repainted. Both the quality of surface cleanliness and the standard of paint application are likely to be lower in such situations and there is a greater probability of coating breakdown leading to further corrosion.

A discussion of protection by paint coatings would not be complete without considering the ageing effects on the coating itself. The physical changes that occur in the paint film may lead to deterioration sufficient to allow the ingress of various corrosive ions and a reduction in adhesion. Although paint coatings are applied to suppress the corrosion of the steel, they are not themselves immune from attack in aggressive environments.

Clearly, the performance of the protective coating will be determined by its properties. The more important of these are considered in the following section.

4.4   Properties of paint films

Specifications and standards for paints and paint coatings are very much concerned with their properties and a range of tests has been devised to check them. Test methods are considered in Chapter 16.

4.4.1   Adhesion

Adhesion is of fundamental importance for any paint film. Good adhesion to the steel surface is essential but sound adhesion between different coats of paint is also important. Some of the many aspects to be considered have been discussed by McGill.[7] Adhesion is influenced by both physical and chemical factors as discussed in Chapter 3.

Blast-cleaning provides a roughened surface to the steel that helps to entrap paint in the surface irregularities. This improves adhesion, particularly with thicker coatings, but mechanical cleaning probably has an even greater influence on the secondary forms of chemical bonding of paints to steel. With increase in the overall surface area, more potential points for bonding arise. Furthermore, the removal of rust and scale provides a clean steel surface which improves bonding with the paint film.

The standard of surface cleanliness has to be higher with two-pack materials than with the oil-type of paint. Red-lead-in-oil paint was particularly good for applying to rusted steel because it was slow drying and penetrating enough to 'wet' the surface. However, the fact that such paints are always brush-applied is probably an important factor. Not only does this assist in the removal and dispersion of small particles of dust and dirt, it also leads to intimate contact between the paint and the surface being painted. Apart from the paint, the surface to be painted will influence the adhesion. All metals have some type of surface film. On steel this is an oxide but this only forms well on a very smooth surface. On blast-cleaned surfaces the oxide becomes broken up. On the other hand, the oxide films on aluminium and stainless steel are much more coherent and may cause difficulties with adhesion. Special etch primers are used on aluminium prior to painting and generally a slight roughening of stainless steel appears to improve adhesion. The adhesion of paints to hot-dip galvanised steel surfaces often leads to problems and this is discussed in Section 7.3.

Apart from oxide films, moisture on the surface to be painted may also lead to poor adhesion. The moisture film may not be visible and may be only a few nanometres in thickness. This may, however, be sufficient to reduce the close contact between the paint and the steel. Dirt and grease on the surface similarly influence adhesion. In practice, the effects of adhesion are not always clear. In the most extreme cases, paints flake off, but even when tests (see below) indicate comparatively poor adhesion the paint film may continue to provide sound protection. There is probably an interaction between adhesion and other properties, such as impact resistance. However, it seems reasonable to assume that high adhesion values are generally advantageous, especially in situations where there has been a local breakdown of the coating and the substrate has begun to corrode. The sideways attack on the substrate will probably be reduced if there is strong bonding between it and the paint.

Adhesion is clearly an important property but it is also a somewhat qualitative matter. It must be sufficient to ensure that there is no detachment of the film from the substrate, but equally there is no experimental evidence of a clear relationship between the degree of adhesion and the practical performance of paint films once the minimum required adhesion has been achieved.

A good deal of maintenance painting arises because of doubts about adhesion of paint films to steelwork rather than from actual breakdown of the coating. This can be an expensive operation with thicker coatings. When an area of coating breakdown is cleaned by blast-cleaning prior to repainting, it is commonly observed that a good deal of apparently sound paintwork can be easily detached from the substrate at the edges of the cleaned area. Consequently, a considerable area of paint may need to be removed to ensure a sound basis for repainting.

Adhesion is still not very well understood and this is reflected in the test methods used to measure it and the lack of precision in determining the significance of the results.[8] Furthermore, the influence of time and ageing of the film on its adhesive properties has not been clearly established. The test methods for determining adhesion are discussed in Chapter 16.

4.4.2 Flexibility

Flexibility, or the ability of a paint film to stretch without cracking, is a measure of the degree to which a dry paint film is able to withstand deformation of the surface to which it is attached. This property is related to the elasticity of the film, which is time dependent. Furthermore, it is affected by the level of adhesion between the paint and its substrate. Good flexibility is an essential property for pre-coated steel sheet which is to be formed into profiles where the elasticity of the films may be important. It is of less importance for paints applied to structural steelwork where the final complete paint system is applied after fabrication.

The tests used to determine the degree of flexibility of paint films are broadly concerned with either bending round a mandrel or measuring extensibility by deformation. These tests measure partly adhesion and partly the brittleness of the film and so, in the hands of experts, provide useful information regarding the overall properties of the film. Brittle films are more likely to crack when subjected to considerable temperature change and are more liable to be affected by abrasion or impact damage.

Flexibility can be introduced into polymers in various ways. In cross-linked films, any loosening of the structure by spacing the cross-links to provide a more open pattern will improve flexibility. The addition of plasticisers, i.e. smaller molecules, into linear polymers also improves flexibility by separating the large polymer chains and allowing movement between them.

4.4.3 Hardness

Although the hardness of the paint films can be measured by various tests, e.g. by indentation, scratching, etc., it is not a property that can easily be defined in relation to the durability or performance of the paint film. It is an important requirement for stoved finishes of domestic appliances such as refrigerators and for car finishes but may have less relevance for paints applied to structural steelwork.

Hardness can first be considered in relation to drying time, where it is an important property so far as the handling and storage of steelwork is concerned. Soft films tend to be easily damaged and deformed during handling, so a reasonably hard film is beneficial. One of the main disadvantages of red-lead-in-oil paints is the very long time required to produce a reasonably hard film. Generally, though, modern polymer coatings dry reasonably quickly to provide acceptably hard films, the hardness of which may increase on prolonged exposure to a point where they become brittle. Although hardness is often related to toughness and durability, considered to be advantageous properties, it may equally be considered in relation to brittleness, which may not be a desirable property.

4.4.4 Abrasion resistance

Resistance to abrasion and erosion, i.e. the wearing away of coats in a paint system by mechanical means, may be of particular importance in some situations. Obvious examples include environments where sand is blown about and industrial works where particles of coke and other substances are wind blown. Other situations include tanks in which abrasive materials are stored and ships' hulls. Abrasion resistance is related to other properties such as hardness and impact resistance. Generally, two-pack paints such as epoxies and hard coatings of the zinc silicate type are reasonably resistant to abrasion.

4.4.5 Permeability

The permeability of a paint film is the property concerned with its ability to absorb and release small molecules or ions. Although this is an important property it is difficult to relate it to practical experience. It seems reasonable to suppose that high degrees of permeability would be disadvantageous and that impermeable films would provide maximum protection. There is evidence to show that the pigment volume concentration (p.v.c.) has a bearing on permeability. Van Loo[9] showed the importance of p.v.c.; in his tests the permeability of the paint film rose markedly beyond what was termed the 'critical pigment volume concentration' (c.p.v.c.). This varies with different paints but may serve to indicate how well the pigment is packed with sufficient binder to maintain a coherent film. If there are unfilled parts of the film, i.e. insufficient binder, then there will be interstices or pores which may provide paths for water to enter the paint film. The rapid escape of volatiles during film formation may also leave pores, as may the release of solvents during drying of the film.

Other work by one of the authors has indicated that the method of paint application may also influence permeability.[10] There is a difference between voids and pores which can be seen, possibly with the aid of an optical microscope, and molecular paths available for ionic diffusion. What might be described as physical paths can be measured by means of various instruments and can be detrimental to coating performance, particularly under immersed conditions. However, the effects of diffusion paths at the molecular level are more difficult to quantify.

The ability of a paint film to absorb moisture or to allow it to diffuse through it does not appear to relate particularly well to its performance as a protective coating. For example, in work carried out by a number of investigators[1,2,3] it has been demonstrated that the amount of water that can diffuse through paint films is greater than the amount that would be consumed by an unpainted steel specimen during the corrosion process. Furthermore, Gay[11] found that under normal conditions paint films may be saturated with water for half their life. It follows that permeability and diffusion of moisture through paint films do not have the effect on their protective properties that might have been anticipated. It must, therefore, be concluded that tests for permeability are of limited value in protective paints. To some extent, this is illustrated by the results of humidity tests, which often cause intense blistering of paint films, although this degree of blistering is not encountered in practice.

Blistering of paint films can occur in two distinct ways: (i) by the local accumulation of corrosion products under the film, which eventually leads to disruption by cracking, and (ii) by the presence of water under the paint film at local sites. There is not complete agreement on the mechanism of water-formed blisters. Some workers[12] consider that they arise mainly from osmotic effects caused by the presence of soluble salts in the film, whereas others[13] consider that it can be explained purely on the basis of water penetration into the film, causing swelling. Generally, water blisters are less of a problem than rust-containing blisters with paints applied to structural steelwork.

4.4.6 Resistance to microorganisms

Attack by bacteria and other microorganisms can cause problems with protective coatings. Microbiological attack may be a problem and certain fungicides may be incorporated into paints. These include organo-tin and organo-mercurial compounds, organic copper compounds and chlorinated phenolics, all of which are toxic. A number of papers have been published on this topic.[14,15] Newer organic materials are available with less potential for severe environmental impact, with engineered polymers seeking to provide a longer term solution to the effects of detrimental microbiological effects.

4.4.7 Ageing of paint films

Paint films have a limited life and require to be maintained. The deterioration may be concerned with decorative requirements, e.g. loss of gloss or colour fading. This must be considered as an inevitable effect of ultraviolet light and general air pollution which does not necessarily seriously reduce the protective properties of the paint film. On the other hand, other forms of deterioration have a marked influence on the protective nature of the coating. These arise from changes in the paint film with time, i.e. ageing, or from corrosion of the substrate which disrupts the film. Although these properties are important it is almost impossible to predict them. Experience gained in using similar coating materials in similar circumstances is the usual way of assessing such properties. Various accelerated tests are used (Section 16.5) but generally they are not effective in determining the overall properties required to determine durability.[15]

This inability to predict performance except in the most general way is undoubtedly one of the main difficulties in both selecting and specifying paints for a particular situation. Most of the properties discussed in this section are capable of measurement or assessment by various test methods. However, these test results may not provide clear data for longer-term performance because of the ageing process.

With the improvement in techniques, apparatus and instruments, it is to be hoped that the ageing properties of films will be more fully investigated and understood, so that reasonable assessments of performance can be made, particularly with new products. Matters such as the changes in the visco-elastic nature of polymer films, their ability to withstand stresses, the effects of such changes on adhesion and the causes of physical disruption all require further investigation.

Paint films fail for a number of reasons and some can be avoided, but others appear to be a natural result of the ageing process. These limit the life of the paint film and often lead to problems in the repainting of the coating, which may have to be completely removed.

4.5   Paint film formation

The protective dry paint film is usually produced from a liquid paint consisting basically of binder, pigment and solvent. The solvent evaporates during drying and the binder is converted to a solid film. The conversion occurs in different ways according to the type of binder. Clearly, the balance of pigment, binder and solvent must be carefully controlled to ensure that the dry protective film achieves its main purpose – to protect the steel from corrosion. There may also be other requirements such as gloss and abrasion resistance. The paint formulator must balance the paint constituents to ensure that all the requirements are achieved. However, even the best liquid paint may not convert to a sound protective coating if the application is poor. It is essential that the paint manufacturers' recommendations are followed to ensure that the conversion from liquid paint occurs in a manner that will provide a good protective film.

4.6   Binders

Most of the organic binders commonly used for protective paints can be classified into one of three broad groups depending on the manner of film formation:

   (i)   oxidation;
  (ii)   solvent evaporation;
 (iii)   chemical reaction.

Inorganic silicate binders fall into a special group and there are also binders which depend upon reaction with atmospheric moisture for their film formation. These do not strictly fall into any of the three groups above, but will be considered under chemical reaction binders.

(i) Oxidation. Natural oils and binders combining oils and resins, i.e. oleoresinous, dry in this way. They combine with oxygen in the air and dry to a hard film.

(ii) Solvent evaporation. Essentially, resins are dissolved in suitable solvents which produces a liquid paint that can be applied to a surface. The

solvent then evaporates leaving a dry film. In practice a number of solvents may be used to ensure that films with suitable properties are produced. The paints that dry in this way include vinyls, chlorinated rubber, acrylated rubber and bitumens. The constrictions of the Environmental Protection Act and similar enforcing Amendment Regulations concerning VOCs will continue to limit the use of this group of paint raw materials.

(iii) Chemical reaction. Although oxidation is strictly a chemical reaction between the binder and atmospheric oxygen, this group covers binders where the reaction occurs within the binder.

Binders can be classified in other ways. A broad grouping covers those that change chemically during drying, called convertible, and those that do not react to produce changes, called non-convertible. These are usually resins dissolved in a solvent, which evaporates during drying, so producing a protective film, which can be re-dissolved with appropriate solvents.

Although paints are usually classified by the main binder(s), e.g. alkyd, the pigmentation may be an important element, so paints with the same binder may vary in their properties. Paints will be considered in Section 4.9; in this section binders will be discussed briefly to indicate their general characteristics.

4.6.1 Oxidation type

Oils such as linseed and tung dry to a solid film under ordinary atmospheric conditions. Other oils such as castor oil do not dry. The drying properties are related to the content and type of fatty acid. These oils, because they are able to form a reasonably coherent dry film, were among the first protective paints or varnishes to be used (a varnish may be considered as an unpigmented paint). The organic chemistry of such oils was not known and was not important to their use some centuries ago. For many years paints were based on these oils extracted from the seeds and fruit of various types of vegetable matter and they are still used. However, oils are now more generally used in *oleo-resinous* paints, in which the oil is mixed with other resins. The resins may react to provide larger molecules or, depending on the resin chosen, they may not actually react with the oil at all but are added to improve the nature of the binder. Phenolic resins are commonly used in oleo-resinous binders but others, such as coumorone-indene and rosin, are also employed. The chemical formulae will not be discussed here.

Whereas oils can be applied without the use of solvents, once resins are added solvents are required to reduce the viscosity of the paint. The addition of resins tends to harden the film and improve the gloss. Provided the resin does not contain ester groups, it adds to the water- and chemical-resistance of the paint and generally improves durability. The ratio of oil to resin weight is called the oil length: 3–5 parts of oil to 1 of resin is called a 'long oil' whereas a ratio of 0.5–1.5 of oil to 1 of resin is called a 'short oil'. The length of oil determines whether the paint will be similar to an oil paint, e.g. long oil, or will be markedly influenced by the resin additions. Many synthetic resins are produced for paint binders and these provide films with considerably improved properties compared with oleo-resinous paints. These resins are called polymers.

A polymer is a molecule composed of many smaller units linked by covalent bonds. The polymer may be built from similar or dissimilar molecules joined together to form a very large molecule. A *monomer* is the small molecule used as the starting point for the production of the polymer. The many molecules are joined together to form long chains and the process by which the monomer molecules react is called *polymerisation*. The most commonly used oleo-resinous binders used for constructional steelwork are considered below.

4.6.1.1   Tung phenolic

These contain up to 70% natural oils with phenolic additions and chemically are the simplest of the oleo-resinous binders.

4.6.1.2   Alkyds

Alkyds are condensation polymers of the reaction between dibasic acids, i.e. having two carboxyl groups, and dihydric alcohols, i.e. having two hydroxyl groups. In fact the name 'alkyd' derives from these two products alcohol and acid, although the spelling has been changed. In practice, alkyds are generally produced from oil by converting it to a monoglyceride by heating with glycerol and then reacting with phthalic anhydride (a derivative of phthalic acid). Other polyhydric acids such as pentaerythritol can be used. The organic formulae can be obtained from books on paint chemistry but are not important for this particular discussion. However, by varying the reactants a range of different resins can be produced.

Alkyds may be formulated with other binders to produce paints with special characteristics, e.g. silicone alkyds and urethane alkyds.

4.6.1.3   Epoxy esters

These are made by esterifying the epoxy and hydroxyl groups with oil fatty acids to produce a binder more comparable with alkyds than two-pack epoxides. A range of properties can be obtained by varying the amount of acid relative to the number of reactive groups in the resin. This can affect the drying time, gloss and chemical resistance. Epoxy esters are best compared with alkyds and should not be considered as epoxides in the usual way the term is used. The epoxy ester binders have some advantages over alkyds because, whereas the alkyd is composed of ester linkages, which are susceptible to attack by alkalis, epoxy esters, despite their name, have a proportion of ether linkages and so are more resistant to alkalis. The binder is used for priming and for undercoats for the protection of steelwork; they are considered to have improved chemical resistance compared with other oleo-resinous paints.

4.6.2 Solvent evaporation type

These are non-convertible binders.

4.6.2.1 Chlorinated rubber

Natural rubber has many properties, including chemical and water resistance, that would make it suitable as a coating for steel. However, it does not provide satisfactory film-forming properties and if vulcanised cannot be dissolved in solvents. Natural rubber will, though, react with chlorine to produce a solid which is called chlorinated rubber and is supplied as a white or cream coloured powder which contains about 65% chlorine.

This powder can be dissolved in a suitable blend of solvents to produce a paint. However, the films produced are brittle, so a plasticiser is added in amounts varying from about 20% to 50%. Suitable plasticisers include chlorinated paraffin waxes and chlorinated diphenyls, both of which have a chemical resistance of the same order as the chlorinated rubber. After suitable pigmentation, the paints are applied and dry by simple solvent evaporation to produce a highly resistant dry film. The original chlorination process used chlorinated hydrocarbons, the use of which was banned under the Montreal Protocol. Manufacture no longer takes place in the UK, but the material may still be obtained for the present time from Germany and Italy. Further restrictions on VOCs have tended to reduce the importance and use of this class of paint raw material.

Chlorinated rubber is soluble in esters, aromatic hydrocarbons, chlorinated hydrocarbons and some ketones. The blend of solvents is determined by the required application method. Often other additions are made and, while these paints are still called chlorinated rubber, their properties are not necessarily the same as the true product. For example, to improve gloss and to allow for brush application, additions of oils and alkyds may be made. Alkyds improve adhesion and have been used to provide what has been termed a 'travel coat' to allow for more rapid handling and storage of chlorinated rubber-coated steelwork.

4.6.2.2   Vinyls

Polyvinyl chloride (PVC) powders are used for plastic coatings, e.g. as a powder in a fluidised bed and as plastisols. However, simple dispersion or dissolution in solvents is not a practicable way to produce air-drying paints at ambient temperatures. To provide a paint it is necessary to polymerise vinyl chloride with about 5–20% of vinyl acetate. A range of products can be produced with different viscosities and solvent characteristics. Little if any plasticiser is required in these binders. Copolymers of vinyl chloride and vinyl acrylate are also blended to provide binders. As with chlorinated rubber, vinyl paints solidify to the protective film by simple evaporation of the solvent but suffer the same volatile organic compounds (VOC) restrictions.

4.6.2.3   Acrylated rubber

This binder is based on styrene butadiene and is produced in several grades, some of which are based on white spirit. Paints based on this binder had been preferred to chlorinated rubber by some, due to a lower requirement on aromaticity of solvent employed. Some acrylated rubber coatings on structural steelwork have proved satisfactory, but it is on masonry and concrete where the majority of this type of paint has been used. With correct formulation, coatings capable of allowing the passage of water molecules, whilst preventing the ingress of the larger carbon dioxide molecule, are possible, offering potential advantage to the protection of reinforced concrete structures. Again, it is presently not possible to provide coatings based on acrylated rubber which are compliant with existing VOC requirements. This restriction due to the environmental need to reduce solvent emissions has caused reduced adoption of this class of material.

4.6.2.4   Bitumen

Bituminous coatings are still used to protect steelwork. As film formers they fall into two distinct groups: naturally occurring, e.g. gilsonite, which usually contain mineral matter and may be called asphalts (particularly in North America); and those distilled from petroleum and coal products. Pitches are produced from the distillation of tars and fall within the same group.

Generally, the materials are used in a suitable solvent for ease of cold application or are hot-applied. They may be 'blown', i.e. oxidised by blowing air through molten bitumen to form materials that provide tougher films. Although black in colour, the use of extenders to reinforce and improve the dry film properties in solvent-based bitumens leads to surfaces exposed to ultraviolet light becoming grey in time. Thermal re-flow and age embrittlement may also prove problematic, hence for these reasons, this inexpensive class of material is mainly confined to below ground or underwater duties.

4.6.3   Chemically reacting type

These binders dry or 'cure' by chemical reaction between different components which are mixed before the paint is applied.

4.6.3.1   Epoxy

Epoxy resins are usually made by reacting di- or polyhydric phenols with epihalohydrins. The phenol used commercially is one called bisphenol-A and the epihalohydrin used is generally epichlorhydrin. The reactions are complex and will not be considered here. However, the properties of epoxides arise from their structure.

(i)   There are no ester groups, so the resin does not saponify.
(ii)   The chains have many polar groups, leading to good adhesion.
(iii) The large number of aromatic rings contributes to their hardness.

The structure also leads to disadvantages because epoxides are not soluble in the cheaper types of solvent and they are not compatible with many other binder materials.

Epoxy resins are produced with a range of melting points and molecular weights, and include those that are liquid at ambient temperatures. These serve as the basis for solvent-free application. When used at ambient temperatures, epoxide resins are cured by the addition of amines or polyamides. The curing agent is stored separately from the epoxide and mixed just before application of the paint in suitable proportions. After mixing there is a limited period during which it can be used, i.e. before the paint hardens in the can. This is called the 'pot life' and varies, usually being between 2 and 8 hours. Instead of using the amine, an amine-adduct may be preferred. The amine and epoxide are partially reacted before-hand and this provides some advantages. The amounts of adduct and epoxide are often similar so the mixing is simpler and more accurate; furthermore, the adduct is less of a skin irritant than the amine and generally the films produced have improved properties. Polyamides are also used as curing agents and have some advantages over amines. They are cheaper and the mixing ratios are less critical than with amines. The films produced, however, tend to exhibit less resistance to chemical attack and to solvents. All liquid-applied epoxies suffer to some extent from 'chalking', hence under external exposure conditions, colour stability is not good.

Isocyanates may be used to cure coal-tar epoxies. These are modified epoxides containing 50–70% of coal-tar pitches. This serves to produce a cheaper coating with many of the advantages of unmodified epoxies with better water-resistance. They are inevitably black or dark brown in colour, so are not suitable where appearance matters. Coal-tar epoxies are widely used for immersed conditions, particularly in seawater.

4.6.3.2   Polyurethanes

Polyurethanes are produced by reacting organic isocyanates with the hydroxyl groups of polyols, usually polyesters. There are a number of different products within the overall term 'polyurethane' and this can lead to some confusion.

There is a range of two-pack materials in which the two components are mixed in the correct proportions before use and are somewhat similar to the two-pack epoxies in their properties, producing tough abrasion-resistant coatings. Variations in the formulation provide a range of finishes from high gloss to semi-matt in a wide choice of colours. The isocyanate component of the two-pack polyurethane may be either aliphatic or aromatic. The aliphatic polymers are generally based on hexamethylene diisocyanate with cycloaliphatic polymers based on isophorone diisocyanate, both imparting good gloss and colour retention, thus differentiating them from the epoxies. Aromatic isocyanates are considerably less expensive and are often based on toluene diisocyanate, but are not colour stable. Two-pack urethane pitches and coal-tar urethanes are also produced and are used for similar purposes to the coal-tar epoxies. Generally, the type of curing agent influences the physical rather than the chemical-resistant properties.

Single-pack materials are also produced in which free isocyanate groups react with moisture in the air to provide a cross-linked polymer structure. These are often termed 'moisture-curing polyurethanes'. Urethane alkyds, sometimes called urethane oils, are similar to alkyds and do not fall within the category of chemical curing binders.

4.6.3.3   Other binders

(a) Silicates. All the binders considered so far have been organic whereas silicates are inorganic. While paints based on silicates are used for painting plaster and similar surfaces, so far as the protection of steelwork is concerned, the only paint to be considered is zinc silicate.

The chemistry of zinc silicate coatings is complex and a number of papers have been written concerning the mechanisms involved.[16,17] In simple terms, the binders consist of zinc dust mixed in a silicate solution which, on curing, cements the zinc particles into a hard dry film. The silicate solution may be produced from a range of silicate materials including sodium, potassium and lithium silicates, with the addition of suitable solvents, either water- or organic-based. The solvent evaporates after application and reactions with moisture and carbon dioxide in the air occur.

As with zinc metal, e.g. hot-dip galvanised coatings, some carbonic acid is formed, which reacts with zinc to form zinc carbonate, and the weakly acidic moisture assists in hydrolysing the organic silicate to silicic acid. The zinc then reacts with the silicic acid to form a somewhat porous coating which, as reactions continue, forms into a cement-like matrix around the zinc particles. The initial pores are filled with zinc hydroxides and carbonates to produce a fairly compact hard coating. In practice, the silicate alkalinity is neutralised either by reaction with the carbonic acid formed from the atmosphere, as already noted, or by using post-curing solutions.

In the process considered above, alkali silicates were the basis of the silicate solution and these are, of course, inorganic. However, similar reactions are possible using alcohol solutions of alkyl silicate esters, e.g. ethyl or isopropyl silicates in organic solvents. Sometimes the two types of coating are referred to as inorganic and organic silicates, but the final film is inorganic because the organic silicate hydrolyses and the alcohols evaporate, leaving the silicic acid to form a cement matrix, as with inorganic solutions.

The above coatings are formed from two-pack materials but single-pack zinc silicates are also produced. These involve various routes including ester exchange reactions with various types of silicates, e.g. alkyl, to produce similar reactions to those already discussed.[18] However, such films are not completely inorganic and generally do not possess the equivalent properties of the two-pack materials.

(b) Silicones. Silicone resins have high-temperature resistance properties and are mainly used in paints for that purpose. With suitable pigmentation, usually aluminium, such paints will withstand temperatures up to about 250°C or above when over inorganic zinc-rich primers or metal sprayed coatings.

Some of the advantages of silicone, which is an expensive resin, can be obtained by modifying it with other binders, particularly alkyd to produce silicone alkyd binders. Provided they contain over 30% silicone, they produce higher gloss retention and improved durability, particularly in resisting intense sunlight, compared with alkyd binders.

4.7 Pigments

Although the binder determines the main characteristics of a paint, pigments also influence durability and other properties. The chemistry of pigments and their properties in relation to the formulation and manufacture of paints is beyond the scope of this book. Apart from certain pigments for primers and micaceous iron oxide in finishing coats, paints are generally categorised and specified by binder. However, so far as the production of suitable and satisfactory paints is concerned, pigments are important. Typical properties include the provision of colour, improvement of the strength of the film, reduction of film permeability and improvements in adhesion. However, there are two prime requirements of pigments. They must be insoluble in the other constituents of the paint and they must not chemically react with them. Furthermore, there must be sufficient pigment in the paint to achieve the required properties. The mere presence of, for example, red lead in a priming coat or micaceous iron oxide in a finishing coat is not sufficient; it must be present in the correct quantity to meet the particular requirements. The relative amount of pigment and binder in the paint is important. Although this relationship was at one time expressed in terms of weight of pigment/volume of paint, it is now recognised that the volume concentration is more important and this is expressed as the pigment volume concentration, or p.v.c.:

$$\text{p.v.c.} = \frac{\text{volume of pigment} \times 100}{\text{volume of pigment} + \text{volume of binder}}$$

A critical p.v.c. (c.p.v.c.) which provides the particular properties for a specific product is an essential element in paint formulation. Many of the pigments used are basically to provide colour, whereas others serve other purposes concerned with the protective qualities of the paint film. The latter class will be considered below.

4.7.1   Inhibitive pigments

Some pigments used in priming coats have inhibitive properties. At one time a range of such pigments was available, but most have some toxic properties and are tending to be barred by many authorities. However, many are still used so will be considered below.

Red lead. Originally the best primer for rusted steel when used in sufficient quantities in an oil-based binder. It was slow drying, but the severe toxicology problems of red lead has restricted its use in present times to a few rare marine applications.

Zinc chromate. The partial solubility in water of zinc chromate was found to provide a good inhibitive effect and was successfully used for many years, the 'zinc chromate red oxide primer' being one of the favourite construction primers. The degree of solubility could be adjusted by blending with zinc tetroxy chromate or strontium chromate and these were widely used in wash or etch primers for non-ferrous metals. Due to these materials now being known carcinogens to man, their use is now scarce, being largely replaced in etch primers by molybdates or similar low toxicity substances.

Metallic lead. This had been used in paints as it had superior drying properties to red lead. Due to toxicity of the metallic lead in fine powder form, manufacture was potentially hazardous, as would be the ultimate removal for future maintenance painting. It has seldom been employed for several years, due to toxicity.

Calcium plumbate. Once used as the main primer in oleo-resinous binders for galvanised window frames, this reactive product was difficult to incorporate into paints and has the same toxicity problems as the lead-based materials detailed above, hence is not now used.

Zinc dust. Historically used with zinc oxide – itself a popular white pigment once, but no longer used in paint – as a primer for iron and steel. Used now almost exclusively for organic and inorganic zinc-rich primers to freshly blasted steel.

Zinc phosphate is considered as an inhibitive pigment and because of problems of toxicity with many other pigments is now the most widely used. Its effectiveness may be influenced by the type of binder used.

4.7.2   Other pigments

Red oxide, or iron oxide, is used in priming paints but only for cheapness as it has no inhibitive properties; because of its colour it may mask rust staining. It strongly absorbs ultraviolet radiation.

Micaceous iron oxide (MIO) is widely used, particularly in Europe, for undercoats and finishing coats. The lamellar form of this pigment tends to form parallel to the surface with overlapping of individual flakes; it is claimed that this provides increased durability to paint films by reducing their permeability and increasing their tensile strength.

The pigment is found naturally in various places around the world and although chemically similar there are significant differences in the pigment shape. From many of the sources the pigments are not lamellar and, therefore, impart different properties to the final paint film (see Figure 4.1). The Austrian source claims to produce the most lamellar form. The pigment is grey-black, so only dark colours are possible in the paint film. Modifications to provide other colours, e.g. green, render the pigment less effective as regards durability. Paints containing MIO tend to weather to produce a metallic lustre and are not considered as decorative coatings.

It has been found technically and commercially possible to produce synthetic MIO, possessing the desired lamella structure and without the

Figure 4.1   SEM photomicrograph of lamella and granular micaceous iron oxide.

possibility of naturally occurring contaminants. The method uses scrap iron as the main raw material. In a single-stage batch process lamellar crystals of very pure iron oxide are formed at elevated temperatures by the oxidation of molten iron complexes. By adjusting reaction parameters, the crystal size and aspect ratio of the particles can be varied. It is possible to produce flakes so thin that they are semi-transparent with a pink pearlescent appearance, or the particle dimensions can be matched closely to the best natural MIO.[19]

Outdoor exposure tests have shown an improved performance when compared with the same paints formulated with the natural material. This is presumably due to the fact that the synthetic MIO is free from such contaminants as sulphur compounds which may be present in the natural material.

Aluminium pigment is produced in two forms: leafing and non-leafing. The leafing property depends upon a thin layer of stearic acid on the pigment surface and this causes the flakes to lie close to the surface of the paint, and by overlapping give an almost continuous bright metallic finish. The pigment also improves the performance of paints under atmospheric conditions, particularly as regards the effects of sunlight. The non-leafing pigment is widely used to lighten the colour of paints made from dark pigments such as micaceous iron oxide. However, aluminium is amphoteric and so is attacked by both acid and alkali and, therefore, paints incorporating either type are not resistant to chemical attack.

Other lamellar pigments such as glass flake and stainless steel are used in protective paints. *Glass flake* has been used in unsaturated polyester, vinyl ester and epoxy tank coatings for some time. High aspect ratio glass flake is used for very heavy duty linings, where application is effected by trowel, due to the size of the flake. Spray grades are available for less chemically onerous lining duties. The good abrasion characteristics imparted by glass flake has found use in marine duties, i.e. for the hulls of ice-breakers, with use of similar formulations gaining acceptance on off-shore rigs and for bridge structures. The material initially caused problems with brush application – for stripe coating welds and edges only – but recent developments have eased this difficulty. Stainless steel pigments do not provide the paint with any of the qualities of the alloy and act purely as barrier pigments. Graphite is also used for heat-resistant paint. White pigments such as titanium dioxide (rutile) are inert and are widely employed in undercoats and finishing coats. Coloured pigments are widely used and these may be split into two categories, inorganic and organic. The inorganic pigments are mainly based on oxides of metals, green chromium oxide being a good example, or stable complexes such as nickel titanium yellow. The large, previously popular range of lead chromates and lead molybdates had excellent opacity, brightness of hue and colour stability. However, North American paint manufacturers first rejected this widely used range on the grounds of their potential toxic nature. The materials still retain some use in Europe, where low-soluble lead types had been commonplace, but the risk and safety phrases necessary on the can label are severe and use in heavy-duty protective paints is diminishing.

There are a vast number of organic pigments covering many chemical classes. Interestingly, it is unusual to find an organic pigment with the same opacity as an inorganic pigment, but they are generally brighter. Certain bright pigments have insufficient light fastness characteristics to be used in paint, i.e. Lake Red C, whilst others, i.e. benzidine yellows, have fallen foul of increased safety awareness. This still leaves a huge range, the most popular of which are azo yellows and reds, phthalocyanine greens and blues and a host of expensive vat dyestuffs covering every colour of the rainbow.

4.7.3 Extenders

These are not strictly pigments but are used to modify the paint properties and/or as a form of cheap pigment. They will not be discussed here but include materials such as barytes ($BaSO_4$), blanc fixe ($BaSO_4$), talc and china clay.

4.8   Solvents

Solvents are not a useful constituent of the dry protective film and so may be considered as adding unnecessarily to the cost of paints. Certainly a 'high solids' content is desirable but solvents are essential for the sound application of most paint coatings. A careful balance between viscosity and volatility is required so that paints can be satisfactorily applied by the required method but at the same time avoid settlement of pigment in the can during application and ensure that the paint, once it reaches the steel surface, quickly achieves stability. For high-performance paint systems a blend of solvents is normally used to give the range of properties required. The blend has to be suitable for the appropriate ambient conditions, and formulations for temperate climates may not be suitable for hotter or colder climates.

The rate of evaporation of solvent from the applied paint film during the drying period may affect the properties of the final coating. Although solvents can evaporate readily in the early stages of film drying, at later stages the rate is controlled by the ability of the solvent to diffuse through the modules in the binder. If solvent evaporates too slowly it may be trapped in the film causing soft films; on the other hand, too rapid evaporation can lead to defects in the final dry film. Some solvent is necessary for ease of processing during the manufacture of most paints. Historically, different solvent blends were used, depending upon whether the paint was to be applied by brush or by spray. Brush coating required a slower blend of solvents, to allow a good 'wet edge time' so that joins would not show. Spray coatings generally required a quicker solvent blend, particularly for shop work, to enable faster handling.

Paint chemists of the twenty-first century are expected to devise a single paint capable of application by brush, roller, conventional and airless spray methods. Generally, only sufficient solvent to ensure satisfactory application should be added to the paint. Only the thinner recommended by the paint manufacturer for that particular paint, and no more than the maximum limit advised, should be employed. It is not often appreciated by contractors that use of wrong thinner or 'gun wash' as a thinner may prejudice the performance of the paint. This is not simply because of possible incompatibility. A thinner intended for an epoxy paint, for instance, added to a polyurethane, may react faster with the urethane hardener component than with the polyurethane resin, causing permanent softness and micropinholes.

Some of the solvents commonly used are listed below.

Aliphatic hydrocarbons, such as white spirit or 'turpentine substitute' are commonly used for oleo-resinous paints and have a relatively high flash point. Even white spirit is not fully aliphatic, containing approximately

14% of an aromatic fraction. White spirit may be used for some alkyd top coats, but this class of solvent is not frequently used in heavy-duty protective paints.

Aromatic hydrocarbons, such as toluene, xylene and trimethyl benzene are stronger solvents, with a faster evaporation rate, lower flash points and a generally more significant odour. Many heavy-duty protective paints contain a portion of such solvents.

Ketones are important solvents and include butan-1-one (colloquially known as methyl ethyl ketone, MEK) and 4-methylpentan-2-one (methyl isobutyl ketone, MiBK). They tend to have characteristic odours with low flash points, posing a potential fire risk.

Esters, such as butyl acetate have a characteristic 'pear drops' odour, are expensive and are not always anhydrous, hence limiting their use in paints for structural steelwork. Higher esters and ethers are, however, employed as a low-volatile solvent, occasionally providing de-gassing properties and enhanced flow characteristics.

Alcohols, such as ethanol, n-butanol or isobutanol are employed more in light industrial paints than in heavy-duty paints. Although they are found in epoxide formulations, they are never used in polyurethanes due to reaction with the curative component.

4.9   Paint classification (see Figure 4.2)

Paint properties are determined mainly by the type of binder. As discussed in Section 4.6, there are three main groups of organic binder based on the method of film formation.

It is convenient to consider the paints in groups based on the type of binder because many of the properties of paints within the group will be similar.

4.9.1   Oil-based and oleo-resinous paints (oxidation drying)

All the paints in this group contain drying oils and this determines many of the properties.

Oil paints. Oil paints, i.e. those without combined resins, are rarely used nowadays. As a group they had advantages, particularly when most steel was merely wire-brushed and painted on-site. However, with modern application methods these are outweighed by the disadvantages. The best

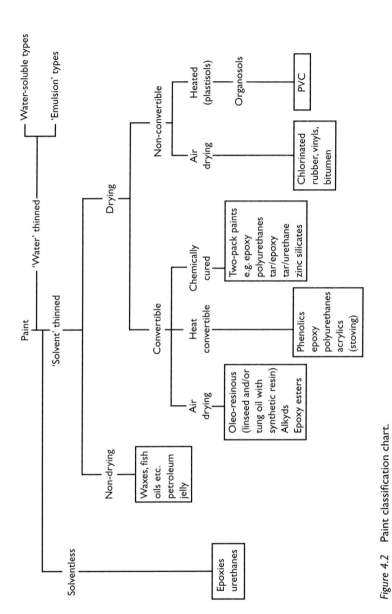

Figure 4.2 Paint classification chart.

known paint in this group is red lead in oil (BS 2523, Type B), which was, at one time, widely used as an effective primer for rusty steelwork.

ADVANTAGES

(i)   Particularly effective as primers for rusty steelwork provided they contain an inhibitive primer.
(ii) Easily overcoated.
(iii) Easily applied, particularly by brush.
(iv) Good wetting properties.
(v)   Relatively cheap.

DISADVANTAGES

(i)   Very slow drying so not really suitable for painting at works.
(ii) Poor resistance to alkalis and acids, and not suitable for cathodic protection.
(iii) Not recommended for immersed conditions.
(iv) Embrittle with age.

Oleo-resinous paints. The combination of oil with synthetic resins improves many of the properties compared with oil paints, but they still suffer many of their disadvantages. Three paints of this class are:

alkyd;
tung oil phenolic;
epoxy ester.

Alkyds tend to be the main product used as a one-pack maintenance paint and these are generally modified to produce silicone alkyds, urethane alkyds or acrylic alkyds. Styrenated alkyds are rarely encountered these days for maintenance painting of steelwork structures.

ADVANTAGES

(i)   Ease of application.
(ii) Many paints in the group provide a range of colours and good gloss.
(iii) Reasonable durability in many atmospheric situations.
(iv) Usually can be overcoated without difficulty.
(v)   Cheaper solvents such as white spirit can be used.
(vi) Generally, reasonably cheap.

DISADVANTAGES

(i)   Because of their oil content they saponify under alkali conditions so cannot be used, for example, where steel is cathodically protected.

(ii)   Not recommended for immersed or very damp conditions.
(iii) Limited film thickness (25–50 μm per coat).
(iv)   Poor resistance to solvents.

Alkyds. For steelwork the long oil type of alkyd is used. Their main advantages are cost and good appearance. Available in a range of colours, they maintain their gloss well and are particularly suitable as decorative coatings. They provide reasonable protection in less aggressive atmospheric situations, but are not particularly resistant to continuously damp conditions and are not recommended for immersion.

Urethane alkyds provide a harder tougher film.

Silicone alkyd paints are a good deal more expensive than ordinary alkyds but more durable, particularly in strong sunlight, and retain their gloss well. They were once used on highway bridges in the UK, but have been superseded by epoxy/polyurethane systems, the latter possessing the necessary superior life expectancy and lower VOC content.

The paints noted above are those commonly used, but it is possible to make an almost unlimited number of modifications to produce different types of binder, e.g. alkyds can be modified with phenolic, vinyl, chlorinated rubber, styrene and acrylic. Generally, the paints tend to provide properties somewhere between those of the component parts. Sometimes they are produced for specific purposes, e.g. chlorinated rubber-alkyd paints have been used as primers for a chlorinated rubber-based system or as 'travel coats' to avoid damage during handling of chlorinated rubber systems, which tend to remain soft for some time after application. Again, this alkyd modification is more easily brushed; often the modified paint dries more quickly, e.g. styrenated alkyds. Recent work on water-based or emulsified alkyd resin binders is showing potential, but these development products are not yet suited to the climatic conditions of northern Europe.

Epoxy ester. These have the same order of resistance as alkyds, although having rather better water resistance. They are considered to have some advantages over alkyds, particularly under adverse drying conditions. Some authorities claim that an advantage of these paints is their quick-drying properties, which makes them eminently suitable for primers. However, this may depend on the nature of the formulation.

Tung phenolics. These have better chemical resistance than either alkyd or oil paints and dry faster. They tend to be rather dark in appearance and become darker on exposure. They produce hard paint films so there is a limit on the overcoating time if intercoat adhesion problems are to be avoided. They can be used for some immersed conditions and are more resistant than most other oleo-resinous paints to humidity; their gloss retention is poorer than that of alkyd paints.

4.9.2   Solvent evaporation paints

The two paints in this group most commonly used for structural steelwork are chlorinated rubber and vinyl. Bituminous coatings fall within this group but they are not comparable in performance with the other paints in the group.

These paints are one-pack materials with a number of advantages over both oleo-resinous and chemically reacting paints. They are sometimes called one-pack chemical-resistant paints.

ADVANTAGES

  (i)   Provide protection in aggressive environments where oleo-resinous paints do not perform particularly well.
 (ii)   Suitable for immersion and damp conditions.
(iii)   Easily re-coated.
 (iv)   Good resistance to many acids and alkalis; suitable for cathodically protected steelwork.
  (v)   Available in a range of colours, but not high gloss.
 (vi)   One-pack materials. This is an advantage compared with two-pack chemical reaction coatings often used as alternatives to these coatings.
(vii)   Although high quality cleaning of the steel surface is always advisable, paints in this group are likely to perform better than two-pack materials on poorly prepared surfaces.

DISADVANTAGES

  (i)   A limit on temperature resistance; generally a maximum at 60°C for continuous heat.
 (ii)   Dark colours not always suitable for high temperatures overseas.
(iii)   Not resistant to most solvents.
 (iv)   High solvent content of paints (typically 60–75%) may cause problems with solvent entrapment or removal of solvent in enclosed spaces.
  (v)   Unable to be formulated to VOC compliance.

4.9.2.1   Chlorinated-rubber paints

Chlorinated-rubber powders are available in a range of viscosities. The higher-viscosity powders are generally considered to provide better film properties but the lower-viscosity ones are used for thicker films, and there are advantages in using paints produced from the two viscosity grades in a chlorinated-rubber paint system. The lower viscosity powders can be used for undercoats to provide sufficient film thickness, with the higher-viscosity materials being used to provide a durable finishing coat.

The paints have excellent chemical resistance and low water permeability. Although they require high-quality surface preparation for exposure to aggressive environments, because of their solvent nature they are somewhat more tolerant to inadequate surface preparation than are the two-pack materials. These paints are more effectively applied by spray than brush, although brushing grades are normally available; these are generally modifications and require special brushing techniques. Ideally, brush application should be confined to small areas.

Suitable non-saponifiable plasticisers are required and, provided the correct quantities are incorporated into the paint, adhesion to the steel is good and a high level of durability can be attained. The addition of too much plasticiser can result in softening at high ambient temperatures. Solvent retained in the film can act as a plasticiser with the same results as adding too much; on the other hand, insufficient plasticiser can cause films to become brittle on ageing. Some plasticisers tend to slowly evaporate so that during exposure films may become brittle. This is particularly so for elevated temperatures and no chlorinated-rubber system should be used on surfaces continuously above 60°C. In hot climates which may not appear to reach such temperatures, dark colours in particular have been known to fail rapidly, sometimes even with the evolution of hydrochloric acid as a degradation product.

Problems can arise when these paints are sprayed; if the viscosity is not right then what is termed 'cobwebbing' may occur. The term is self-descriptive, indicating the difficulty of applying a sound paint film. Solvent entrapment may also take place leading to what is called 'vacuoles' in the dry paint film, i.e. small pores or holes.

Although chlorinated rubber dries to the touch fairly quickly, retained solvent and the comparatively soft nature of the films in the early stages after application can lead to problems with handling and stacking. The soft films tend to stick together, which seriously damages the coating. This can be overcome by ensuring that the paint is fully dried, and by handling and stacking sections with some care.

Maintenance is comparatively straightforward because the solvent in the paint allows for excellent adhesion between coatings. Chlorinated-rubber paints are not particularly resistant to many solvents or oils.

Care must be taken when welding steelwork that has been coated with chlorinated rubber because, if the temperature reaches a certain level, acid decomposition products can cause attack on the steel at some distance from the actual weld.

Chlorinated rubber paints are still used for minor maintenance and for certain marine work, but are becoming obsolete due to improvements in epoxy and polyurethane technology and the strengthening grip of the Environment Protection Act and similar regulations.

4.9.2.2   Acrylated-rubber paints

These paints are somewhat similar to chlorinated-rubber paints but they are claimed to have improved properties and do not, for example, suffer from 'vacuoles'. As with chlorinated rubbers, it has not been found possible to provide acrylated-rubber paints which comply with the VOC restrictions, hence their use is limited.

4.9.2.3   Vinyl paints

These have something in common with polyvinyl chloride (PVC). These paints have similarities to chlorinated rubber and also dry by solvent evaporation. They are superior to chlorinated rubber in their ability to resist solvents but are still attacked by many of the stronger ones, including esters, ketones and aromatic hydrocarbons such as xylene. They are less tolerant of inadequate surface preparation, especially surface moisture, than chlorinated rubbers but are more tolerant than the two-pack materials. When the paints are blended in appropriate proportions, little, if any, plasticiser is required and this can be advantageous because plasticisers can cause problems (cf. chlorinated rubber).

Generally, vinyl systems have a higher gloss than chlorinated rubber systems. However, they are more sensitive to surface moisture which can affect their adhesion to steel. To improve adhesion a maleic-modified primer is commonly used. The pigments used for vinyl paints are similar to those for chlorinated rubber.

Provided attention is paid to application (which may require somewhat more control than with chlorinated-rubber paints), vinyls are usually more durable. Generally, they are used for similar applications as chlorinated rubber and are often preferred to them in North America.

For seawater immersion, some authorities consider that a wash primer should be applied before application of the full vinyl system. Wash primers, such as those based on polyvinyl butyral resins combined with phosphoric acid and some phenolic to improve moisture resistance, fall into the category of vinyl coatings. They are also called etch primers and pre-treatment primers. Apart from their use with paint systems, they are also used to prepare a galvanised surface for painting. However, they are susceptible to the presence of moisture during drying. Vinyls can be modified with other binders such as alkyds and such paints are reasonably durable in all but the most aggressive environments and can be applied by brush. Furthermore, they are not so sensitive to surface conditions during application. With the exception of wash or etch primers, which currently have a generous VOC limit of 720 grammes per litre, vinyls suffer from similar problems to the rest of the group regarding solvent content.

Vinyl tars. Combinations of selected coal tars and vinyl resin can be used in many of the situations where coal-tar epoxies are used (Section 4.9.3.2) but are cheaper and easier to apply. However, they cannot be applied to the same thickness as epoxies so a multi-coat system is required. They are generally not so durable as the epoxies.

4.9.2.4 Bituminous coatings

These can be considered as comparatively cheap one-pack chemical-resistant paints, although they can only be used in a limited number of situations. They should not be confused with the hot-applied coatings used for buried and submerged pipelines, which are usually very thick and are not normally considered as paints. There are, however, other thinner coatings that are used in the same way as paints and are sometimes called black paints (tar-based) as in BS 1070. They are solvent based and, after application, the solvent evaporates leaving a bituminous coating.

They are quite useful as coatings for immersed conditions, although coal-tar epoxies provide a greater durability. Under atmospheric exposure conditions, bituminous coatings are liable to crack and sag in hot weather. They also exhibit a form of coating defect called 'alligatoring' where the bituminous coating contracts from the steel surface. Bituminous paints are not usually pigmented and so are black or dark brown, but the addition of aluminium flake improves both the durability and appearance under atmospheric exposure conditions. In tests, aluminium-pigmented bituminous coatings have performed well and it is somewhat surprising that they have not been used to a greater extent for industrial situations. One possible reason is that many paint manufacturers are reluctant to manufacture such materials because of the possibility of contaminating other paints being made at the same time. They therefore tend to buy-in items from the specialist manufacturer.

4.9.3 Chemical reaction paints

These are two-pack materials based on epoxies and urethanes. They have now become the most commonly used materials for new and maintenance painting of steel structures. They are the most durable of all site-applied coatings but application requires careful control if the full potential of such coatings is to be attained.

ADVANTAGES

(i) Durable coatings, resistant to seawater, many chemicals, solvents and oils.
(ii) Can be applied to provide a much thicker coating than is possible with other types of paint.

(iii) Do not saponify under alkaline conditions.
(iv) Most now have modified formulas allowing infinite overcoatability, have pot lives which last a full shift and cure down to +5°C or below for epoxies, or minus 10°C for polyurethanes.
(v) May be formulated to comply with current European and North American solvent content (VOC) requirements.

DISADVANTAGES

(i) More expensive than other paints.
(ii) The majority are two-component materials, requiring more control over mixing and application.
(iii) A high standard of surface preparation is required.
(iv) Maintenance painting is more difficult than with most other paints. Generally, old surfaces need to be abraded.
(v) After mixing, there is limited time during which paint can be applied.
(vi) Unmodified epoxies still suffer from requiring a substrate temperature in double figures to ensure cure.

Despite these disadvantages, the paints in this group prove to be the most suitable coatings for very aggressive situations. However, these coatings will only realise their full potential in situations where surface preparation, application, etc. are of a high standard. If these standards cannot be met, then alternative coatings should be considered. Because two components are mixed to produce these materials, the term 'pot life' is used to indicate the length of time available before the paint is no longer suitable for application. The component added to polymerise the paint is sometimes called a catalyst, although this is not a correct description. More suitable terms are curing agent, hardener or additive.

4.9.3.1 Epoxies

These are two-component materials and require mixing prior to application. The reaction to produce the dry film is temperature and time dependent.

Epoxies are available in a range of materials from liquids to solids, depending on their molecular weight. The chemistry involved in polymerisation is fairly complex but basically the hardening agent reacts with groups in the resin to produce a film suitable for its purpose as a protective coating. As various curing agents can be used, so a range of different epoxy coatings can be produced.

For coatings to protect steelwork, two main types of curing agents are used: polyamides and polyamines. The former are more commonly used and their properties and uses can be summarised as follows:

Polyamide curing

(i)   Generally, the two components are of similar volume, making mixing less critical.
(ii) Slow curing at low temperatures; does not cure below 5°C.
(iii) Longer pot life than with polyamines.
(iv) More flexible and durable than polyamine cured epoxies; take longer to cure.
(v)   Excellent alkali resistance but poorer acid resistance than polyamines.
(vi) Less tendency to 'bloom'.

Polyamine-cured epoxies are generally used for specialist purposes such as tank linings. Their overall properties are summarised below:

Polyamine curing

(i)   Excellent chemical- and corrosion-resistant properties.
(ii)   Good solvent resistance; better than polyamides.
(iii) Excellent alkali resistance.
(iv)   Good water resistance.
(v)   'Chalk' badly on atmospheric exposure.
(vi)   Can form amine 'bloom' when coatings applied at high humidities; if this is not removed, poor intercoat adhesion may occur.
(vii) Amines may cause skin irritation during handling.
(viii) The mixing ratios are more critical than with polyamides.

Amine-adduct curing. The amine is partially reacted with the epoxy beforehand. This provides some advantages because the mixing ratio is simpler so less control is required. Furthermore, the adduct is less of a skin irritant.

Isocyanate curing. This is sometimes used for curing at low temperatures, down to 0°C. A short pot life of 1.5–2 hours and a tendency for film embrittlement on ageing are disadvantages.

Epoxies provide hard, durable films, although they tend to chalk fairly readily. They must be applied to steel that has been blast-cleaned to at least ISO Standard Sa2$\frac{1}{2}$. Primers may be pigmented with zinc phosphate or zinc dust. Provided the latter gives a suitable concentration of pigment in the dry film, then the paint is a zinc-rich type. In many situations, primers as such are not used, undercoats sufficing for application to the steel. Often solvent additions are made to the undercoat, producing a coating material that will 'wet' the steel surface when it is used as a primer. Undercoats and finishing coats are pigmented with titanium dioxide (rutile) and colouring pigments to provide a full range of colours. Micaceous iron oxide pigments can also be used. Aluminium is sometimes added as a pigment but this is not recommended for exposure to chemical environments.

Epoxy paints are widely used for offshore structures and for other aggressive and chemically polluted situations. They are sometimes employed in less aggressive environments for long-term protection, but maintenance is not always a straightforward matter with these paints. They produce hard films which tend to harden further on ageing. This generally results in difficulties with the adhesion of maintenance coatings, unless the old epoxy coating is roughened, e.g. by light blast-cleaning before repainting.

Epoxies generally contain suitable solvents and are applied by spraying. Solventless materials are also used and are sprayed using a special two-component hot airless spray. Solventless products are available for spraying at ambient temperatures. High-film-thickness coatings can be applied by solventless application, e.g. $300 \mu m$ to over 1 mm.

A greater measure of skill is required for epoxy coating than for one-pack materials. Mixing of the two components must be carried out correctly. The requirements regarding 'pot life' must be carefully observed and if it is exceeded, paints must be discarded. Some epoxy materials cure slowly below certain temperatures, or cure very quickly at higher ambient temperatures, so careful attention must be paid to manufacturers' data sheets. Incidentally, some data sheets do not contain sufficient information on the effects of temperature.

Developments in the 1970s and 1980s led to a new class of surface-tolerant epoxies coming onto the market. Originally from the USA and smelling heavily of phenols, the new range of aluminium flake-filled epoxies had a volume solids content of 90% and were capable of brush application to vertical and horizontal surfaces at a minimum of 100 microns dry film thickness. The aluminium content was not so high as to support the thermite reaction and was suitably enclosed in the epoxy matrix. The term 'surface tolerant' meant that the materials could be successfully applied over adherent rust and adherent residues of other coatings, no matter what the generic type.

Initial scepticism was overcome when good case histories had been developed. At this time, both low-odour and low-temperature curing versions capable of spray application appeared, as well as conventionally pigmented derivatives. These 'epoxy mastic' paints have become established maintenance paints, where the rather slow rate of curing is acceptable. Indeed, where grit blasting to a minimum cleanliness standard of ISO Sa2$\frac{1}{2}$ is achieved, the aluminium mastic may even be employed sub-sea. See also Section 4.9.7.

4.9.3.2 Coal-tar epoxies

Coal-tar epoxies, sometimes called epoxy pitches, are epoxies that have been modified with coal tar to provide a cheaper but very useful coating material with excellent water-resistant properties; this makes such coatings particularly useful for immersion in seawater and because of their non-saponifiable character they are suitable for use with cathodic protection systems.

Various agents are used to cure these materials, but the polyamides provide coatings of high moisture resistance. Isocyanate-cured coal-tar epoxies are also used and have been reported as being particularly suitable for sheet steel piling, and have the considerable advantage for the UK of curing below 5°C. It is claimed that such materials can cure satisfactorily to temperatures well below 0°C but, of course, if ice is present on the surface to be painted this will be of no avail.

Coal-tar epoxy coatings can also be employed for atmospheric exposure and do not suffer from the forms of breakdown typical of bituminous coatings. If applied under damp conditions they may 'blush', i.e. produce rather unsightly deposits on the coating.

Coal-tar epoxies can be applied by all the conventional methods, although spray application is generally recommended. There is a limit on the time between the application of successive coats to ensure good inter-coat adhesion; 7 days at 20°C is often considered to be a maximum. Generally, these paints are not pigmented, although micaceous iron oxide or carbon black can be added. A disadvantage of coal-tar epoxides is their dark colour, black or dark brown, which is not suitable for decorative purposes. Primers such as zinc-rich/epoxy and zinc silicate paints are sometimes used as part of a protective system but authorities differ in their views on the use of zinc-containing primers for immersed conditions. Often, the standard coal-tar epoxy, with added solvent to improve application, is used as a priming coat. Coal-tar epoxies are almost as corrosion resistant as epoxy-polyamides, but are less resistant to solvents. Coal-tar epoxies are widely used for ships' hulls, immersed parts of offshore platforms, dock gates, sewage systems, tanks and many similar purposes. Despite the known risk caused by components of tar having carcinogenic properties and the practical banning from use in mainland Europe, coal-tar epoxies continue to provide the most cost-effective solution to many UK effluent and water immersion duties.

4.9.3.3 Two-pack polyurethanes

The organic chemistry of urethanes is fairly complex and will not be considered here. However, in simple terms the coatings are based on the reaction of isocyanates with hydroxyl groups present in a variety of compounds, most commonly polyols. Two types of isocyanate are used for coating materials: aliphatic and aromatic. These produce different properties. The aliphatic type are used for atmospheric exposure and have excellent abrasive and chemical resistance. They are hard and have good gloss and colour retention. The aromatic coatings are cheaper and have poorer gloss and colour retention, but unlike aliphatic urethanes they are suitable for some immersion service.

Urethanes may be used as a finishing coat for epoxy systems because they retain their gloss for a longer period, but complete urethane systems can also be employed. The aromatic types are often used for the primer and undercoats, with the aliphatic type as a finishing coat.

Since the range of products falls within the overall 'urethane' classification and problems may arise if they are used without the required background knowledge of particular products, it is advisable to seek advice before using these materials. The National Association of Corrosion Engineers (NACE) has published a useful technical committee report on urethane coatings.[20]

Two-pack urethanes are similar to epoxies in their properties such as resistance to water and solvents but with improved atmospheric durability, particularly to intense sunlight and with lower curing temperatures. Moisture must be excluded during application because the isocyanate, an essential component, reacts with the hydroxyl group in water. Ultra heavy-duty polyurethanes have found use in linings and coatings on pump internals, whilst the lighter acrylic modified urethanes are now widely used over epoxies as the finishing coat of many specifications for exposed structural steelwork.

4.9.3.4   Urethane pitches

These are comparable with coal-tar epoxies with similar properties and are used for the same type of service conditions. The two-pack urethanes are normally modified with asphaltic pitch or coal tar to produce suitable coatings, which can be applied by all the usual methods although spraying is generally recommended. The coatings are black or brown and so are not suitable for decorative purposes.

4.9.3.5   Moisture-curing urethanes

These are not two-pack materials, but produce coatings that fall into this category. Basically a polyol is pre-reacted with excess isocyanate, so producing a one-pack material that will react with atmospheric moisture. Problems can arise in the production of these materials, not least that of excluding moisture from the material. As most pigments contain moisture, it must be removed to prevent reaction in the can. Provided these paints are manufactured and used properly they are a useful addition to the range of paints available for steelwork. The advantages claimed are as follows:

(i)   One-pack material that provides a high level of protection to steel-work.
(ii) Will cure at low temperatures, below 0°C.
(iii) Can be applied under adverse weather conditions, which may make them particularly useful for maintenance work.
(iv) Because of the nature of the coating it is useful for anti-graffiti purposes.
(v)   The very short overcoating time makes it possible to apply a complete three-coat system in a single day.

However, the very rapid cure also has its disadvantages, in that an undercoat can become too hard and solvent-resistant to provide a suitable key for subsequent coats. Therefore, these materials should be overcoated as soon as is practicable and particularly in conditions of high humidity, partial immersion and exposure to direct strong sunlight. Another cause of intercoat adhesion weakness is undue delay in the use of the paint after the can has been opened.[21]

4.9.4   Zinc-rich coatings

These are mainly used as primers but can be used at suitable thickness for a complete paint system. Unlike all the other paint coatings discussed above, which are based on the binder, certain zinc pigmented paints with a high proportion of zinc in the dried film have special properties which can reasonably be considered as relating mainly to the pigment (although the binders are important) and can be considered as a special group of coatings.

Although comparatively small amounts of zinc powder and zinc oxide can be used to pigment paints, those in the class under consideration all contain about 80–90% zinc in the paint. There are two types of zinc paint in this category and they are variously described, but are formulated (i) with organic binders and generally called *zinc-rich paints*, and (ii) with inorganic binders based on silicates and called *zinc silicate paints*. These paints contain sufficient zinc to provide a measure of cathodic protection in low-resistivity electrolytes but should not be considered as metal coatings because the binder, albeit in small amounts, effectively alters their properties as compared with, for example, hot-dip galvanised coatings. These coatings are sometimes called 'cold galvanised', but that can be misleading.

4.9.4.1   Organic zinc-rich paints

The most resistant paint coatings are based on two-pack epoxies, although a range of single-pack products has been produced based on chlorinated rubber, styrene butadiene, polystyrene and high-molecular-weight epoxy. The single-pack products are easier to apply but are not generally used for structural steelwork, with the exception of the epoxy type. A high level of surface preparation is necessary to ensure the highest performance from zinc-rich coatings. Although coatings of sufficient thickness may be used without further protective coatings, e.g. to patch damaged or weld areas on hot-dip galvanised steelwork, they are more frequently used as primers for other coating systems, e.g. epoxies.

While these primers are generally excellent in the atmosphere, some authorities have reservations regarding their use under immersed conditions because of the possibility of attack on the zinc by moisture or water permeating through the paint system. Such attack could cause adhesion problems. However, a more serious problem may arise if the zinc in the primer corrodes and is not properly treated before it is overcoated, as this can lead to breakdown of the system.

4.9.4.2   Inorganic zinc-rich paints

Zinc silicate coatings are, unlike all the other paints considered in this section, based on an inorganic binder. Silicate binders are either aqueous solutions of alkali silicate, usually sodium, potassium or lithium silicates, or alcohol solutions of alkyl esters of silicate. The former are commonly termed 'alkali silicates' and the latter 'alkyl silicates' or 'organic zinc silicates' because they are based on organic silicates; however, the organic silicate hydrolyses, releasing volatile alcohols, leaving an inorganic dry film (see Section 4.6.3.3 for a discussion of the mechanism of formation).

Zinc silicate coatings are cement-like, being based on polysilicic acid which cements the zinc particles into a hard film. Although the amount of zinc may vary, it must be of the required high concentration if the coating is to fall within the class of zinc-rich paints. However, films with lower zinc contents also provide useful properties. The cement-like nature of the coating confers a number of useful properties; the films are resistant to solvents and the alkyl silicates provide heat-resistance up to about 400°C. They are more resistant to abrasion than paints based on organic binders and, because the coefficient of friction of the dry film is about 0.5 compared with about 20% of that for most paints, they can be used for surfaces to be connected by high-friction grip bolts. Surface preparation of the steelwork must be of the highest standard, i.e. Sa3 or the equivalent,*

---

*Some American authorities recommend a much lower standard.

and a comparatively low profile height is generally recommended (25–50 μm). Zinc silicates will not adhere properly to surfaces carrying the remnants of old organic coatings, i.e. most conventional paints. Unlike most paints, a reasonably high humidity is advantageous because the film must remain 'wet' at least for a few seconds to ensure curing of the binder. Otherwise 'dry spray' will cause poor adhesion of subsequent coats or lack of durability of the final coat.

The paints are usually sprayed to provide a thickness of about 75 μm; only small areas should be coated by brushing. 'Mud-cracking', a self-descriptive term, only occurs if coatings are applied at thicknesses in excess of the paint manufacturer's recommendations.

Zinc silicates are widely used with no further protection, e.g. for tank linings. However, where they are used as primers for other systems, care must be exercised to ensure that zinc salts are not present on the silicate coating as this may impair adhesion of coats applied over it. If they are present then removal either by vigorous use of a bristle brush or by light blast-cleaning is recommended. Brushing is preferred because of the loss of coating with blast-cleaning which, if not carefully controlled, may be significant. Most of the chemically resistant paints can be applied to zinc silicate without difficulty, but its porous nature can trap air or solvent and then cause pinholing of subsequent coats. The application of a thin sealer coat generally eliminates this problem. Paint manufacturers' lists may contain a considerable number of silicate paints, some single-pack; many of them will not compare in properties on durability with those considered above, so advice should be sought to ensure that suitable silicate paints are chosen.

Inorganic zinc-rich paints have found a niche as pre-fabrication primers. The decision to use pre- or post-fabrication primers is highlighted later in Section 5.2.1. Automatic shot blasting followed by a preheat and automatic spraying of inorganic zinc pre-fabrication primers to just 20 μ ($\pm 2\mu$) is widely used on sheet steel in Europe and North America and Korea. After application, the coated sheets pass through a tunnel heated to circa 50°C, allowing the sheet steel to be stacked immediately. So tough is the dry film, that even shackling with chains will not remove all of the coating. It has been found that sheet steel coated with inorganic zinc pre-fabrication primers to 20 μm will cut by oxy-acetylene torch as quickly as uncoated sheet, but with a cleaner edge. The fume is considered acceptable in normally ventilated shop conditions.

Additionally, all normal welding techniques provide welds with full integrity and without any weld spatter remaining on the periphery of the weld. Indeed, the inorganic zinc pre-fabrication primer prevents adhesion of the weld spatter, such falling from the surface of the sheet of coated steel. The 'burn-back' of the coating after cutting or welding is minimal and easily cleaned by pencil blast. Recent trials with laser cutting has confirmed the rate of cutting is not diminished by the inorganic zinc pre-fabrication primer, but no coating has yet been found to be compatible to present laser welding techniques. Unlike other forms of pre-fabrication primer, it is not generally necessary to remove all of the inorganic zinc pre-fabrication primer before continuing fabrication.

Epoxy pre-fabrication shop primers are less expensive, but do not possess the unique range of benefits exhibited by inorganic zinc pre-fabrication primers. Water-based modified alkyd pre-fabrication shop primers have recently been introduced in Europe, but such require nearly double the thickness to provide 6 months' protection, approach double the price and have none of the attributes of the inorganic zinc pre-fabrication primers. The paint and construction industry awaits the first water-based, self-curing inorganic zinc pre-fabrication primer.

4.9.5   Water-borne coatings

There are four main types: water solubles, water reducibles, latex and emulsions.

Water solubles include alkyds treated to be water soluble. They dry by water evaporation and cure by auto-oxidation polymerisation. Examples include, household paints and electro coats.

Water reducibles dry by water and solvent evaporation. Two pack types cure by cross linking polymerisation. They include epoxies where the water acts as a dilutent and reduces the need for strong solvents, but does not dissolve the resin. Examples of uses include structural steel coating.

Latex may be a copolymer blend of polyvinyl acetate, polyvinyl chloride and acrylics. A solid phase disperses into a liquid phase. Examples of uses include interior and exterior (acrylic only) 'emulsion paints'. They dry by water evaporation and cure by coalescing.

True emulsions consist of an immiscible liquid phase dispersed into another liquid phase. Examples are epoxy emulsions used on structural steel.

4.9.6   Compatibility of different paints

Caution should be exercised when mixing different types of binder in a single paint system. Table 4.1 gives a broad indication of the effect of applying one type of binder over another. However, it is advisable to consult paint suppliers before mixing binders in the system.

4.9.7   Heat-resistant coatings

Surfaces subject to temperatures up to 95°C. A normal paint system of primer and gloss finish is usually satisfactory. For surfaces subject to

Table 4.1   Compatibility of different paints

Applied second coat	First coat						
	Bitumen	Vinyl	Chlor-rubber	Alkyd	Epoxide	C/T epoxide	Poly-urethane
Bitumen	*	×	×	×	×	×	×
Vinyl	×	*	×	×	*	*B	*
Chlor-rubber	×	*	*	○	*	*B	*
Alkyd	×	×	×	*	*	*B	*
Epoxide	×	×	×	×	*	*B	×
C/T epoxide	×	×	×	×	*	*	×
Polyurethane	×	×	×	×	*	*B	*

* Probably all right.
× Not recommended.
○ Results in crazing.
B Adhesion may be satisfactory but appearance affected by 'bleeding'.

temperatures approaching 95°C, better results are obtained by omitting undercoats and applying one or more coats of finishing paint direct to the primer.

Surfaces subject to temperatures between 95 and 130°C. It is generally satisfactory to apply gloss finishing coats over a modified alkyd zinc phosphate primer, although discoloration of the lighter shades must be expected. Aluminium (general-purpose) paint can be used over a modified alkyd zinc phosphate primer.

Surfaces subject to temperatures between 130 and 150°C. It is preferable to use alkyd gloss finishing paint only, without undercoats or primers.

Surfaces subject to temperatures between 130 and 260°C. Application direct to bare metal of an aluminium (general-purpose) paint often gives better results than the heat-resisting type based solely or partly on silicones. Where acid resistance is also required (in steel chimneys, ducts, etc.) a zinc-dust–graphite paint applied in one single heavy coat can be effective for the short term. Inorganic zinc followed by silicone, silicate or siloxane, pigmented with aluminium flake, have been used with success if the substrate is constantly operating above 130°C.

Surfaces subject to temperatures between 120 and 950°C. Wherever possible, sprayed aluminium metal coatings are to be preferred to paint coatings for heat resistance. A very high standard of surface preparation is required and the coatings must comply strictly with BS 2569, Part 2 or similar.

Surfaces subject to temperatures between 260 and 540°C. Silicone-based aluminium paints can be used for this temperature range, but it is essential if good results are to be obtained to ensure that the metal surfaces are really clean and free from rust. These paints also require a minimum temperature for curing (usually about 260°C). Another difficulty with these paints is that, although they have good resistance to high temperatures, they have poor corrosion and weather resistance properties.

4.9.8   Other coating materials and fillers

Intumescent paints are used in providing a degree of fire protection to steel structures.

These materials swell up and char in fire situations, protecting the substrate from intense heat. The lighter duty materials are often applied by a similar autoblast/autospray technique as for inorganic zinc pre-fabrication primers, except that steel grit is used instead of shot, to provide a deeper, sharper profile to hold the thicker coating of intumescent paint.

Heavy-duty intumescents, the type used on off-shore rigs, are generally epoxy based and are applied over an inorganic zinc-rich or organic-zinc phosphate primer, at several millimetres in thickness, but operate in a similar fashion. A further type of passive fireproofing is the encapsulation of the steel members in a low-weight concrete, such as magnesium oxychloride, which evolves water of crystallisation as steam in fire situations, so cooling the surface (see also Section 6.5).

A number of materials based on waxes, oils and greases are used either for temporary protection of steel components or for protection of steelwork, particularly in fairly inaccessible areas. These materials are not paints, but for convenience they are considered here.

Waxes and greases, sometimes containing inhibitors, are generally applied as thick, reasonably soft, films to provide barrier protection. Some of the compositions are hot-applied and are useful for injection into spaces where steel may be in contact with damp materials. Others are cold-applied to produce either hard wax-like or soft greasy coatings. Although these coatings are often effective and do not require a high degree of surface preparation, their use on structures is limited to areas which personnel are not likely to visit very frequently, e.g. interiors of box girders, or virtually inaccessible areas. They are difficult to inspect for corrosion because often a considerable amount of rust may form before it becomes noticeable. They are cheap and easily applied but authorities differ in their views on their overall effectiveness. It is advisable to consult the manufacturers of these products before using them because they cover a wide range of properties. For example, some harden to a marked degree and can be overpainted with conventional paints to provide a coloured finish.

Generally, the same types of paint are used for initial and maintenance painting, even though the conditions under which the two types of painting are carried out is often vastly different. A few special types of coating have been developed. These are sometimes called 'surface-tolerant coatings'. These are designed to be applied to a range of coatings, to rusty steel and to damp surfaces. High-build epoxy aluminium mastics were introduced in the late 1970s. They are high in solids (over 80% by volume), so thick films can be obtained with a few coats using brush application. They are formulated to give good wetting properties and to reduce solvent attack on the old paint coatings to which they are applied. They have good resistance to solvents and to saponification. Generally they provide protection equal or better than many of the conventional paints, but are not particularly decorative.

Temporary protectives are so-called because the coatings can easily be removed by simple solvents such as white spirit, and this term is not necessarily an indication of the time of protection afforded by them. Generally, they fall into a number of clearly defined groups, as classified in BS 1133, 'The Packaging Code', Section 6.

(i)   Solvent-deposited type: consisting of film-forming ingredients dissolved in solvents to provide low-viscosity liquids. On evaporation of the solvents a thin, tough, abrasion-resistant film is formed. The materials are usually applied by dipping or spraying. They are not particularly suited to batches of items that might stick together but are used for components of fairly high surface finish.

(ii) A soft-film type similar to (i), but less resistant to abrasion, and mainly used for the interior protection of assemblies.

(iii) A type rather similar to (i) above but applied by hot-dipping to produce thicker films, which are more protective.

(iv) Types based on grease, sometimes containing an inhibitor, which are applied by smearing or brushing, unlike (i)–(iii) which are applied by dipping.

(v)   More expensive strippable coatings are also available; generally, they are fairly expensive but can easily be removed without solvents.

Mastics are thick coatings used to fill gaps, crevices, overlaps, etc., in a manner similar to wax and grease paints. They are generally based on synthetic rubbers, synthetic resins and bitumen. Polysulphides are frequently used, as are butyl rubbers. Most mastics tend to harden to some extent with ageing and some shrinkage may occur. Consequently their effectiveness should be checked regularly, e.g. during maintenance, to ensure that wide gaps are not converted to crevice situations which may be worse from the corrosion standpoint.

4.10   Health and safety matters

Most heavy-duty protective paints must be regarded as hazardous and harmful to humans, dangerous to the environment and even marine pollutants. Solvents used are frequently flammable, have a narcotic effect and have a de-fatting action to human skin. Epoxies are potentially liable to cause dermatitis on prolonged contact with unprotected skin and the hardeners used in polyurethanes may give rise to pulmonary problems. It is therefore extremely important for users of paint to have an up-to-date copy of the paint manufacturers' Material Safety Data Sheet (M.S.D.S.) available at point of use. Additionally, in the UK obtaining and making reference to a current copy of EH 40, available from H.M. Stationery Office, is both a wise and necessary precaution.

Personal protection and ensuring extremities are well protected is the responsibility of all concerned. If natural ventilation is suspect, forced ventilation to safe areas should be employed. Consideration must be given to other trades working in the area, particularly if welding is being carried out in close proximity to application of flammable paints. If there is any doubt, advice from the site Safety Officer and paint manufacturer should be sought. Breathing solvent vapour should be avoided. For short spray runs, a carbon filter mask – never a simple dust mask – may suffice, but for longer spraying times, a full face mask with independent air supply passing through suitable filters is mandatory. The use of lip balm is recommended to avoid discomfort caused by long-term use of such masks.

It should also be remembered that the operative mechanically mixing the paint will require protection. Not only is there the probability of particles of paint of respirable size being present, but with the operator's head being over the mixing drum, solvent inhalation is certain without taking adequate precautions. Separate considerations apply to all areas which may be regarded as confined spaces, including low-voltage, non-incandescent lighting and absolute earthing of spray equipment, with safety harnesses employed for heights over 2 metres.

Under the Control of Substances Hazardous to Health Regulations, for every painting job, a risk and safety analysis should be completed. All concerned in the contract, not simply the operatives, must be made aware of safety precautions and procedures, plus actions to be taken in case of accident or emergency.

References

1. Edwards, J. D. and Wray, R. I., *Industrial Engineering Chem.*, **28** (1936) 549.
2. Guruviah, S., *J. Oil and Colour Chemists' Assoc.*, **53** (1970) 669.
3. Mayne, J. E. O., *Corrosion*, Vol. 2, Newnes Butterworths, London, 1976, p. 15.26.
4. Mayne, J. E. O., *J. Iron Steel Inst.*, **176** (1954) 143.

5. Mayne, J. E. O., *J. Appl. Chem.*, **9** (1959) 673.
6. Chandler, K. A. and Stanners, J. F., *Second International Congress on Metallic Corrosion*, 1966, NACE, Houston, pp. 325–33.
7. McGill, W. J., *J. Oil and Colour Chemists' Assoc.*, **60** (1977) 121.
8. Walker, P., *Paint Technology*, **31**(8) (1967) 28.
9. Van Loo, M., *Official Digest*, **28** (1956) 144.
10. Bayliss, D. A. and Bray, H., *Materials Performance*, **20** (1981) 11.
11. Gay, P. J., *J. Oil and Colour Chemists' Assoc.*, **31** (1948) 481.
12. Bullett, T. R. and Rudram, A. T. S., *J. Oil and Colour Chemists' Assoc.*, **44** (1961) 787.
13. Brunt, N. A., *J. Oil and Colour Chemists' Assoc.*, **47** (1964) 31.
14. Klens, P. F. and Lang, J. F., *J. Oil and Colour Chemists' Assoc.*, **39** (1956) 887.
15. Shapiro, S., *Official Digest*, **30** (1958) 414.
16. Munger, C. G., *Material Performance*, **14**(5) (1975) 25.
17. Ginsberg, T. *et al.*, *J. Oil and Colour Chemists' Assoc.*, **59** (1976) 315.
18. *Steel Structures Painting Manual*, Vol. 1, Steel Structures Painting Council, Pittsburgh, 1982, p. 13.
19. Carter, E. V., A new synthetic process for the manufacture of lamellar iron oxide pigment for use in anti-corrosive coatings, *J. Oil and Colour Chemists' Assoc.*, **5** (1988) 132–8.
20. Urethane protective coatings for atmospheric exposure, 6B273, NACE, Houston.
21. Thompson, I. and Temple, C., The relationship between cure and performance of single pack moisture cured urethanes. *Industrial Corrosion*, **8** (1990) 11–16.

Further reading

Boxall, J. and Von Fraunhofer, J. A. (1980). *Paint Formulation*. George Godwin, London.
Introduction to Paint Technology. Oil and Colour Chemists Association, London, 1967.
Paint Technology Manuals, Services published for the Oil and Colour Chemists' Association by Chapman and Hall, London, including Part 1 'Non Convertible Coatings' and Part 3 'Convertible Coatings'.
Turner, G. P. S. (1967). *Introduction to Paint Chemistry*. Chapman and Hall, London. (A simple account of the organic chemistry of paint binders, drying mechanisms and paint composition and formulation.)

Chapter 5

Paint application

Paint application has an important influence on the life of a paint system and must be properly controlled to ensure that coatings attain their full potential. This is achieved by employing skilled operatives and ensuring that paint is applied correctly.

Despite the improvements that have taken place which allow for more economic application, the older methods are still widely used. Methods such as brush application are basically simple but still require to be properly controlled if good coatings are to be obtained. Furthermore, well-trained painters are far more likely to produce sound coatings than those with little knowledge of paint application.

In this chapter, only the application methods commonly used to paint constructional steelwork will be considered in detail. Methods used to apply paints to domestic equipment, cars and consumer goods are not used to any extent for constructional work. Furthermore, they are applied under factory-controlled conditions to the manufacturers' specifications, which cannot easily be varied by the purchaser.

5.1   Methods of application

The methods used for paint application fall into the following general categories:

(a) manual application – brush and roller;
(b) spray – airspray and airless, hot or cold;
(c) dip coating;
(d) flow coating;
(e) roller coating;
(f) electrostatic;
(g) powder coating.

Additionally, paint coatings may be air-dried or stoved.

For constructional steelwork, methods (a) and (b) are generally used for applying paint. Method (e) is sometimes used and (f) and (g) may be used, but generally are only for special components or for plastic coatings.

5.1.1 Brush application

Although centuries old, application of paint by brushing is still widely used, e.g. for:

(a) Sensitive areas, i.e. close to property, equipment, etc. where overspray could cause a problem.
(b) Small or irregular surfaces, such as lattice structures, pipe racks, etc.
(c) Interior areas with relatively poor ventilation.
(d) Application to surfaces with an unavoidable low standard of preparation. The brush will tend to penetrate into pits and, to a certain extent, displace dust and even moisture.
(e) Small areas of touch-up work.

Other important advantages of the use of brushes are their relative cheapness and ease of cleaning.

The brush serves two purposes. It acts as the container for the paint as it is transferred from the can or kettle and also as the actual method of transfer. This means that the bristles must be capable of holding sufficient paint to allow for reasonable coverage at each transfer and must also be of material capable of transferring the paint in an even way through the points of contact at the steel surface. It is important to choose the right type of brush for the job in hand. For example: square-ended brushes for large flat surfaces, angular-cut brushes for narrow surfaces and oval shaped brushes for small surfaces such as nuts and bolts. Top quality brush bristles are made from animal hairs and are flagged, i.e. have split ends. Brushes made from polyester or nylon have better wearing properties but give a lower standard of finish. They should be used, however, for waterborne paints. Cheap brushes of any type often end up costing more when one takes into consideration productivity, brush life and the possible need to rework areas of poor quality.

Apart from the conventional brush described above, another type made from short synthetic fibres about $1\frac{1}{4}$ in (3.2 cm) long attached to a flat base about 4 in (10.2 cm) × 6 in (15.2 cm) is also used. This holds more paint and covers a greater area. It is particularly useful for forcing paint into crevices and for use where it is difficult to apply the paint in the conventional manner.

Various attempts have been made to mechanise the process by using methods such as hoses to supply the paint to the brush. Recently, equipment of this type has been made available for the application of emulsion paints for the domestic market. Generally, such methods are not widely used for the painting of steel. In part this arises from the additional cleaning of equipment required, so eliminating one of the main advantages of brush application. Additionally, skilled painters used to application with conventional brushes appear to find that they lose some of the control over the operation if the paint is automatically fed to the brush. The advantages of brush application are considered to be as follows:

(i)   Cheap, requiring no expensive equipment.
(ii)   Relatively clean, requiring no masking of adjacent areas.
(iii) Can be used in restricted areas.
(iv)   Particularly suitable for displacing dust and moisture on imperfectly prepared surfaces.
(v)   Allows paint to be worked into crevices and other difficult areas.
(vi)   Brushes are easily cleaned and can be easily maintained in good condition.
(vii) The only practicable method of applying certain types of paint, e.g. lead-based.

The main disadvantage is the slow rate of application, which can be less than 10% of the rate achieved by spray methods. However, taking into account the time saved on masking, cleaning, etc., the overall rate of application is better than this figure would indicate. It is also generally used for stripe coats along edges, corners, etc., even where other methods may be used for the application of paint to the main areas.

In general, literature and paint manufacturers' data sheets do not provide guidance concerning the application of stripe coats. Some consider that brushing is most suitable for small complex shapes (such as lattice members and bolted connections), whereas spray application is appropriate for the edge of large structural shapes. Guidance should be obtained from the paint manufacturer, at the specification stage, for suitable methods of application of their material as a stripe coat.

Not all modern paints are capable of being applied by brush. For example, high-build epoxies or urethanes, or quick drying materials such as zinc-rich primers, do not usually lend themselves to brush or roller application, except possibly for small areas of touch-up. This should be checked by consulting the materials data sheet, or the coating manufacturer. Even where application is possible, they will be best applied at a lower film thickness with spray application.

Brushing may not be suitable for paints drying by solvent evaporation such as chlorinated-rubber paints. This is because the solvent evaporates quickly, leaving a very viscous polymer, and to apply the paint satisfactorily requires a low viscosity material which may cause sagging of thicker films. Problems also arise with applying subsequent coats by brush because the solvent tends to dissolve the previous coat. However, suitably formulated solvent-type paints are produced for brush application.

Generally, brushing is used to apply oleo-resinous types of paint (including long oil alkyds) for which solvents of the white-spirit type are used. Rapidly drying paints, e.g. those drying in under 1 hour, are difficult to apply satisfactorily by brush. Although other types of paint may be applied by brush, special formulations are usually required.

There is little justification for the belief that one should always apply primers by brush. However, for maintenance work where a high quality of surface preparation is not proposed, brush application gives better wetting of rusty and pitted surfaces.

5.1.2 Roller application

This method should not be confused with roller coating application, which is a means of applying paint to flat steel sheet under factory conditions. In the context of painting structural steelwork, it is similar to the method commonly used to apply emulsion paints in domestic situations.

Roller application is particularly suitable for painting large flat areas and does not require the same skill as for spray application. Rollers also allow for application rates up to four times faster than those achieved with brushes. With this method, large trays or containers are used for the paint and the roller is dipped in to transfer the paint to the steel surface. Although useful for rapid covering of large areas, the method does not have the advantages of brush application for priming coats on hand-cleaned steel surfaces, and generally requires the additional use of a brush for edges, corners and other areas not accessible to the roller. In addition, aeration has often been shown to induce small pinpoint voids into roller applied films.

In general, rollers are not recommended for the application of primers or high-build coatings because it is difficult to control film thicknesses with such a method of application. There is also the tendency to 'roll out' the coating, i.e. to apply it too thinly, especially with high-opacity coatings containing aluminium pigment. In these cases it is necessary to apply an increased number of coats to obtain the correct film build. Also, with metallic pigmented paints it is necessary to finish off in one direction to avoid showing the path of each rolling.

The size of the roller affects the speed of application; a $230\,mm \times 65\,mm$ roller will hold $2\frac{1}{2}$ times as much paint as a $180\,mm \times 30\,mm$ roller. Extension handles can help to gain access and reduce scaffolding but it is doubtful whether the practice sometimes seen of a roller fixed to the end of a long pole leads to a uniform paint application or increased speed of application.

The type of roller surface is important. Roller naps of $5\,mm$ to $10\,mm$

are best for smooth surfaces and 10 mm to 20 mm for rough, pitted surfaces. Special rollers are available for pipe coating work. Care must be
taken with some modern high-performance coatings since their solvents
may attack the adhesive within the roller.

Because of the inherent difficulty in getting even films with brush or
roller, it is advisable in the interests of overall practicability to give the
contractor some tolerance regarding film thickness. A reasonable figure is
that for roller application not more than 10% of the dry film thickness
readings may be up to 10% below the specified minimum.

Cleaning is not usually a problem because the roller covers are often
discarded at the end of the day's work. The most commonly used cover
materials are mohair, lambswool, nylon and polyester. The material and
the length of nap are usually chosen in relation to the paints being applied.
The roller itself must be made of a material resistant to the solvents used
for particular paints. Special rollers consisting of a series of narrow rolls
are often used for application to complex-shaped surfaces and pipes.
Although rollers are generally used by dipping the roller into the paint and
then applying the paint, mechanised methods are utilised. In these, the
paint is supplied from a pressurised tank into the roller head. Other
methods such as paint pads or special lambswool gloves or paint mitts are
also sometimes used to apply paint. Generally, these methods are more
suitable for general coverage rather than for controlled painting. They are
useful for painting areas of difficult access and restricted space, e.g. back-
to-back angles.

5.1.3   Airspray application

Airspray was the first type of spray equipment developed to provide
speedier application of paints, particularly when the quicker-drying types
were developed, as these could not easily be applied by brush. With airspray application, sometimes known as 'conventional' spray application,
compressed air is used both to atomise the paint and to carry it to the
surface to be painted (see Figure 5.1). Paint is fed to the spray gun by one
of the following methods.

(i) Gravity feed: the paint is contained in a cup on top of the spray gun
and is fed by gravity to the gun nozzle. This action is also helped by a
vacuum created at the nozzle by the flow of compressed air. Gravity feed
guns are normally used for paints with low or intermediate viscosities, and
are very popular for painting small areas because of the ease with which
they can be refilled and cleaned. They cannot be used for spraying surfaces
overhead.

(ii) Suction feed: the paint is contained in a cup fixed to the underneath

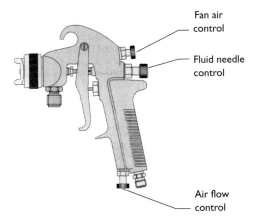

Fan air
control

Fluid needle
control

Air flow
control

Figure 5.1   Spray gun controls.

of the gun and is sucked to the paint nozzle by a vacuum obtained with compressed air.

Both gravity feed and suction feed are relatively low speed application methods.

(iii) Pressure feed container: the paint under pressure is supplied to the gun by a displacement pump operated by air pressure. This method is the most efficient for large applications and is particularly effective for applying mastic or coatings containing large or abrasive pigments.

Important considerations in air (conventional) spray painting are:

1   Proper gun distance from the work (150–200 mm).
2   Gun should be held perpendicular to the surface to be painted.
3   Paint should be sprayed at the lowest possible atomisation pressure that will adequately atomise the paint.
4   Since it is possible to adjust the controls of the gun to alter the application volume, pattern, etc., the operator should be skilled in its use.
5   Material losses are likely to be 25–50% depending upon numerous factors.
6   The air to the gun should be passed through suitable filters or traps to remove moisture. Airlines to the gun can be extremely moist and this can cause problems with paint materials, particularly urethanes and some types of epoxy.

In recent years, new types of spray gun have been designed specifically for the application of high-solids and water-borne coatings. These are available in pressure-, gravity-, and suction-fed versions.

Equipment for the application of water-borne coatings generally has all-metallic parts that come into contact with the paint, made of stainless steel. Chromium plated steel, bronze or other alloys may cause water-borne coatings to coagulate[1] and, before spraying with equipment that has been used previously for solvent-based coatings, it should be cleaned out with a water-miscible solvent, followed by clean water.

5.1.4 Airless spraying

In conventional air spraying the volume of air required to atomise a given volume of paint is considerable. This arises from the large differences in the specific gravities of air and paint and the requirement to impart high velocity to the paint. This leads to an inbuilt inefficiency in the system because of the large volumes of air needed and the high overspray that occurs with this method. Air spraying can be efficiently used in automatic plants and in properly designed spray booths, but is less efficient for painting a wide variety of structural-steel shapes. Airless spraying is generally more suited to this type of application (see Figure 5.2).

In airless spraying, the paint is forced through a small jet so that it reaches the velocity required for atomisation and, as with a garden hose, a spray of droplets is produced. As there is no expanding compressed air, as in air spraying, to disperse the fluid particles, most of the paint adheres to the work surface, thus eliminating to a great extent spray mist and paint wastage. This results in a faster rate of spraying (up to twice that of air spraying) and less loss of paint by overspray.

The spray gun used looks similar to that used for air spraying, but does not have a compressed-air hose connected. However, the operating parts inside are different. There is a ball valve and seat, a sieve and a tip. The gun is connected by a suitable hose to an air-powered pump which forces the paint to the tip at a pressure some 20–50 times that of the compressed air used to operate the pump. Consequently, although the air pressure is of the same order as that used for air spraying, the actual spray pressure is 12–35 MN/m^2 (1800–5000 psi).

Very thin fluids such as sealers can be successfully atomised at pressures around 400–800 psi. Most protective coatings require 1500 to 2000 psi while some high-build paints need 2500 psi. Mastics and high-solids materials may have to be applied at pressures up to 5000 psi. The high pressures involved with airless spray present a safety hazard.

Since airless spray does not require compressed air to atomise and deliver the paint, it is not essential for the operation of the airless spray pump, and electric or petrol motors can be used instead. This means that the units can be made extremely portable. Agitators are often built into spray units to ensure adequate mixing during operation. The tip of the gun is made of tungsten carbide to reduce wear from pigments in the paint and

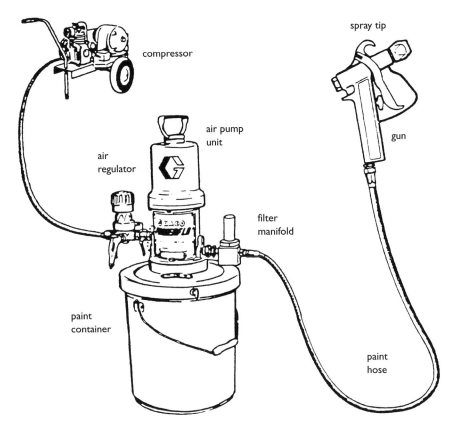

Figure 5.2   The airless spray system.

is an essential element in the control of the process. A range of tip shapes is available for different spray patterns and different types of paint and the paint suppliers' recommendations must be followed regarding the tip to be employed. Apart from the choice of tip, the only other variables in the process are the paint viscosity and the input pressure. Extension units or 'pole guns' may be employed with airless spraying to reduce or avoid scaffolding. These may be up to 3 m in length.

The main advantages of airless spray over airspray are:

(a) Higher output.
(b) Less paint fog, or rebound.
(c) Gives the painter the ability to apply thick films in a single pass.
(d) Coatings often require no thinning before application.

(e) Gives good penetration into pits and crevices due to the high kinetic energy involved.
(f) Extensions can be used to reach some inaccessible areas.
(g) An operator can install the pump and feed at floor level and in dry conditions and work at a height with only a single fluid line.

The main disadvantages are:

(a) Higher purchasing cost.
(b) Greater safety hazard.
(c) Cannot be used for all types of paint.
(d) Cannot be used with pigments with large particle sizes.
(e) Not suitable for fine, decorative finishes, since the spray edges are not sufficiently 'feathered'.
(f) Since the spraying characteristics, i.e. both volume and spray pattern are fixed with the airless spray tip or nozzle, a selection is required to meet a variety of spraying conditions. Over 90 types of airless spray nozzle are currently available.
(g) Inorganic zinc-rich primers and other highly abrasive paints are not generally suitable for airless spray application.
(h) Frequent clogging of the gun and wear on the orifice causes problems.
(i) It requires a greater distance from the work surface, normally 400–500 mm, than conventional spray.
(j) Although it requires less skill than the use of air spray, it still requires experience and care in handling and regular maintenance of the equipment is important.

5.1.5 Application of plural-component paints by spray

Two-component paints, such as epoxies, have a separate base and curing agent, which are mixed before application. Immediately they are mixed a chemical reaction begins which proceeds over a period of time to produce a solid product. This occurs at ambient temperature and the speed of the reaction is influenced by the nature of the two components. Clearly, once the components have been mixed there is a limit to the time they can be used in the spraying equipment because they will solidify reasonably quickly. Because the final product is soluble only in strong solvents, problems can arise if these materials are not used during the comparatively short period during which application is practicable. This period is called the 'pot life' of the paint. To overcome these problems, spraying equipment is available in which the two components are mixed at, or immediately before, the spray gun, so there is no possibility of chemical reaction between the two components in the rest of the equipment (see Figure 5.3). The correct mixing ratios are of paramount importance to achieve the

Figure 5.3   Plural system equipment for application of heated two-component coatings.
Source: Covercat Spray Systems Ltd.

designed stoichiometric ratio. New equipment features pressure detectors
to ensure against 'air bubbles' spoiling this ratio.

The majority of machine mixing systems use reciprocating positive displacement pumps to automatically supply each component at the required
ratio. They are then combined in a mix manifold and then a mixer, which
as its name implies, thoroughly mixes the base and the curing agent. The
mixer uses baffles, whirling impellers or impingement of the fluid streams
under pressure, to fold and refold the two components into a homogeneous mixture. Without this key component the two materials would
flow through the hose as separate streams and the quality of the final
coating would be very poor. The mixed material can then be applied by
spray, airless spray, manual or automatic devices, conventional or electrostatic. This method is not suitable for the rapid curing types of urethane,
which require mix-at-nozzle, plural spray units.

Of particular interest in recent years has been the airless spray application of heated, plural-component materials. The heating process, for

example the base to 55–65°C and the curing agent 35–45°C, reduces the viscosity and permits the application of very high solids materials, such as solvent-free epoxies polyurea systems or urethanes. Apart from the very rapid application of very thick coatings, these also meet the environmental requirements of reduced solvent emissions. Correctly applied these materials can give very long-term performance, such as with the internal and external coatings on the UK's Thames Barrier, where such a system has lasted at least twenty-two years and, apart from some mechanical damage with ship collisions, is still in excellent condition.

The obvious difficulty with all dual-component equipment is that it is more complicated than single feed units. Because of its high output, any malfunctioning or operator error can result in a large area of defective coating that is inevitably both difficult and costly to rectify. All operators of such equipment, regardless of how experienced they are with normal airless spray application, should have special training.

5.1.6   Electrostatic spray

This method is suitable for spraying either liquid or powder coatings. Unlike other spraying processes, it has the ability to paint all parts of an object from a fixed position, e.g. the whole of a pipe can be coated without moving the spray from the front of the pipe to the back (see Figure 5.4). This is achieved by passing paint droplets through a powerful electrostatic field, during which they receive a high charge, typically up to 75 kV for

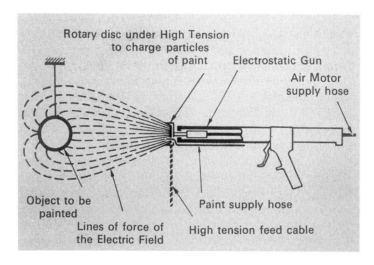

Figure 5.4 Details of electrostatic spray gun.

hand-held guns and up to 180 kV for some automatic equipment and so become attracted to the item to be coated, which is earthed.

This method is widely used for tubular goods, e.g. pipes, garden furniture, bicycle frames and wire fences. Electrostatic spray painting is generally more economical than other types. It is universally used for the factory application of fusion-bonded epoxy powder coatings to the external surfaces of pipelines. The method is not used for the painting of general constructional steelwork, but on sites it is well suited for open steel work such as gratings or railings.

The advantages of electrostatic spraying include:

(a) It can give a fairly complete overall coverage simultaneously to both back and front of small sections of simple shape.
(b) There is less spray loss than with other methods of application, typically by 50–75% compared with air spray.
(c) It gives a uniform film thickness and edges and corners are well coated.
(d) It is quicker than conventional spraying techniques and, when operated in works, no complex exhaust systems or paint booths are required.

Its disadvantages include:

(a) Only comparatively thin coatings can be applied.
(b) It can only be used on the bare, conductive substrates.
(c) Generally, only one coat can be applied by this method since the applied coating insulates the surface.
(d) The paint or powder has to be specifically formulated for the purpose.
(e) Paints containing conductive pigments, for example zinc-rich primers, cannot be used.
(f) The high voltages used present a safety hazard.
(g) The equipment is more expensive and more cumbersome to use than a normal spray gun.
(h) It is not possible to spray into some crevices or internal surfaces due to the Faraday Cage effect.

5.1.7   Other application methods

Most structural steelwork is painted by one of the methods discussed above. However, other application methods are also used to a limited extent and these will be briefly discussed in this section.

5.1.7.1   Dipping

Application by dipping is sometimes carried out for small fabrications difficult to paint by other methods; pipes are also coated by dipping. The method is basically simple. Either the steel is dipped into a tank containing paint or, sometimes, the tank is raised to cover the stationary steelwork. Paints have to be specially formulated and suitably thinned for the dipping method. Owing to the large exposed surface of the paint, excessive evaporation of solvent can occur and the viscosity must be regularly monitored and corrected with addition of more solvent. The whole process is generally less easily controlled than the other methods and is not suited to all types of paint. Clearly, there is a limitation on the size of steelwork that can be treated in this way, partly because of the size of the tank and partly because of the amount of handling required. Provided the design of the process allows for flow of the paint to all parts, internal and external surfaces can be coated in one operation. The method is rapid and skilled operators are not required. However, 'drips' and 'tears' tend to occur on the lower parts of the painted steelwork and there is a tendency for edges to be coated to a lower thickness than the main surfaces.

This method is widely used for comparatively small fabrications, particularly where final stoving is carried out and where the process can be automated. However, it is not generally used for structural steelwork, because it is not suited to many of the paints commonly used for such structures and the shear bulk of the steelwork would make handling difficult.

5.1.7.2   Flow coating

As a variation of the dipping process, items on-site are occasionally flow coated. This consists of building a watertight reservoir, generally in plastic around the base of the item to be coated, and pumping the paint through a hose to the top of the item. It is most useful for complicated structures such as transformer radiator tanks where an operator directs the hose from a point above so that the paint covers the inner surfaces. The only alternative to reach such surfaces is by spray nozzles on the end of extension tubes and in most such cases the spray operator has to work 'blind'. Such a process still requires the cleaning and preparation of the surface before painting, and if this cannot be achieved by solvent or steam washing, the paint application is unlikely to achieve its object.

If flow coating is carried out the paint must be carefully selected for its purpose, and during the flow coating operation the paint viscosity must be checked and rectified at frequent intervals.

5.1.7.3   Miscellaneous methods

Methods such as trowelling on thick coatings of high viscosity are some-times used. Barrelling, similar to the method used for sherardising zinc coatings, is also used for small components. Other methods widely used for industrial painting such as roller coating and curtain coating are not used for structural steelwork.

5.1.8   Comparison of application methods

The main comparison between the four common methods of application must be on the speed of operation. The following figures indicate the average area of steelwork that an operator might be expected to cover with one coat of paint per day: brush, $100\,m^2$; roller, $200–400\,m^2$; airspray, $400–800\,m^2$; and airless spray, $800–1200\,m^2$.

The rate of application is not the only matter to be taken into account, but the advantages to be gained by spraying are clearly demonstrated. Brush application is the slowest but for many situations the most effective, and masking and loss of paint by overspray are avoided. Roller coating is generally limited to large flat surfaces, although special rollers have been developed for other shapes. Spraying is the fastest method and in most situations airless spray is preferred.

5.2   Application conditions

Only paint application will be considered here; surface preparation is con-sidered in Chapter 3.

To obtain a satisfactory job with all methods of coating application the following are important:

1   There is access by an unrestricted, safe, working platform.
2   A cleaning team is ahead of the applicators. This particularly applies to airless spray, which pumps out a large volume of paint so that there is a tendency for applicators to spray everything in the path of the spray gun. Most experienced painting inspectors can recall such items as mud, piles of abrasive, cigarette packets, broken light bulbs, etc., that have all received a coat of paint in lieu of the sub-strate.
3   Correct lighting so that the operative can move about safely and have sufficient light on the work surface to carry out the job correctly. This is particularly important for high-build coatings, such as coal-tar epoxies where succeeding coats are of the same or very similar colour.
4   For spray operators on large work there should be a paint loader to keep up with the sprayers' output.

5   With spray operation it is preferable to have another operator to follow up the sprayer and touch-in and remedy visual defects.

5.2.1   Pre- or post-fabrication

For new structural steelwork there are four main procedures for carrying out the coating operations:

1   Surface preparation and priming at the fabricators. Coating system completed on site after erection.
2   Surface preparation and priming, plus the main part of the coating system completed at fabricators. Damage repaired and final coat only applied on site after erection.
3   Surface preparation, priming and coating on the ground at site and *before* erection.
4   Surface preparation, priming and coating at site and *after* erection.

Obviously there are further variations possible but one of the above is normally the main choice. It is not possible to state that any one method is preferable, there are advantages and disadvantages to each one. A NACE publication[2] comprehensively sets out the factors that have to be considered. Briefly the choice depends upon, amongst other factors:

1   The size and complexity of the fabrication.
2   The extent to which welding has to be completed after surface preparation and priming.
3   Availability of facilities at the fabricators, such as, convenience of access to items on the ground, adequate and suitable storage, etc.
4   Access on site.
5   Type of coating materials to be used, e.g. quick-drying, stacking resistance, etc.
6   Ambient conditions on site.

Pre-fabrication surface preparation and priming is generally the most economic option. In such cases, the user is seldom able to specify the paint manufacturer to be used since it is not practicable for the fabricator to change materials frequently. Ideally, specially formulated primers should be used, such as those based on polyamide-cured epoxy and pigmented with zinc phosphate. These dry rapidly and can be overcoated with most other types of paint for periods up to at least one year, without the need for special preparation other than the removal of dirt and repair of damage. The addition of fine-grade micaceous pigment to such primers also has advantages since it can provide a matt surface that aids overcoatability.

If primers are to be on the surface during welding operations, it is important that they should have two, quite separate, certificates of approval: (i) to indicate a satisfactory level of toxicity during the welding operation and (ii) to confirm that it will not adversely affect the strength of the weld. The latter is dependent upon paint film thickness and manufacturers' recommendations to this effect should be followed. If the area to be welded is relatively large, then preparation and priming after fabrication may well be the most efficient and economical option.

In almost all cases there are considerable advantages in stripe coating, that is, applying an extra coat of primer along edges, welds, etc. This is not popular with many painters because of the amount of effort involved for such a relatively small area of coverage. It must also add to the cost of application but for the end-user there is ample economic justification in increased durability. Paint systems commonly fail first at edges, welds, etc. because of the reduced paint thickness. T.N.O. Paint Research Institute in the Netherlands has studied the rheology of coatings applied to edges and has found that for normal, solvent-based paint systems there is a reduction of 40–70% of film thickness compared with the adjacent flat surfaces.[3]

5.2.2 The painting shop

The layout, equipment and environmental controls are important factors in determining both the efficiency of the painting process and the standard of the coating, i.e. its overall durability. Generally, insufficient attention is paid to the ability of a fabricator to achieve the necessary standards required and specified. It may well be advantageous to carry out a check or audit of the works before contracting the painting of steelwork. Apart from surface preparation, which is now capable of fairly strict controls, poor painting procedures are the most likely causes of inadequate coating performance. Against the obvious advantages of shop-controlled work must be weighed the occasional situation where numerous small, loose and separate items have to be painted, or where access has to be obtained to all faces of a large fabricated structure close to, or on, the ground.

Requirements such as cranes suitable for handling the steelwork, sufficient weatherproof area to paint and store the steelwork, properly maintained equipment for painting, sufficient trained operators and a proper division between blast-cleaning and painting operations are very important and should be checked. However, the particular factors that affect paint application are those concerned with environmental control, temperature and lighting, and these will be considered in this section. These will often be considered for health and safety reasons rather than for purely technical ones. Nevertheless, these environmental factors do influence the efficiency of paint application and should be properly controlled.

5.2.3 Ambient conditions

Although it would seem obvious that site painting should be scheduled for the most favourable months of the year, in practice it is a common experience to observe paint being applied under virtually any weather conditions, short of heavy rainfall. Even with the latter it is also a common experience to see work re-start immediately after the rain has stopped and when the surfaces are still, at least, damp. This is perhaps not surprising in countries like the UK with unpredictable summers. McKelvie[4] has produced evidence on the influence of weather conditions in the UK. This shows that the period of December and January cannot be expected to be favourable for outdoor painting unless special precautions such as enclosure, heating and/or special paint systems are used. February is also generally unsatisfactory and November and March to a lesser extent. The months of May and June provide the best conditions with a greater likelihood of efficient use of working time, when the painting is started later in the day. The work applied to London, Manchester and the Cardiff area; further north, for example in the Shetland Isles, the 'weather window' is generally even shorter and in some years ideal conditions have been non-existent. It is more important in these circumstances to select paint systems on the basis of their tolerance to application conditions rather than standard systems that really only give high performance under close to ideal application conditions.

5.2.3.1 Temperature

Temperature influences coating application in a number of ways, e.g. drying time, curing time of two-pack materials, solvent evaporation and paint viscosity. Temperature has a marked effect on the viscosity of paints; the actual effect depends on the paint but a typical gloss paint may become twice as viscous when temperatures drop from 30 to 20°C. The temperature at which paint is stored is also important. Depending on the type of paint and temperature, it may be necessary to warm or cool the paint before use.

Two-pack materials which cure by chemical reaction are extremely sensitive to temperature. Low temperatures will slow the reaction and high temperatures will cause it to accelerate.

Most epoxies will not cure below 5°C and ideally should be applied to surfaces above 10°C. Any paints applied to a cold surface will instantaneously cool to the temperature of the substrate and this will impair the flow into the innumerable interstices and crevices of, for example, a blast-cleaned surface.

Conversely, at high temperatures, the viscosity of any type of paint will drop, possibly to the point where it will have undesirable rheological characteristics such as runs and sags on vertical surfaces.

In extremes of temperature, brush and roller application may require the addition of extra solvent, with consequent loss of film thickness. Spray operation, both air or airless, will also need modification to spraying pressures, tip sizes and possibly the use of special solvent to avoid dry spray and other film defects.

Temperature is also critical for the application of water-borne coatings since there is a minimum temperature at which a proper film will form by coalescence.

Many paint manufacturers only provide data about their products, such as two-pack epoxies, at standard temperatures of 20 or 23°C, whereas variations from the standard can make significant practical differences to such properties as pot life, minimum and maximum allowable overcoating periods, and drying and curing times. Very approximately, a difference of 10°C can double or halve the time, as appropriate. The effect varies with the quantity of solvent present, the type of curing agent and mass of paint. This should be checked for each type and make of paint used. Apart from their effect on paints, low temperatures may lead to moisture condensation on cold steel surfaces. This can lead to problems with paint application. When temperatures have to be raised, only indirect heating methods should be used. Heaters that produce combustion products inside the shop should be avoided. In situations where it is impracticable to heat the whole shop, local heating in the working area may be employed. Where the steel surface has to be heated, this must be carried out cautiously to avoid contamination or overheating of the steel itself.

The outer surface of a paint film should never be force heated in order to speed drying. This is because with most materials, particularly two-pack, that cure by chemical reaction, this will cause the top surface of the paint to cure and leave the underlying paint film uncured: this can occur in hot sunlight on a newly applied paint and the effect is magnified with black or dark-coloured coatings.

5.2.3.2 Relative humidity

The moisture content of the air has an influence on painting operations and is indicated by the relative humidity at a particular temperature. Relative humidity is important during paint application for two reasons. Firstly, depending on the amount of moisture in the air and the temperature of the steel surface, some condensation may occur on the steel and this may lead to adhesion problems with some paints. Secondly, some paints are sensitive to humidity during their curing, regardless of whether condensation occurs or not.

It is normal practice to specify that paints shall not be applied to surfaces that are within 3°C of the dewpoint. The higher the relative humidity, the greater the risk of condensation, for example, in the temperature range 0–10°C, if the RH is 80% the temperature needs to drop 4 degrees before condensation occurs, but at 98% RH it only needs to drop a half a degree. Therefore, if the RH is 85% or higher, painting operations should be critically reviewed since the dewpoint is only 1 or 2°C away at the most. These degrees of cooling may be reached during application solely by solvent evaporation effect, particularly when there is rapid evaporation of solvents, such as with low-boiling types with high evaporation rates. Fast evaporation of solvents can, therefore, reduce the temperature of the paint droplets below the dew point during the application. When this occurs, moisture from the atmosphere condenses on the surface of the paint particles and is then entrapped in the paint film. This may prevent proper film formation.

An important factor is the trend of humidity conditions. If weather conditions can confidently be expected to remain static or improve in, say, the following 6 hours, then application could start when the surface to be painted is less than 3°C above dewpoint.

Some paint systems, such as moisture-cured urethanes and zinc silicate primers, require a degree of moisture before they will cure, but most high-performance coatings are adversely affected by moisture during application and curing.

The relative humidity conditions applicable to various coatings should be but are not always, given in the paint manufacturer's data sheets. Water-borne coatings are particularly sensitive to conditions of high relative humidity. The moisture in the coating film will evaporate more slowly and increase the time the coating remains water sensitive, i.e. can be damaged or even washed off the surface by rain or condensation. Prolonged exposure of a wet coating on a bare steel substrate can also cause flash rusting.

5.2.3.3 Ventilation

Ventilation is necessary in a paint shop or confined space to keep the concentration of fumes and vapours to an acceptable level. This requirement is determined by the threshold limit value (TLV) for solvents. This is the total, in parts per million, of the solvent which over an 8-hour day will give no ill-effects. Tables providing this information have been prepared and for some solvents, including most ketones, it is low, e.g. 50–100 ppm. Generally, in modern painting shops overspray is collected using either down-draught ventilation near where the painting is carried out or by using spray arrestors. However, ventilation is also required in other parts of the shop to reduce the possibilities of explosion of low-flash-point solvent and also to reduce the amount of dust and contamination in the shop.

5.2.3.4 Lighting

Adequate lighting might appear to be a fairly obvious requirement, but in some paint shops and under some site conditions, the amount of light is in fact inadequate for proper painting of the steel and effective inspection of the applied paint coating. The intensity will be determined by the type of work but generally requires to be between 500 and 1000 lux.

5.2.3.5 Wind and weather

Exposure of surfaces before paint application or of uncured paint films to rain, frost or dew will generally have a damaging affect both on the inherent adhesion and on the ultimate durability.

Painting in high winds should be avoided, especially when spraying as, apart from risk of overspray, the wind blowing across the atomised paint can result in a powdery, porous, dry-sprayed film. High winds also increase the possibility of dirt, sand, debris, etc. falling on the cleaned surface of the wet paint.

5.2.4 Storage of paint

Paint must be stored in proper store rooms under proper temperature control with adequate fire precautions and should be open only to authorised personnel. All paint and solvent containers should be clearly marked and recorded in a book so that stock is used in the correct order. If the shelf life is exceeded, then the particular paints should be disposed of or, depending on the contract terms, returned to the supplier.

5.2.5 Preparation of paint before use

Even where paints are properly stored, it is necessary to prepare them correctly before use. Pigments tend to settle during storage so thorough mixing is required to ensure homogeneity. This is particularly important with heavy pigments, e.g. in zinc-rich paints. If these are not thoroughly mixed, the dry film may not meet the requirements of the specification. Before mixing, the paint should be brought to the approximate temperature of use. Mixing is usually carried out mechanically in special equipment by means of revolving stirrers. Alternatively, up to about 20-litre cans can be prepared by shaking in suitable units. If manual mixing is to be carried out, this should not be attempted with larger containers of paint; a can size of about 10 litres is the maximum that can be mixed in this way. Two-component paints, e.g. epoxies, must be mixed carefully. The 'hardener' should be added in the correct ratio as specified by the paint manufacturer. Two-pack materials should only be used in the complete 'pack' size. Attempts to use fractions inevitably lead to incorrect proportions. Apart from the inaccuracies in measurement, there is also the difficulty of making an allowance for the amount of viscous material left on the sides of the measuring container.

The pack size should always be chosen in order to allow for application within the 'pot life' of the paint. Thinning of paint before and during application must be carried out in accordance with the paint manufacturer's recommendations. Only thinners of the recommended type must be used and should be added to a well-mixed paint. If, after mixing, there is any skin or contamination of the paint, then it should be strained to remove all foreign particles. It is probably necessary to strain all paints that have been used and left standing for some time.

5.3   The painter

Paradoxically, in many countries, including the UK, a house painter and decorator is more likely to be trained and experienced in the methods of paint application than an industrial painter working on a major industrial project. To rectify this the National Association of Corrosion Engineers (NACE) in the USA is starting a painters' qualification programme. Painters will be required to demonstrate their skill in or knowledge of safety regulations, degreasing, surface preparation of both steel and concrete, methods of application and use of quality control instruments such as wet film thickness gauges (see Figure 5.5), dry film thickness gauges and wet sponge holiday detectors.

It is to be hoped that other countries will follow this excellent lead. However, in the past, painting contractors have been reluctant to finance such schemes because of the transient nature of the employment. Such ideas can only work if the end-user insists on, and pays a premium for, qualified painters and also makes allowance for the interim period at the start of the scheme. Fortunately, in many countries, including the UK and the USA, there are schemes being implemented for the pre-qualification or approval under a quality assurance scheme of painting contractors. This may well redress this deficiency.

5.4   Paint manufacturers' data sheets

Data sheets produced by the paint manufacturers provide the major source of information concerning their materials. Very often compliance

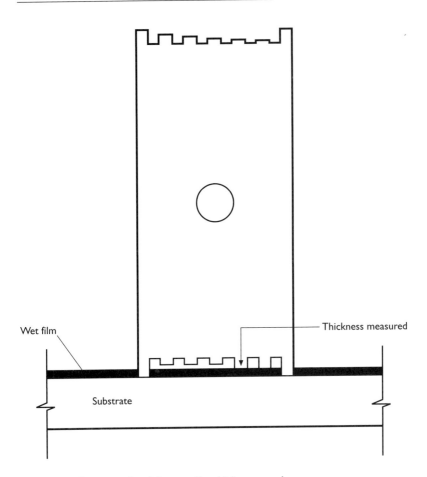

Figure 5.5   The principle of the wet film thickness comb gauge.

with the manufacturer's data sheet forms part of the requirements of the contract. Disregarding these recommendations could relieve the paint manufacturer of any responsibility in the event of subsequent problems. It is essential to possess the latest version of the relevant data sheets which are available to both painters and quality control inspectors.

Of necessity, a paint material has to be a compromise of properties, some of which are in direct conflict with each other. Unfortunately, paint manufacturers are reluctant to mention any aspect that requires special care and attention in case it implies a weakness in their own proprietary material rather than an inherent weakness of the particular type of paint. The following information is the minimum required:

(a) General descriptive name and colour.

(b) Principal characteristics and recommended uses.
(c) General description of binder and, if appropriate, main pigmentation.
(d) Any special requirements in application and use.
(e) Mass density.
(f) Solids content by volume and tolerance allowable.
(g) Recommended dry film thickness with a minimum and maximum permitted.
(h) Touch-dry time.
(i) Minimum safe period for the stacking of painted articles.
(j) Minimum interval before overcoating.
(k) Maximum interval before overcoating.
(l) Full cure.
(m) Pot life, if applicable.
(n) Properties (h)–(m), as above, at a range of temperatures.
(o) Shelf life and recommended storage conditions.
(p) Recommended methods of application.
(q) If appropriate, precautions necessary for any particular form of application.
(r) Recommendations for suitable spray set-ups, where appropriate.
(s) Recommended thinners.
(t) Cleaning solvent required.
(u) Compatibility with primers or topcoats, as applicable.
(v) Mixing instructions.
(w) Induction time, if applicable.
(x) Safety precautions. For most modern paint systems there should be separate data sheets on safety requirements giving full details of threshold level values and lower explosive limits and any other hazards applicable.

5.5 Health and safety matters

5.5.1 Airless spray

The pressures involved in airless spray may be up to 6000 psi. There are obvious dangers of handling equipment at such pressures. The fluid hose and fittings must be kept in good condition and replaced if there are any signs of damage. The material from which the hose is made must be fully resistant to solvents and suitably earthed to prevent a build-up of static electricity. All hose connections should be tightened securely and checked before use. The airless spray gun should be fitted with a safety catch to prevent accidental operation of the trigger and fitted with safety 'horns' or 'tips' that protrude in front of the orifice. These are often coloured yellow and their purpose is to prevent anyone getting part of their body too close to the orifice. With such high pressures forcing material through a small orifice there is a danger that paint could be injected into the skin, resulting in a loss of a limb or even a fatality.

5.5.2 Paint materials

5.5.2.1 General

The solvents, resins, pigments and other ingredients in paint coatings can all affect health by inhalation, ingestion and absorption in the body. Some possible hazards are highlighted below but they do not represent all that might occur. It is essential that a paint manufacturers' Material Safety Data Sheet (MSDS) or, in some cases, the Technical Department of the paint manufacturer, is consulted before using any specific material, or if in doubt about any risk or safety phrases on the label of the containers.

5.5.2.2 Flammability

All coatings that contain organic solvents are flammable and so also may be the resins. The greatest danger lies in application when coatings in the liquid state are present.

5.5.2.3 Explosive hazard

Most coatings are not explosive in the liquid state. Even if the surface of the liquid is alight it will not explode. However, in a fire the paint container can expand and the lid can be blown off the container owing to expansion. In that case flammable material could be spread over a wider area.

In an enclosed space or stagnant pocket, the possibility of explosion depends upon the concentration of solvent vapours in the atmosphere. With very high concentrations, no explosion can occur because there will be insufficient oxygen. In low concentrations there may be insufficient solvent vapour present to ignite and explode. Painters are only concerned with this lower explosive limit (LEL). Table 5.1 gives typical LEL figures by volume in air for common solvents. Table 5.2 gives typical LEL figures for different types of paint.

Adequate ventilation is the key requirement for painting in confined spaces, but even with adequate ventilation there are other necessary precautions:

(i)   No smoking, welding or flame cutting within at least 15 m from the painting operation.
(ii) All electrical equipment should be explosion-proof and no commutator-type electric motors should be used in the vicinity.

(iii)   All tools, equipment, footwear, etc. should be of the type that does not generate sparks.
(iv)   Nylon overalls or other plastic items liable to cause static discharge should not be worn.
(v)   Equipment liable to generate static electricity, such as blast or paint spray hoses, should be adequately earthed.
(vi)   Solvents or paints should not be applied to hot surfaces.

5.5.2.4   Flash point

The flash point is the temperature of a solvent at which it releases sufficient vapour to ignite in the presence of a flame. In this respect, the higher the flash point, the safer the solvent. The closed-cup method of flash point determination gives lower results than the open-cup method and is therefore normally the figure quoted. Table 5.1 gives typical closed-cup flash points for common solvents.

5.5.2.5   Evaporation rate

The relative evaporation rate for a solvent is based on an arbitrary value of 1 for ethyl ether. The higher the evaporation rate the longer it will take for the solvent to evaporate from the paint film and form solvent vapour in the atmosphere. However, the slow evaporation means that the paint film stays wet longer and presents a greater flammability hazard. Table 5.1 gives the relative evaporation rates for common solvents.

5.5.2.6   Solvent vapour density

The greater the solvent density, the more likely that the vapours will accumulate in the lower portions of a confined space. Localised pockets may

Table 5.1   Solvent properties

Solvent	Evaporation relative rate	Closed-cup flash point, 0°C	LEL vol % in air	TLV (ppm in air)	
				TWA	STEL
Acetone	4	32	2.15	230	1000
Ethyl alcohol	20	14	2.23	1000	—
Methyl ethyl ketone (MEK)	8	−1	1.81	200	300
White spirit	150	38–43	1.10	100	—
Naphtha	105	38–43	1.20	300	400
Toluene	15	4	1.27	100	150
Xylene	35	16	1.00	100	150

then reach the lower explosive limit even though the bulk of the volume is satisfactory.

5.5.2.7 Reactivity

Many of the two-pack, chemically cured coatings, such as polyesters, epoxies and urethanes, and particularly those with 100% solids, generate a substantial heat if left in the container for a long period after the curing agent has been added. For example, if left overnight the exothermic reaction could build up to a flammable level.

5.5.2.8 Hazards from solid components of paint

Many high-performance coatings can cause dermatitis. Epoxies containing solvents are particularly hazardous, since they can penetrate the skin more easily. The effect appears to be cumulative in that once a person is sensitised, even limited contact can cause a reaction all over the body. The use of protective ointments and creams is a worthwhile precaution. Coatings may contain metals or metallic compounds; cobalt can cause pneumonia and asthma;[5] chromium can irritate or damage the nose, lungs, stomach and intestines and can increase the risk of cancer;[5] arsenic can increase the risk of cancer;[5] zinc chromate increases the risk of cancer and has been banned in many European countries. Isocyanates are used in urethane coatings. If present as free monomeric isocyanates they can result in chest tightness, vomiting, abdominal pain and redness, swelling and blistering of the skin. Chronic exposure may result in flu-like symptoms such as fever, chills, aching and nausea with reduced lung function and possible lung damage. Once workers are sensitised, even exposure to airborne quantities below the occupational exposure limits may result in serious asthmatic reaction.

Epoxy resins and curing agents are used in both water- and solvent-borne coatings. Materials such as glycidyl ethers are used to modify the resins and these are irritants to the respiratory tract as well as the eyes and skin.[5] Aliphatic polyamines are sometimes used as curing agents for these plural component materials and are strong irritants and sensitisers.

Coatings may contain forms of crystalline silica. Types that may be listed in a safety data sheet include: quartz, cristobalite and trydimite. The dry dust from these could cause silicosis.

5.5.2.9 Solvent hazard

Overexposure to solvents can lead to severe health problems. Typical solvents found in industrial paint systems and their possible effect on health are as follows:

Ketones: for example, methyl ethyl ketone (butan-2-one or MEK) and methyl isobutyl ketone (4-methylpentan-2-one or MIBK) can cause irritation to the eyes, nose and throat. In high concentrations, exposure can result in narcosis, with symptoms of headache, nausea, light-headedness, vomiting, dizziness, loss of coordination and loss of consciousness; prolonged exposure can be fatal.

Aromatic hydrocarbons: for example, toluene, xylene, dimethyl benzene and other solvents similar in structure to benzene. Contact with these solvents can cause skin irritation, but the respiratory problems are severe. Acute exposure can result in narcosis or can damage the lungs. Chronic exposure can damage the liver, kidneys and bone marrow. High concentrations can be fatal.

Alcohols: for example, methyl alcohol (methanol) may be reported as toxic and other alcohols tend to irritate the skin, eyes and respiratory system. Acute exposure can result in depression of the nervous system, thus slowing the activity of the brain and the spinal cord. A sufficiently high exposure can be fatal.

5.5.2.10   Injury by penetration into the skin

Solvents or solvent vapours that enter the pores of the skin will give symptoms similar to those of solvent inhalation. Skin contact with solvents should be avoided wherever possible.

5.5.2.11   Injury by swallowing

Solvents that remain as liquid in the stomach should be rendered harmless by being metabolised. However, lung irritations can occur because much of the dose will leave the body as vapour in the expired air. Fortunately, a frequent result of swallowing solvent is the protective reflex of vomiting.

5.5.2.12   Eye injury

Solvents in the eye can cause corneal necrosis. Afflicted eyes should be flushed with water and medical attention should be obtained at once.

5.5.2.13   Toxicity

The maximum allowable concentrations of solvent vapours are known as threshold limit values (TLV) and can be listed in two forms: TLV-TWA (time-weighted average) and TLV-STEL (short-term exposure limit); the latter is more commonly quoted in North America than in Europe. The definition of TWA is the time-weighted average concentration for a normal 8-hour workday and a 40-hour work week, to which nearly all workers may be repeatedly exposed, day after day without adverse effect. STEL is the maximum concentration to which workers can be exposed for a period up to 15 minutes continuously without suffering from: irritation to the lungs; chronic or irreversible tissue changes; or narcosis of sufficient degree to increase accident-proneness, impair self-rescue or materially reduce work efficiency; provided that no more than four exposures per day are permitted and provided that the TLV-TWA is also not exceeded.

Ventilation is the key to safe application of coatings in enclosed areas. Its significance cannot be overemphasised from the standpoint of fire, explosion and health. Suction fans should always be used and positioned so that they draw from the lower areas of the enclosed space. Ventilation should be continued until the coating is sufficiently dry to ensure that, with the ventilation removed, no area of the tank will build up vapours to the explosive limit.

The minimum ventilation air in cubic metres per minute may be calculated from the formula:

$$\frac{(P \times A) + (Q \times B)}{t}$$

where P = volume of paint applied in litres in time t; Q = volume of added solvent used in the paint applied in time t; A = ventilation air quantity for 1 litre of paint to reach 10% LEL (obtain this information from paint supplier); B = ventilation air quantity for 1 litre of solvent to reach 10% LEL (obtain this information from paint supplier); t = time of application in minutes, of volume P of paint.

EXAMPLE

100 litres of paint (P) plus 5 litres of thinner (Q) are used within 45 minutes (t). $A = 60\,\text{m}^3$, $B = 130\,\text{m}^3$.

Ventilation air quantity to reach 10% LEL is then

$$\frac{(100 \times 60) + (5 \times 130)}{45} = 147.7\,\text{m}^3/\text{min}$$

Table 5.1 gives the TLV-TWA and TLV-STEL for common solvents. Table 5.2 gives the TLV-TWA and TLV-STEL for common paints. Solvent vapour meters which monitor toxic or combustible gas levels should always be used for the application of coatings in enclosed areas. Visitors and workers in such areas should wear air-fed masks.

Today, there is particular concern about the spraying of paints containing isocyanates, for example urethanes, isocyanate-cured epoxies, etc. Their wide use is due to their considerable weather-resistant and

Table 5.2 Coating properties

Coating	Flammability	Solvent	LEL vol. % in air	TLV-TWA (ppm)	Toxicity
Oil-based	Flammable	Aliphatic hydrocarbon may contain turpentine	0.8 if contains turpentine; 1.1 if white spirit	100 if turpentine; 500 if white spirit	Non-toxic
Alkyd	Flammable	Aliphatic or aromatic petroleum	1.1	200–500	Non-irritating
Chlorinated rubber	Non-flammable	Aromatics	1–1.3	100	Possible skin irritation due to solvent
Vinyls	Self-extinguishing	Ketones and aromatics	1.3–1.8	100	Non-toxic
Epoxies, solvent-based	Will support combustion	Ketones and aromatics	1.3–1.8	100	Dermatitis
Coal-tar epoxy	Will support combustion	Ketones and aromatics	1.3–1.8	100	Fumes irritating Skin irritation Dermatitis
Polyurethane	Flammable	Ketones and aromatics	1–1.3	100	Toxic fumes Dermatitis
Coal-tar	Flammable	Aromatics	1.1–1.27	200	Severe skin irritation
Inorganic zinc silicate	Non-flammable	Ethyl alcohol	3.2	1000	Possible mild skin irritation

abrasion-resistant properties. Their use by operators possibly unfamiliar with the health hazards has prompted the UK Health and Safety Executive to produce a guidance leaflet for the motor repair trade.[6]

5.5.2.14   Instrumentation

The percentage of solvent vapour present in a confined atmosphere can be determined by the use of an apparatus that draws a sample of the air through a calibrated tube containing a chemical reagent. The chemical in each type of tube is sensitive to a particular gas and changes colour appropriately. Advice should be obtained from the paint manufacturers as to which type of tube is the most suitable for their product. Although the method is not suitable for continuous sampling, its cheapness and ease of use make it suitable for applications where accuracy is not of prime importance.

Solvent vapour meters are also available that can detect toxic or flammable gases or oxygen deficiency either continuously or as a portable instrument. Such instruments generally work by the use of a semiconductor metal oxide detector. Semiconductor devices (for example, transistors) depend upon the modification of their electrical conductivity properties by the addition of certain chemical impurities, known as doping. When metal oxides are 'doped' with small concentrations of suitable metals or rare earths, they can be made to function as gas detectors. The semiconductor element of such detectors consists of a small pellet of 'doped' oxide which is heated by a minute platinum/rhodium filament. When a flammable gas passes over the pellet, the gas is adsorbed by the oxide surface and alters the electrical conductivity. This change in conductivity produces a small voltage which, after amplification is recorded on a meter. These instruments can be very sensitive and selective and can measure low concentrations. It is reported, however,[7] that some makes of instrument on the world market lack selectivity in that they will respond equally well to a number of gases with different explosive ranges. This may produce misleading and possibly dangerous readings if the user is unaware that the gas sampled is not that for which the instrument is calibrated.

The instruments also measure the oxygen content of the atmosphere by the use of a galvanic cell which produces current in proportion to the oxygen content. When the current reads a pre-set level, an alarm sounds.

5.5.2.15   Water-borne coatings

It is commonly thought that application of water-borne coatings eliminates all the hazards associated with the application of solvent-borne coatings. It is certainly true that they present much less risk of fire and explosion since they have a much higher flash point. It is however important to remember that most water-borne coatings, and this especially applies to the water-reducible types (see 4.9.5), contain some organic solvent, and so personal safety precautions are still important and often necessary. For instance, dizziness, watery eyes or headaches while applying water-borne coatings may indicate inadequate ventilation in the work space and the need for a respirator. This is particularly the case when water-reducible types are being spray applied. It may not be necessary with brush or roller application but much depends upon the degree of ventilation.

The amount of solvent in a water-borne coating should be determined from the Materials Safety Data Sheet (MSDS) and this should be consulted to determine the degree of protection required for any particular product.

Different people have different sensitivities to materials and some people are more sensitive to water-borne coatings than others. The use of gloves and/or protective cream is always an important safeguard.

5.5.2.16   Polyurea spray coatings

Because they have no VOC's and other advantages over their traditional counterparts, such as polyurethane, the relatively new polyurea elastomer coatings are now increasingly used for pipelines, tank interiors etc. The potential health and safety hazards associated with the handling and spraying of polyurea arise both from the toxicological properties of the chemical components and the mechanical aspects of the dual-component spray application necessary for these fast curing, high-build materials.

Under normal equipment operating conditions, the isocyanate component and the resin component react instantaneously so that there is not an excess of either unreacted isocyanate or amine at the spray nozzle. However, even under ideal operating conditions, very small amounts may be present as an aerosol or vapour. These can cause irritation to the eyes and respiratory tract. Repeated inhalation of such an aerosol or vapour may produce a hypersensitivity reaction of the respiratory tract similar to an asthma-like response. Contact with the skin can also cause similar irritation and a hypersensitivity reaction.

References

1. Bosklopper, R., *J. Protective Coatings and Linings* (Sept. 2000) 17.
2. *Structural Steel Painting, A Comparison of Various Work Sites and Procedures.* NACE Publication 6H 184, 1984.
3. Winkeler, M. and Van Der Poel, H., Improved service life expectancy through better edge coverage. *J. Protective Coatings and Linings* (Dec. 1989) 16–20.

4. McKelvie, A. N., Prediction of the suitability of the weather conditions for external painting, *Br. Corros. J.*, **10**(2) (1975) 107–10.
5. O'Mally, D., *J. Protective Coatings and Linings* (Aug. 2000) 40–7.
6. UK Health and Safety Executive. Guidance Leaflet on Two Pack Spray Paints.
7. Anon. Detection and Monitoring Part 1. Flammable Gas Detectors. Safety Surveyor. Vol. 2, No. 3, Sept. 1974.
8. Applicator Training Bulletin. *Protective Coatings Europe*, (Sept. 2001) 13–15.
9. Knight, W. J., Polyurea Spray Coatings: an introduction. *Protective Coatings Europe* (Sept. 2001) 48–52.

Chapter 6

Specialist coatings and applications

Modern paint systems can provide very corrosion-resistant coatings and are used successfully in many aggressive environments. There are, however, some situations, such as the lining of tanks, where, although the materials may be similar to those used for structural steel, there are special requirements.

There are also specialist coating materials, such as powder coatings, tapes and fire-proofing coatings that are also associated with structures.

6.1   Coating or lining of tanks

Coatings for tank internals are often called linings and they may be applied directly to the steel as coatings, or as sheets cut to shape and cemented to the steel to provide a barrier to the liquids in the tank. The choice of coating is generally determined by the nature of the liquid and its corrosivity to the steel. However, with some solutions, tainting with even a small amount of iron corrosion products is not acceptable. Consequently, even a limited amount of corrosion which might not seriously affect the integrity of the tank has to be avoided by the choice of suitable lining materials and application methods. Lining of tanks, therefore, may be to protect the tank from corrosion caused by the contents, but also to protect the contents, or cargo, from the metal of the tank.

6.1.1   Corrosion protection

From a corrosion protection point of view, a tank can be divided into three areas, bottom, walls and roof, both internally and externally.

6.1.1.1   Internal tank bottom corrosion

Providing it is supported, the bottom plate is not considered a structural member and is therefore often made of thinner plate than the walls. Since water, sometimes seawater, can collect at this point and differential aeration cells can form due to deposits on the surface, bottom plates can suffer severe pitting and metal loss, for example, 0.5–2.0 mm per year.

Sulphate-reducing bacteria have been detected in oil-storage tanks and this could add to the problem, but Delahunt[1] considers that pitting from this cause is minimal.

6.1.1.2   External tank bottom corrosion

Tank bottoms resting on the ground are susceptible to corrosion attack. Ideally the foundations should consist of an inert and compact surface such as a layer of fine aggregate covered with oiled sand or asphalt or bitumen. The use of coal cinders or other acidic materials, or large rocks that allow differential aeration cells to form, can greatly accelerate corrosion.

Corrosion can also be caused by stray currents. An external source of DC, such as from electrified railways or nearby impressed-current cathodic protection, may cause current to flow through the ground to the tank bottom and cause pitting. Measurements of potential, and particularly fluctuations in the potential, can indicate stray currents.

6.1.1.3   Internal wall corrosion

Attack on the shell occurs at the junction with the bottom plate owing to water collection, but this normally extends no higher than 300 mm. Corrosion can also occur in the upper courses of the shell because of condensation and corrosive atmospheres entering the tank owing to expansion and contraction of the stored product.

6.1.1.4   External wall corrosion

The external wall is subject to normal atmospheric corrosion attack with two added problems. Firstly, the temperature of the stored product, if it is different from the ambient, can cause condensation. Such condensation contaminated with, for example, salt spray from a marine atmosphere, can cause severe corrosion. Secondly, if the tanks are lagged then, regardless of the type of insulation used, corrosion under the lagging can be very rapid and severe.

Hot tanks experience severe attack of the walls at the bottom where the insulation can become saturated with water. Refrigerated tanks can corrode under the insulation on the roof and upper strakes. Pitting rates of up to 1.5 mm per year have been reported and such corrosion is impossible to locate by visual inspection without removal of the insulation.

Methods of prevention for such corrosion usually include the provision of external cladding of either metal or a mastic coating to prevent water penetration. Mastic systems based on butyl or Hypalon rubber are becoming increasingly popular because they can be more easily and effectively sealed at points where the insulation is penetrated, for example, at ladder fixings and the like. In addition, it is advisable to paint the steel, before applying the insulation, with a suitable barrier coat, such as polyamide-cured epoxy. Zinc-rich systems should not be used for this purpose.

6.1.1.5   Internal corrosion of tank roofs

In roof areas, corrosion can occur owing to condensation and the corrosion rate may be greatly accelerated by chlorides from the atmosphere or hydrogen sulphide from the product under store, such as oil. Overlapping roof plates that are not sealed underneath also provide crevices that cannot be painted and will promote corrosion. If permissible in the design of the structure, such areas should be sealed by welding.

Severe corrosion can occur with floating roof construction because the movement of the roof can dislodge corrosion products and reveal a fresh surface for corrosion attack.

6.1.1.6   External corrosion of tank roofs

This is similar to that occurring on the tank walls except that buckling of plates, depressions on flat surfaces, etc. can cause ponding of water and accelerate corrosion.

6.1.2   Lining materials

A wide range of coating materials is used for tank linings. The choice will depend upon the nature of the solutions to be contained or, in the case of solid materials, the amount of abrasion and wear likely to occur.

The materials considered below are those most commonly used. Some general indication of their chemical resistance is given but detailed advice should be sought before selecting linings for highly corrosive liquids. Useful tables on the chemical resistance of various materials have been published,[2] but it may be necessary to carry out tests before finalising the choice of lining. Variations from published data, such as differences in concentration and temperature or impurities in the solutions, may affect the performance of lining materials.

6.1.2.1   Organic coatings

It is unlikely that any one type of coating would be suitable for all tank linings. They are required to withstand a multitude of environments and be applied under many different circumstances. For exposure to water immersion, zinc silicates and epoxies have been used successfully. For exposure to alcohols and phenols, zinc silicates have been used successfully. For chemical resistance, there are often suitable formulations based on epoxy resin, where stoved phenolic linings were used in the past. To reduce the problems arising from solvent emissions there is an increased use of solvent-free coating and dual-component spraying equipment. The US Navy has reported[3] that, as a result of a five-year test programme and five years' subsequent use in their ships, the use of solvent-free epoxies has significantly improved the maintenance-free life of their tank coatings. In particular, formulations based on bisphenol-A epoxy resin and amine-based curing agent for seawater ballast tanks and the lower viscosity bisphenol-F epoxy resin for chemical and fuel tanks. It was also appreciated that the choice of the right coating material was not sufficient on its own and that they also had to improve their application techniques and quality control to realise the coatings, potential. Sometimes tank linings have to be applied in less than ideal conditions, for example: cold ambient temperatures, high humidity and often the need to re-commission the tank as soon as possible. There are epoxy resin systems which can fulfil many, if not all, of these requirements.[4] However, there is no miracle coating and again it must be emphasised that even with judicious coating selection this must be accompanied by a well-written specification, conscientious quality control and qualified applicators using suitable application equipment, with de-humidifiers as required.

For very thick coating systems, and particularly for the repair of corroded tank bottoms, hand lay-up of glass-reinforced polyester coatings over an epoxy primer, is frequently used. Glass fibre or glass flake-reinforced epoxy resins are also used and have excellent adhesion and good chemical resistance, particularly to alkalis, which is a weak point for the polyesters. The epoxy systems, however, are generally more expensive, and because only solventless types can be used at the thickness required, they tend not to wet out the glass fibre or flake quite so satisfactorily.

In the petroleum industry there has been a move towards vinyl esters because they have superior resistance to unleaded fuels. The vinyl esters are based on an epoxy resin reacted with an acrylate monomer and dissolved in styrene. Like polyesters, they cure by the addition of a peroxide catalyst.

The polyesters and vinyl esters in particular require special care in handling and application. Nill, of Phillips Petroleum USA, has published a comprehensive proposal for the method of installing fibre-reinforced plastic (FRP) coatings in petroleum storage tanks.[5] It is particularly concerned with the safety requirements, method of repairing holes, surface preparation and quality control.

Other organic-type lining systems include thermoplastic resins such as polyvinyl chloride, polypropylene, polyvinylidene fluoride,[6] chlorinated polyether and chlorosulphonated polyethylene. Although in some cases these plastics can be applied as dispersions, for tank linings they are normally in sheet form. The sheets are often provided with a woven glass fibre cloth backing in order to assist adhesion to the steel. The seams are then heat-welded to provide a continuous lining. These materials or their adhesives have temperature limitations but have the advantage of being able to apply a thick film in one application. Sheet linings are usable in thicknesses varying from 1.5 to 6 mm.

A number of sheet lining systems are based on elastomeric materials such as semi-hard or soft rubber, neoprene, or Hypalon. The rubber systems have excellent ageing properties, particularly when protected from heat or oxidising solutions. Neoprene is superior to rubber in its resistance to oils, sunlight, heat, etc. Hypalon is used where resistance to sunlight, ozone, abrasion, aliphatic oils and oxidising chemicals is required.

6.1.2.2   Cementitious

Cementitious linings are composed of inert aggregate in an inorganic binder, such as potassium or sodium silicate. These materials can be applied by trowel, but for linings, normally a special high-pressure spray system is used. To assist adhesion, special anchor points are generally welded to the substrate.

The actual lining material has good heat, abrasion and corrosion resistance but *in situ* tends to crack due to thermal cycling and has poor resistance to steam, alkalis and weak acid solutions. They are therefore normally applied over an acid-resistant membrane.

In the majority of cases, cementitious linings are applied at site because transport of a lined tank would involve extra weight and the danger of cracking of the lining due to the flexing of the tank walls.

Acid-resistant brick linings are also used for the protection of certain types of tanks or process vessels. The cement mortar between the bricks can be based upon polyester, phenolic, epoxy or Furan cements depending upon the service conditions.

6.1.3   Application of linings

6.1.3.1   Design requirements

Tank linings require an exact specification, careful application and rigorous inspection. Surface preparation is normally required to a higher standard than for structural steel.

Correct design of the tank is important. For example, it has to be realised that there must be sufficient and suitably sized access holes for all the equipment, ventilation hoses, etc.

All holes, cracks and the like need to be repaired before work starts. It is also generally necessary to fill all lap seams, depressions, sharp corners, etc. with epoxy putty to provide a smooth surface. All edges and welds must be radiused, with all welds free from undercutting, porosity and inclusions. Porous or rough welds must be treated. In riveted tanks it is desirable to seal-weld the rivets and install water stops. Contrary to what may appear obvious, many authorities believe that coating application should start at the bottom of the tank and work upwards. This is because it is exceptionally difficult to rid the tank floor of all traces of abrasive. Treating that surface first gives the minimum amount of abrasive that has to be removed before the coating is applied. Only non-metallic abrasives should be used. Obviously the applicator must take special care to avoid damage to the coating already applied. Rubber-soled shoes or overshoes should be worn by any person inside the tank after the primer has been applied and ideally these shoes should be changed and left inside the tank on entering and leaving.

6.1.3.2   Rigging or scaffolding

Tanks up to 10m to the higher point in the roof are often painted using movable platforms. The height of the work platform should not exceed four times the main base dimension.

Beebe[7] suggests that higher tanks and those up to 9m in diameter can be painted with a painters' boom or 'merry-go-round'. This consists of a tight vertical cable between a roof opening and a fixing lug on the tank bottom. The painter's stage is controlled by ropes attached to the vertical cable, and manipulated by the painter and a helper on the floor of the tank.

Even larger tanks can be painted using a swing stage. This requires two attachment points: one at or near the top centre of the tank and one on or near the tank's walls.

All this points to the need to consult experts on rigging at the design stage of the tank so that attachment points can be included to provide safe, effective and economic access.

6.1.3.3   Ventilation

Toxicity of solvent vapours and risk of explosion are the hazards most likely to be present when painting inside a tank or confined space. However, ventilation is not only necessary for reasons of safety: the performance of a coating system can be greatly affected by the type and amount of solvent trapped in a coating when it cures or dries. Adhesion, water resistance, mechanical or chemical properties can all be adversely affected. Very slow evaporation of trapped solvents can also develop internal stresses. Condensation of solvent onto adjacent coated surfaces can also cause problems.

Many solvents are heavier than air and if the ventilation is inadequate can remain in stagnant pockets at the base of the tank. Ideally, air should be exhausted from the bottom of such tanks and clean air drawn in from the top of the tank. In addition suction fans tend to reduce the air pressure within the tank slightly and this helps with solvent evaporation.

Exhaust fans must be capable of achieving the required volume of air movement. Table 6.1 gives typical ventilation volumes for maintaining solvent vapour concentrations below 10% of the Lower Explosive Limit.[8]

6.1.3.4   Dehumidification

In cold climates, or at night-time in hot climates, the temperature of the air inside the tank is often higher than outside. This is particularly so if the temperature is raised in order to speed cure and the walls are not insulated. Under these conditions, the hot air impinging on the tank walls can cause copious condensation. The solution is dehumidification of the air to reduce the moisture content to 50% RH or less. However, some applicators claim that very dry conditions, for example below 20% RH, make the atmosphere uncomfortable to work in and causes respiration problems.

In selecting a dehumidifier for the job, emphasis should be put on its low-temperature performance, since this is when it is mainly required. Refrigeration dehumidifiers are commonly used in industry and are capable of removing large amounts of water at high temperatures and during winter conditions when the water has to be frozen out of the air, rather than condensed as a liquid. Absorption or desiccant dehumidifiers, on the other hand, have nearly the same performance in summer or winter

Table 6.1   Typical ventilation volumes required to maintain solvent vapour concentrations below 10% of the L.E.L.

Tank volume (ft^3)	Ventilation volume (ft^3/min)
670	1 000
2 800	2 000
8 400	3 000
11 200	4 000
14 000	5 000
28 000	6 000
56 000	10 000
112 000	20 000
168 000	30 000

From Ref. 8.

because the affinity of the chemical absorbent for moisture is not greatly affected by temperature.

The benefits of dehumidification to the contractor are considered[9] to be that:

1 Crews can start work earlier in the morning.
2 It generally enables large areas to be prepared and completed in one go.
3 It avoids possible delays between application of coats.
4 It enables a contractor to predict accurately when the work will be completed rather than being at the mercy of the weather.

6.1.3.5 Inspection

Inspection of the application of tank linings generally needs to be even more rigorous than for structural steel. This should include a continuity check of the whole surface. NACE have prepared a recommendation for the inspection of linings on steel and concrete.[10]

6.1.3.6 Health and safety matters

Since most tank lining operations take place in enclosed spaces, correct ventilation is extremely important to avoid health and safety hazards. NACE International publishes 'Manual for Painters Safety' and 'Coatings and Linings for Immersion Service' which provide an in-depth review of many safety issues when lining tanks, some of which are given below.

Steps should be taken to detect any hazards such as concentrations of noxious fumes, or any remaining splashes or spillage of harmful liquids, conditions of excessive heat and/or areas where the oxygen content of the air may be dangerously low.

To avoid human error, all valves on pipelines feeding to or from the tank should be locked in the off position and further protected from ingress of spent abrasive. Adequate clothing and respiratory equipment should be worn by anyone entering the tank during the lining operation, according to the relevant safety regulations. Some lining operations take place in potentially explosive atmospheres and there must be no source of ignition that would trigger an explosion. Precautions include the use of explosion-proof lighting equipment and approved safety, non-sparking type hand and power tools.

The chemicals and resins used in fibreglass-reinforced polyester coating (FRP) linings present a very real fire and explosive hazard if used incorrectly. It is important that all personnel involved should receive and obey the instructions in the material safety data specification. Persons carrying out the application of any coating should follow safety rules as set out in the specification or by the safety engineer or person responsible for safety on a particular job. Because of the hazards associated with tank lining, this becomes even more important.

6.2   Powder coatings

The use of powder coatings has increased significantly in recent years. They offer films with excellent corrosion and chemical resistance, high resistance to abrasion and excellent adhesion to metal, and being totally free from solvents they are more acceptable in this age of heightened awareness of environmental issues.

There are two broad groups of resins or polymers used for powders to produce what are commonly called plastic coatings. These are *thermoplastic polymers* which can be heated without chemical change, the process can be repeated any number of times, and *thermosetting polymers* which, when heated, produce a chemical change which is permanent. Some examples of thermoplastic powders used for corrosion protection are: nylon 11, nylon 12, low-density or high-density polyethylene, polypropylene, ethylene vinyl acetate (EVA), polyvinyl chloride (PVC), chlorinated polyether (Penton), acrylic and methylacrylic resins, cellulose aceto-butyrate, polystyrene, chlorinated polyethers and polyesters.

In general, the nylons have excellent resistance to oils, solvents and abrasion; the polyethylenes can resist strong acids and other corrosive fluids; EVA resists mild acids and alkalis and is particularly suitable for marine conditions; PVC has good resistance to impact and moisture and is comparatively cheap; Penton is a relatively expensive material which is used for the coating of smaller items such as pumps and valve bodies. Cellulose aceto-butyrate is a relatively low-cost thermoplastic with good decorative properties and is available in a wide range of colours. It has excellent colour and gloss retention under normal conditions. Chlorinated polyether is a relatively expensive material but possesses outstanding chemical and corrosion resistance. It is used to protect chemical equipment and in particular to replace glass linings. It can also be machined which makes it useful for valves, etc. Polyesters are available both as thermoplastic and thermosetting polymers and are moderate in cost. They can be applied in relatively thin films. They are available in a wide range of colours and have superior decorative properties as well as outstanding electrical properties. Polyethylene is available as two types, high and low density. Low density (LD) is generally used for wire goods and high density (HD) for chemical resistance. Polyvinylchloride is a low-cost thermoplastic with superior decorative properties, outstanding flexibility and resistance to impact, water and corrosion. It has some of the qualities of rubber and is occasionally used to replace rubber linings at a lower cost.

All thermoplastic materials have limited resistance to heat.

Some examples of thermoset powder coating systems used for corrosion protection are epoxy, epoxy/polyester hybrids, polyester/triglycidal iso-cyanate (TGIC) and polyester/isocyanate (PUR). In general, these materials harden rapidly, have good adhesion after curing and excellent chemical and corrosion resistance. The epoxy and epoxy/polyester systems have outstanding chemical resistance especially against aqueous acids and alkalis. The epoxy/polyester powder systems are preferred for decorative interior use but have limited gloss retention and resistance to yellowing on exterior exposure. Where weather resistance is required, the TGIC and PUR systems are preferred. The TGIC system has the advantage of lower curing temperatures and shorter curing times, plus better film coverage on edges.[11]

6.2.1 Application methods

The following methods are used to apply these coatings.

(i) *Flock spraying*: with this process the powder is mixed with low pressure compressed air and blown onto the surface through a wide-orifice nozzle.

(ii) *Flame spraying*: in which the powder is heated as it passes through the nozzle of a special spray gun and the molten powder falls onto the steel in the form of globules which are then fused together with gentle use of a flame. The method is most suitable for the application of thermoplastic powders.

(iii) *Electrostatic spraying* has been discussed in Chapter 5 and is unusual in that the whole surface of components can be sprayed from one position. This is the most widely used method of applying thermoset and thermoplastic powders, as for example in pipe coating (see Section 6.6).

(iv) *Vacuum coating* is a specialised process used mainly for pipes and chemical vessels to be coated with Penton powder. A high vacuum is used to draw powder from a fluidised bed onto the heated object to be coated. The vacuum is maintained during the fusion of the powder, so eliminating the possibility of air entrapment.

(v) *Dipping* covers two different processes. The *plastisol method* consists of dipping a primed and heated steel object into a tank of cold plastisol, which is a colloidal dispersion of the resin in a plasticiser. The action of the heat results in the formation of a gelatinous deposit arising from the cross-linking of the main components. Heating at about 175°C results in fusion into a solid film. This method is used to coat tanks, pipes and other components with PVC, in one coating operation, which can produce thicknesses up to 12 mm. The technique is widely employed to coat steel strip (see Section 6.3).

The other method of dipping is based on the *fluidised bed* technique. In this method a tank with perforations at the bottom is used. Air is introduced through the perforations to maintain the powder in the tank as a suspension. The powder in suspension acts as a fluid. When a steel object, heated to a temperature just above the melting point of the powder in the bath, is dipped into the fluidised bed, the powder in contact with the steel fuses to the surface and flows to produce a coating. The steel object is then removed from the bath and chilled in cold water. The thickness of the coating will be determined by the temperature of steel article when dipped and by its heat capacity, effectively its mass.

A number of components can be treated at one time in the fluidised bed, and in some situations the powder may become agglomerated because of the temperature and heat capacities of the objects being processed. One method of overcoming this problem is to cause the powder to adhere to the article by electrostatic action, so eliminating the need to heat the articles before dipping. The electrostatic charge is applied from a grid mounted just above the tank bottom which is maintained at a suitable potential to impart the charge to the powder. As more of the powder particles adhere to the article, the force of attraction between the article and the charged particles in suspension is decreased due to the insulating effect of the coating build-up and the consequent decrease in electrical conductivity. This allows the thickness of coating to be controlled by altering the applied potential. The article with adherent powder is finally fused by heating after removal from the bath.

Compared with paint and metal coatings, plastic coatings are used to a limited extent on steel structures. They are being increasingly employed for components, cladding and subsidiary parts of structures, e.g. balustrades, and for submarine pipelines. They are comparatively expensive to apply so tend to be utilised where appearance is important or for specialised purposes where plastic-coated steel can be used as a replacement for more expensive alloys, or where coatings of high integrity are required. The application of plastic coatings must be carefully controlled to ensure sound films with full fusion of powders to provide reasonably pore-free coatings. When plastic-coated steelwork is damaged or where it has to be welded after coating application, some form of 'touch-up' is required. Generally, this is an air-drying paint, which is unlikely to provide protection equivalent to the original plastic coating; even special fusion sticks are unlikely to meet this requirement. Also, when maintenance is required then air-drying paints are usually used.

Special methods have been adopted for field welds on fusion-bonded epoxy coatings which provide coatings of a high standard. However, such methods are not generally used for plastic coatings.

6.3   Coil coatings

Plastic coatings are applied to thin gauge steel, which is used generally in profiled form for cladding and roofing for buildings and also, when formed, for some structural purposes. Many of the coatings are primarily decorative with only limited corrosion resistance and are used internally for domestic appliances, office furniture and shelving. These will not be discussed in this section, which is concerned with those coatings intended for exterior application.

Coatings are generally applied to steel that has been zinc coated or hot-dip galvanised for exterior exposure. This prevents or delays corrosion at the edges and damaged areas. The plastic coatings are applied in one of two ways; liquid coatings, similar to those discussed in earlier sections of this chapter, are more commonly used, but some coatings are produced using a film of the plastic material which is bonded to the steel surface. Both methods are used in a continuous process: applying the films to coils of steel which, after coating, are cut into sheets of suitable size, hence the term 'coil coating'.

The general method of applying liquid coatings is as follows: the coil is fed through the line and is either pickled beforehand or in some cases is pickled in-line. After cleaning of the surface, a primary coat is usually applied, then baked in an oven and followed by the application of the final coat. Different companies vary the process but it follows these general lines.

A number of different organic finishes have been developed to provide different levels of performance. The major generic types are as follows:

Polyesters, at a nominal 25 μm thickness, are low-cost coatings with reasonable flexibility but relatively poor performance outdoors. Life to first maintenance is claimed as 5–7 years.[12]

Silicone modified polyesters, at a nominal 25 μm, have improved colour fastness and resistance to sunlight than polyesters but less flexibility and are more expensive. Life to first maintenance is claimed as 7–10 years.[12]

Polyurethanes, at 25 μm, have a durability similar to the silicone modified polyesters but are more flexible and cheaper.

Fluorocarbons can be a 35-μm thick film bonded to the substrate or a dispersion in an acrylic resin, to a nominal 25 μm. These materials have good colour stability and chalk resistance but the relatively thin films are more easily damaged during installation than the thicker plastisols. It is claimed that these materials will give a life to first maintenance of 10–12 years in non-aggressive environments.

PVC plastisols at 100–200 μm have the maximum flexibility and resistance to damage and abrasion of all the coated cladding systems. However, these materials can degrade under the influence of sunlight and rainfall so that the film is eroded, becomes porous and eventually can detach from substrate. Therefore, its durability is least in clean, sunny environments and greatest in shaded, or dirty industrial atmospheres. It is claimed that life to first maintenance should be between 10 and 25 years. If the system is caught before it starts to lose adhesion it can be overpainted, otherwise it is necessary to strip off the coating and replace it with a new coating.

There are also special proprietary cladding sheets available which consist of felt impregnated with bitumen on a sheet which is generally galvanised. The sheet is then top-coated with a coloured finish, such as a styrenated alkyd. The aesthetic durability therefore tends to be governed by the performance of the top coat.

Most of the pre-coated sheets are embossed for decorative purposes. These tend to collect dirt and grime. This can often be removed by suitable cleaning with warm water containing a detergent, with careful brushing where required. The use of solvents to clean the coatings should be carried out with caution. Both chlorinated hydrocarbons and ketones attack PVC, so a check should be made before using organic solvents.

With all types of coated cladding it is advisable that the purchaser fully understands the implications of the manufacturers' claims for durability. In some cases this might mean a relatively simple overcoating and in others complete stripping of the original material and replacement with a new and different material. In some cases this latter option may be economically or operationally unacceptable. Advice should be sought on this aspect.

Premature failure of cladding sheets may also occur due to poor fixing or inadequate design of details.

Pre-coated sheets are supplied as standard products and reliance must be placed on the quality control systems operated by the suppliers. As the whole process is capable of a much higher level of control than would be anticipated with coatings applied to structural steel, there should be little or no requirement for further inspection by the user. Any shortcomings in the process are not likely to be immediately evident but will result in premature failure of the coating or loss of decorative aspects, such as colour fading. There are no obvious ways by which users can check the properties likely to cause such failures before use. The thickness of the coating, its freedom from pinholes and forming properties can be checked by standard test methods.

Materials used for repainting coated cladding sheets need to be specially formulated to allow for the required flexibility, adhesion, etc. Currently in the UK, a single-pack moisture-cured urethane primer and a high-build single-pack moisture-cured urethane finishing coat, is the main system being used for this purpose. These materials have the added advantage for site application that they can be applied to slightly damp surfaces and will cure at temperatures as low as −9°C.[13]

6.4   Wrapping tapes

Wrapping tapes are utilised mainly for tubular sections and pipes. They can be used for angles and I-beams but the wrapping is more difficult and the protection is therefore often less effective than on round sections. If properly applied with suitable overlapping, tapes provide sound protection in many situations. They are widely used to protect bare pipes and pipe joints even where other forms of protection are employed for the main surface. Sometimes they are used to provide additional protection where pipes are buried in rocky soils.

Tapes can be applied by hand over comparatively short lengths, but special wrapping machines are generally used; these may be hand or power operated (see Figure 6.1). Although tapes are sometimes applied directly to the steel surface, it is more usual to apply a suitable primer and often a mastic or heavy grease to improve the contours for wrapping, particularly at joints. The wrapping may consist of a number of different tapes to provide the required protection, with the outer tape chosen to suit the environmental condition.

A range of tapes is available and some of those most commonly used include the following:

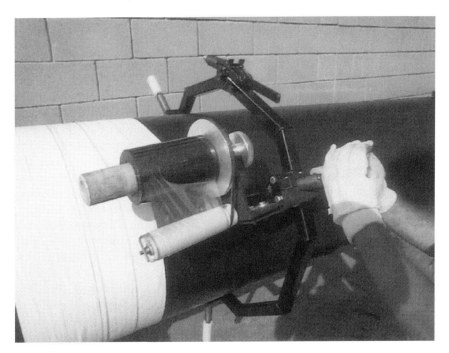

Figure 6.1   Hand tape wrapping machine.

Source: Winn and Coles (Denso) Ltd.

(i) Petrolatum compound tapes consisting of a fabric of natural or synthetic fibre impregnated with a jelly-like petrolatum compound, usually with neutral fillers. A petrolatum-type primer is generally recommended. Such tapes are highly conformable, do not harden and can be applied to irregular profiles. They can be wrapped externally with bitumen-type or plastic tapes to resist damage, particularly when used under atmospheric conditions.

(ii) Petrolatum compound reinforced with woven polypropylene laminate is used with special machines for wrapping pipelines. A bitumen-type primer is used.

(iii) Plastic materials such as PVC or polyethylene are also used for tapes: these are usually coated on one side with a rubber-bitumen or a butyl rubber compound.

(iv) Cold-applied, two-pack, cementitious coatings combined with reinforcing fabric are used for the protection and sealing of petrolatum tapes.

(v) Bitumen and coal-tar compounds are also used with fabric reinforcements. A range of products is used: some involve heating the material with a flame, and some authorities consider this method to be safer to apply than hot-poured flood coatings. These tapes are generally used for buried pipes and a suitable primer is applied before wrapping. Other materials are also used as tapes, e.g. polyethylene may be extruded as a wrap on pipes (see Section 6.6.1.3).

Although most generally used for pipelines, tapes are also used for other purposes, e.g. for insulating bimetallic joints. Logan[14] has discussed the use of tapes for protecting steel used in marine environments and an earlier paper gave examples of the application of petrolatum-coated tapes in such situations. Generally, only tubular or box-type sections have been protected as these allow spiral or circumferential wrapping. Jetty piles have been protected with tapes in many parts of the world. In all cases the primary protection has been by petrolatum tape over a specially formulated petrolatum primer paste. Outer wraps are chosen on the basis of the aggressivity of the environment and the degree of damage anticipated; they have included a cold applied bitumen tape using a nylon carrier with a PVC backing, PVC and polyethylene tapes and even a preformed glass reinforced jacket bolted round the pile, where severe mechanical effects were considered to be likely.

Although tapes are usually comparatively easy to apply, the manufacturer's instructions must be followed to ensure a high degree of protection to the steel. If tapes are not properly applied and corrosion occurs at the steel surface, it may be some considerable time before this is detected. Unlike conventional coatings where signs of rusting can be rapidly detected either by the visual appearance of rust staining or changes in the coating itself, e.g. blistering, there may be no immediate changes in the appearance of tapes. Recommendations should cover the following:

(i) Method of application to ensure close contact with the steel and a tight fit without folds.
(ii) Overlap required when wrapping.
(iii) Mastic to be used to contour projections and protuberances.
(iv) Primer to be used.
(v) Requirements for outer wrap.

6.5 Fire protection

Increasingly there is a need to give some fire protection to steel structures. This is particularly so for petroleum and petrochemical installations both on- and off-shore. The materials traditionally used, such as cementitious products, have limited physical strength and flexibility and are easily damaged in an operational environment. This has led in recent years to the development of fire-retardant coatings.

The majority of paint binders are organic in nature. Their organic molecules contain carbon which, under heat, will react with atmospheric oxygen and so burn quite readily. The function of a fire-retardant coating is to minimise or delay the conflagration and, even more important, to delay the flames or hot gases coming into contact with the substrate.

Fire-retardant coatings fall into three different types:

(i) Those that work by forming a heat barrier and giving off non-inflammable gases when subjected to a fire. Early types were silicate based but nowadays there are more durable systems based on phenol formaldehyde and polyamide resins. In a fire these are gradually destroyed to form a heat-insulating carbonaceous char and give off carbon dioxide and water.

(ii) Those that are generally based on halogenated compounds, such as chlorinated alkyds or chlorinated paraffin and evolve non-combustible gases such as hydrogen chloride or bromine when heated. These paints usually contain antimony oxide pigments and can be formulated on water- or solvent-based resins. They are very effective in retarding fire but the corrosive nature of the gases evolved can be a problem.

(iii) Intumescent coatings, which are becoming increasingly popular. At elevated temperatures they swell up and form a thick, insulating layer of char or foam. Non-combustible gases are also trapped in this layer and add to the insulation. Originally these coatings were based on thermoplastic binders, either solvent- or water-based, but there are proprietary materials based on epoxy resins which claim improved durability and faster use.[15]

6.6   Pipelines

There are a number of forms of pipelines: (i) laid above the ground, (ii) buried, and (iii) submarine. Some pipelines may appear in all three forms. From the standpoint of corrosion protection, buried and submarine pipelines are likely to cause most problems. The selection of coatings is a specialised matter and often sophisticated application techniques are used. NACE International Standard RPO 1692 lists the desirable performance criteria for pipeline coatings as follows

(a) Effective electrical insulation.
(b) Effective moisture barrier.
(c) Application to piping by a method that will not adversely affect the properties of the pipe.
(d) Application to piping with minimum of defects.
(e) Good adhesion to the pipe surface.
(f) Ability to resist development of holidays with time.
(g) Ability to resist damage during handling, storage and installation.
(h) Ability to maintain substantially constant electrical resistivity with time.
(i) Resistant to disbonding.
(j) Resistant to chemical degradation.
(k) Ease of repair.
(l) Retention of physical characteristics.
(m) Non-toxic to the environment.
(n) Resistant to changes and deterioration during above-ground storage and long-distance transportation.

The important requirements for factory application of pipe coatings are:

(i) All incoming steel must be stored to prevent localised corrosion and in the factory the metal surfaces must be suitably dressed and cleaned from contamination such as oil, grease, water, paint, waxes and soluble iron corrosion products.
(ii) The surface preparation of the metal must produce a chemically clean surface with the required blast profile overall (including dressed areas).
(iii) Application must produce homogeneous coatings and where heat is required for curing this must be uniformly applied and the cure carefully controlled.
(iv) Quality control of the preparation and application must be continuous and to a high standard.
(v) Coated objects must be suitably handled and stored to minimise mechanical damage.

In this section some of the more common forms of protection are considered but there are continuing developments, particularly on submarine pipelines, and those concerned with such matters should keep abreast with them.

6.6.1 Subsea pipelines

A number of factors must be reviewed when selecting coatings for pipelines immersed in water. However, one aspect specific to such pipelines is the requirement for concrete weight coatings to ensure negative buoyancy. The problems of applying a concrete weight coating are:

  (i) The method of application and the aggregate causes damage and penetration of the corrosion protection coating.
  (ii) In the laying operation there can be slippage between the concrete weight coating and the corrosion protection coating due to poor bonding between the two.

These problems particularly apply to relatively thin, hard coatings such as the fusion-bonded epoxy. In some cases a polymer cement barrier coating is used, which, it is claimed, both reduces damage from the concrete weight coat and prevents slippage. However, such a coating requires a special machine for application and special curing conditions. Another solution is to increase the coating thickness of materials such as fusion-bonded epoxies to a minimum of $625\,\mu$m, although apart from the extra cost there is no assurance that this is sufficient to eliminate damage. Another method to reduce slippage is to use a 1-m band of two-pack epoxy heavily filled with aggregate and installed near the trailing end of each pipe, before the weight coat is applied.

  Furthermore, cathodic protection is generally used, which may involve the removal of part of the coating to allow for the welding of small steel plates to hold the anode. The method chosen for the laying of the pipeline may also influence the coating selection. The four methods commonly employed are:

  (i)   pulling method, in which the pipeline is constructed in lengths and pulled into position by winches fixed on ships anchored at sea;
  (ii) lay barge method, where the pipes are assembled and jointed on a special barge and then lowered to the sea bed from the barge;
  (iii) float and sink method, in which the pipeline is prepared on the shore, then floated using buoyancy tanks. It is then lowered into place from pontoons;
  (iv) reel barge, in which the pipeline is reeled in long lengths from special craft.

The ability of the coatings to resist bending and the effectiveness of field joints will be typical of the factors to be taken into account when selecting coatings. Others include:

(i)   durability in seawater;
(ii)   adhesion to the pipe;
(iii) resistance to marine organisms and bacteria present in seawater;
(iv)   abrasion resistance; and
(v)   resistance to cathodic disbondment.

The coatings typically used for submarine pipelines are considered below, although, as noted above, there are continuing developments in this field.

6.6.1.1   Coat and wrap

'Coat and wrap' was at one time extensively used but other types of coating tend to be preferred nowadays. This method covers various types of application both hot and cold but, generally, for hot application the steel is blast-cleaned and pre-heated to about 90°C. A coal-tar primer is then applied to the hot pipe followed by a flood coating of a suitable enamel. An inner wrap such as glass fibre is pulled into the hot enamel and the coating is finally covered with a coal-tar impregnated wrapping. Because of the possible carcinogenic properties of coal tar, some authorities insist on replacement by bitumen. This, however, can raise problems of reduced durability and increased susceptibility to bacterial attack.

Welded joints on-site are often coated by the use of shrink wraps or sleeves. These are composed of an irradiated polyolefin backing which, upon brief exposure to a temperature in excess of 128°C, for example from a propane torch, will shrink from the expanded diameter as supplied, to a predetermined recovered diameter. The sleeves are precoated internally with a controlled thickness of thermally activated adhesive which bonds to either metal or existing coatings. Some problems have arisen due to the fact that the heat application, since it depends upon the operator, can be non-uniform and cause disbonding at the overlap. A further problem arises with overlapping onto bitumen or coal tar in that careless operation of the flame can cause damage.

In the USA and the UK there has been a move away from both coal-tar and bitumen enamel coatings, because of limitations in their performance. In particular they are:

(i)   easily damaged during construction;
(ii)   prone to cracking on bending or hydrostatic testing;
(iii) cracked and torn by stresses;
(iv)   liable to soften at the higher operating temperatures;

(v) affected by loss of adhesion between the coating and the steel sub-
     strate which, combined with the flexibility of the material, allows
     appreciable corrosion to proceed without detection; and
(vi) subject to disbonding between the enamel and the concrete weight
     coating.

However, there is an equally strong school of thought that still considers
that they have an important role for buried and sub-sea pipelines, because
the coatings are:

(i)   generally more foolproof in application;
(ii) sufficiently thick to accommodate flaws in the steel surface;
(iii) generally cheaper than other types; and
(iv) the only ones with genuinely long-term experience.

6.6.1.2 Fusion-bonded epoxy coatings

Fusion-bonded epoxy coatings (as compared with coat and wrap coatings)
are thin films, about 0.3 mm thick, so their effectiveness depends very
much upon the quality of the coating procedures. The coating processes
can vary but generally follow a pattern, such as:

(i)   Degrease, then pre-heat to remove moisture.
(ii)   Initial blast-clean to reveal steel surface defects.
(iii) Remedy defects by grinding.
(iv)   Final blast-clean to required standard.
(v)   Heat pipe.
(vi)   Apply epoxy powder by electrostatic methods while pipe rotates.
(vii) Quench in water.

The steel surface must be of a high standard and the blast-cleaning is
carried out to International Standard Sa2½ or Sa3.

Inspection of a high standard is essential with such a comparatively thin
coating and full 'pinhole' surveys are generally carried out (see Figure
6.2). The pinholes are repaired before the pipeline is immersed in the sea.
Large areas of damage are repaired by the application of a two-pack
epoxy, and small pinholes by means of nylon 'melt sticks' (see Figure 6.3).
Lengths of coated pipe have to be welded together as there is a limit to the
length that can be reeled at one time. Such areas are often coated in the
field with the same material using portable blast-cleaning, induction
heating and application equipment. Problems that have arisen include:

(i) Non-uniform heat sink at the weld causing non-uniform heating and
     therefore inadequate curing of the epoxy.

Figure 6.2   Pinhole testing of coated pipeline.

(ii)   Water absorption by existing coating at the overlap causing blister-
       ing on heating.
(iii) Damage by burning of previous repairs, using melt sticks which are
       too close to the weld area.

Fusion-bonded epoxy coating cannot be applied in all field situations, e.g.
tie-in welds. Alternative coating materials must be used. Multi-component
urethane coatings applied by hot airless spray have been used successfully
for this purpose.

6.6.1.3   Polyethylene coatings

In general, early experience with these materials, mainly applied by the
sintered process, was very unsatisfactory. There were pinholes or 'holidays' in the coatings and large-scale failure from disbonding of the coating
from the steel. Considerable improvements have now been made:

(i)   The application method is by flat extrusion.
(ii)   High-strength polyethylene with suitable elongation properties is
       used.
(iii) The materials are checked to ensure that they have high resistance
       to environmental stress cracking (tested to ASTM D 1693–70).

Figure 6.3 Repair with nylon 'melt stick'.

(iv)   By attainment of correct profile on the blast-cleaned surface.
(v)   Use of hot thermal bonded primer.

German Standard DIN 30670 'Polyethylene coating for steel pipes for gas and water supply' and British Gas Draft Eng. Std. PS/CW4 'Standard for specification for polythene cladding on steel pipe' are useful references. Some of the problems remaining are:

(i) Although the thicker and more flexible nature of the polyethylene coating is considered to be more resistant to damage from the concrete weight coat application, there remains the problem of bonding between the two materials. The surface of the polyethylene needs to be roughened; either mechanically or by addition of a surface dressing of small aggregate during the coating process while the coating is still soft.
(ii) Polyethylene coatings must be protected from damage by weld spatter. The joints have to be coated by cold wrapping tapes or heat shrink wrap-arounds.

6.6.1.4   Other coating materials

Neoprene and solventless epoxies may also be considered for the protection of submarine pipelines.

Generally, the pipelines are regularly monitored *in situ* and cathodic protection is also normally employed to ensure overall protection of a high quality. The satisfactory protection of submarine pipelines is a matter requiring considerable expertise and advice should normally be sought from specialists before selecting coatings.

6.6.2   Buried pipelines

Most buried pipelines are coated and, provided the soil is of a suitable resistivity, are also cathodically protected. The requirements for the coatings vary, depending upon the conditions of burial, but the following are typical:

(i)   Good adhesion to the steel pipe.
(ii)   The coating is reasonably impervious to moisture.
(iii) Resistance to mechanical damage both in handling and during burial.
(iv)   Stability over the temperature range encountered in service.
(v)   Where appropriate, suitability for use with cathodic protection and resistance to cathodic disbondment.
(vi)   Resistance to bacteria in the soil.
(vii) Suitable electrical resistivity.

A range of coatings is employed and selection will be determined by factors such as accessibility of the pipe for repairs, type of soil, operating temperatures, availability of application methods and, of course, cost.

Sometimes a thermal insulation coating is applied to ensure that the material flowing inside the pipe does not solidify. Such coatings are usually applied over those applied for corrosion protection. Some of the coatings used for pipelines immersed in the sea are also used for those buried in the ground, but the requirements are not the same. Buried pipes may be of much greater diameter and can be laid directly into trenches. The coatings commonly used are discussed below.

(i) Galvanised coatings are rarely used without additional protection, but provide some cathodic protection at scratches.

(ii) Bituminous materials produced from naturally occurring bitumens or from petroleum and gas by-products have been the most widely used coatings for buried pipelines, often with glass reinforcements and felt wraps. A number of applications can be made to build up the overall thickness of the coating. The application may be by cold or hot methods and for some purposes flame-treated tapes of bitumen or coal tar are applied in preference to hot-poured coatings.

There has been a good deal of satisfactory experience with these coatings and they are comparatively cheap. Furthermore, they can be applied in the field using special machines. Damaged areas can also be repaired fairly easily. On the other hand, they have a limited temperature range, somewhere in the region of 0–80°C.

Coal-tar coatings are used in a manner similar to the bituminous ones. They also have a limited temperature range but somewhat better than for bituminous coatings. Asphalt mastics containing sand, limestone and reinforcing fibres may be extruded onto pipes to provide thick seamless coatings.

(iii) Fusion-bonded coatings, similar to those used for submarine pipelines, can also be used in underground situations (Section 6.6.1.2).

(iv) Extruded polyethylene coatings (see Section 6.6.1.3) are also used.

(v) Tapes are widely used for protecting joints on pipelines, and they can also be utilised for the main protection. Special wrapping equipment is used for cold application of the tapes either in works or in the field. Generally, a number of tapes are applied to produce a coating of sufficient thickness.

(vi) Epoxies, coal-tar epoxies and phenolics can also be used and are applied by spray methods. The materials may be solventless and are usually applied with specially designed equipment, where the steel pipe is first blast-cleaned, then coated, and often treated with hot air to assist in curing.

To avoid damage to pipe coatings which could affect their corrosion resistance, perhaps resulting subsequently in expensive repair or even catastrophic failure, it is essential that coated pipes are handled and transported in the correct manner so as to avoid or at least minimise damage.[16]

Inevitably coatings on pipelines in service will need maintenance and repair. Coatings used for repair obviously require similar properties to those applied at works but have the extra problems of site application where poor accessibility and the working environment play an important part. Solvent-free epoxies or polyurethanes applied by plural component hot air spray equipment (see Section 5.1.5) possess the advantage that with suitable formulation they can be applied in extremes of climate. Polyurethanes can be applied down to sub-zero temperatures, providing the surface to be coated is ice-free. Epoxy resin blends can be formulated to be applied by plural component or conventional hot airless spray, to give greater tolerance to application and cure in cold damp conditions.[17]

References

1. Delahunt, J. F., Coating and lining applications to control storage tank corrosion. *J. Protective Coatings and Linings* (Feb. 1987) 22–31.
2. *Coatings and Linings for Immersion Service*, TPC Publication No. 2, NACE, Houston, 1972.
3. Thomas, D. and Webb, A. A. *J. Protective Coatings and Linings* (Feb. 2000) 29–41.
4. O'Donoghue, M. and Garrett, R. *Protective Coatings Europe* (Jan. 1999) 16.
5. Nill, K., A proposed standard for installing F R P linings in petroleum storage tanks. *J Protective Coatings and Linings* (Dec. 1984) 32–42.
6. Johnson, M. E., *Chem-Therm Linings: PVF₂ Impregnated Glass Laminate Corrosion Resistant Linings*. Managing Corrosion with Plastics. NACE, Houston, 1977.
7. Beebe, D. G., Rigging and ventilation in tank painting. *J Protective Coatings and Linings* (March 1985) 18–24.
8. *Procedure Handbook: Surface Preparation and Painting of Tanks and Closed Areas*. Prepared by Complete Abrasives Blasting Systems Inc. in co-operation with Avondale Shipyards Inc. USA, Sept. 1981.
9. LaCasse, G. A. and Bechtol, D., Using dessicant dehumidification during surface preparation and coating operations. *Material Performance* (Aug. 89) 32–6.
10. RP 0288–88 *Standard Recommended Practice: Inspection of Linings on Steel and Concrete*. NACE, Houston.
11. Hoppe, M., Powder coatings. *J. Oil and Colour Chemists' Assoc.*, **71**(8) (1988) 237–40.
12. Vyse, R. G., Pre-printed strips for the building and construction industry – products and performance. *Proceedings, UK Corrosion 89*. Institute of Corrosion Science and Technology, London, 1989, pp. 2/1–2/18.
13. Asher, J. M. B. and Pratt, E. N., *Repaint Systems for Cladding. Proceedings, UK Corrosion 89*. Institute of Corrosion Science and Technology, London, 1989, pp. 2/27–2/37.

14. Logan, A. G. T., Splash zone protection: anti corrosion tape systems. *UK National Corrosion Conference*, Institute of Corrosion Science and Technology, London, 1982, p. 45.
15. Dunk, J. V., *Fire-proofing of Offshore Structures with Epoxy Intumescent Materials*. JOCCA 89.10, pp. 413–21.
16. Dept. of Industry Guide, No. 5. The Handling and Storage of Coated and Wrapped Steel Pipes.
17. Norman, D., Swinburne, R. *Corrosion Management* (March/April 1999) 6–9.

Chapter 7

Metal coatings

Under most conditions, non-ferrous metals are more corrosion resistant than
carbon steels. However, they are more expensive and, for many purposes, do
not have the required mechanical properties. The advantages to be gained by
combining the properties of the two groups of metals by coating steel with a
non-ferrous metal are clear. Zinc has been used to coat steel articles for over
a hundred years and many other examples of metal coatings are commonplace, e.g. tinplate for food cans. Metal coatings can be applied for decorative
and engineering purposes, e.g. for hard facings. However, as these are not
primarily for steel protection, they will not be considered here. The metals
most commonly applied to impart corrosion-resistance properties to constructional steelwork are zinc and aluminium. Of these, zinc is by far the
most widely used. Other metals and alloys are also used to coat steel in
certain critical situations, e.g. 'Monel'. Although they will be discussed, such
alloys are not used as general protective coatings for steelwork.

The choice of metal coating for steel is determined by the cost of the
particular metal, its corrosion resistance and the ability to apply it economically to the steel. The more commonly used application methods will
be considered below.

7.1 Application methods

Although metals can be applied to steel in a variety of ways, including
vacuum evaporation and plasma spraying, four methods are commonly
used for structural steel, and the components used for construction:

   (i) hot-dipping;
  (ii) spraying;
 (iii) electrodeposition;
  (iv) diffusion.

(i) and (ii) are used for structural sections and (i), (iii) and (iv) for components and smaller items. Other methods are also used for protecting

steelwork in critical areas, in particular cladding and weld overlays. Although they are not as widely used as the four noted above, they are sometimes employed for aggressive situations, so will also be considered.

7.1.1 Hot-dipping

Zinc is the metal most widely applied by this method, and for heavy steel sections is the only one. Tin and lead are commonly applied by hot-dipping but not for structural steelwork. Aluminium is also applied by this method, particularly to sheet steel, which is marketed as a pre-coated product, often as Zn–Al-coated steel. Although aluminium can be applied to heavier sections of steel by hot-dipping, it is a more difficult and expensive process than for zinc and is rarely, if ever, used. However, if an economic form of this process could be developed it might well prove to be a suitable method of coating with aluminium, which, in many situations, provides a higher degree of corrosion resistance than does zinc.

Zinc is particularly suited to hot-dipping because of its low melting point (420°C) and the nature of the alloy layer formed during the process. For many years hot-dipped galvanising has been specified by BS 729:1971 (1986). A new International and European Standard BS EN ISO 1461:1999 has now been published. Aluminium is by no means as easy to apply by dipping techniques. It has a higher melting point than zinc (660°C) and this means that the bath usually has to be operated at a temperature over 700°C. At this temperature the reaction between aluminium and steel is rapid, resulting in high dross formation. Furthermore, at this temperature, because of the reaction with steel, it is necessary to use ceramic-lined tanks, which are more expensive than the standard steel type used for zinc. Aluminium oxidises readily to produce an oxide (Al_2O_3) and this makes fluxing more difficult than with zinc. Oxide particles may also become entrapped in the coating.

7.1.1.1 Hot-dip galvanising

The procedures for hot-dip galvanising are as follows:

(a) Steel is cleaned of grease and oil by dipping into an alkaline or acidic degreasing solution. Welding slag, paint and heavy grease requires additional manual cleaning methods for their removal.

(b) The steel is then immersed in a bath containing acid to remove rust and scale. This is termed 'pickling' and is usually carried out in a dilute solution of hydrochloric acid. Blast-cleaning may be employed before pickling, either to obtain thicker zinc coatings on steel or to remove all sand from iron castings.

(c) After a further rinsing operation the steel will then undergo a fluxing procedure. This is normally applied by dipping in a flux solution. Alternatively the steel can be immersed through a flux blanket floating on top of the molten zinc. The fluxing operation removes any last traces of oxide from the surface and allows the molten zinc to wet the steel.

(d) The steel is then dipped into the molten zinc. Large sections of steel are dipped directly into the bath. If the section to be coated is longer than the bath it may be possible to 'double dip' it, provided the overall length of the steel is somewhat less than twice that of the bath. Small components are either fixed in a jig or placed in a suitable perforated container, then, on removal, centrifuged to remove excess zinc and to eliminate irregularities in the coating. This reduces the coating thickness and tends to produce a matt surface; on the other hand, it removes excess zinc from threads on nuts and bolts.

(e) Sometimes further treatments may be carried out immediately after the steel has been removed from the bath. These include wiping and centrifuging to remove some of the molten metal from the steel.

7.1.1.2   Continuous galvanising

Heavy steel sections are individually dipped but wire, tube and sheet can be coated by continuous processes. To produce coated sheet, coils of thin gauge steel are degreased then treated to remove scale and rust before passing through a protective gas atmosphere into the zinc bath. The strip from the bath passes through fine jets of air or steam which remove zinc to provide the thickness of coating required. The strip emerging from the process may be cut into sheets or coiled for storage and transport prior to further coating treatments: hence the process is often called 'coil coating'.

The continuous process is not used for heavy sections but components made from sheet may be used in structures and buildings.

7.1.1.3   Reactions occurring in hot-dip galvanising

When the clean iron or steel component is dipped into the molten zinc, a series of iron–zinc alloy layers are formed. The rate of reaction between the steel and the zinc is normally parabolic with time[1] and so the initial rate of reaction is very rapid and considerable agitation can be seen in the zinc bath. The main thickness of the coating is formed during this period. Subsequently the reaction slows down and the coating thickness is not increased significantly even if the article is in the bath for a longer period. A typical time of immersion is about four or five minutes but it can be longer for heavy articles that have high thermal inertia or where the zinc is required to penetrate internal spaces. Upon withdrawal from the galvanising bath, a proportion of molten zinc will be taken out on top of the alloy layer and generally cooled to exhibit the characteristic bright, spangle finish associated with galvanising.

The exception to this occurs when the steel contains in excess of 0.25% silicon. Silicon may be present because of its use as a deoxidant in the manufacturing process of steel (silicon-killed steels) or may be present accidentally, and in these instances the coating thickness continues to increase with the time of immersion in the molten zinc. For steels without relatively high silicon contents, the most important influence on zinc coating thickness formed is the mass of the iron or steel component. The higher the mass the greater the coating thickness. The thickness can be increased by grit blasting before the initial pickling process, since this provides a greater surface area with which the molten zinc can react. Galvanised coating thickness cannot be increased to any significant extent by double dipping, i.e. dipping the article more than once.

7.1.1.4   Thickness of hot-dipped galvanised coatings

The corrosion protection of galvanising is proportional to its thickness. For steel articles 1–3 mm thick, the localised coating thickness will be 60 μm and the average 70 μm. For steel articles 3–6 mm thick the localised coating will be 85 μm and the average 95 μm. For steel in excess of 6 mm, the localised thickness will be 100 μm and the average 115 μm.

Where grit blasting is specified, for example using grade G24 chilled iron grit, the minimum thickness can be raised to 140 μm. In practice, for structural steel, the minimum thickness is normally much higher than this without the need for grit blasting. When the steel contains a silicon content in excess of 0.25%, then coating thickness of 210 μm or greater can be obtained. Excessively thick coatings of galvanising can be brittle and flake from the surface when the steel is under stress, so the specification and implementation of thick coatings by the use of high silicon content steel should only be used by agreement between purchaser and galvaniser and is seldom justified. It is more likely that the high silicon content, and the consequent high film thickness, will be there by accident. Threaded articles such as nuts and bolts are normally centrifuged on drawing from the bath and while the zinc is in a molten state. This reduces the zinc coating thickness and ensures that the threads remain free running. These are normally required to have a minimum coating thickness of 43 μm.

7.1.1.5   Design of fabrications for hot-dip galvanising

Although fabrications can usually be galvanised without serious problems, care must be exercised in the design of the steelwork to ensure that all parts are adequately coated and that distortion is avoided.

Experienced galvanisers can advise on the requirements and BS EN ISO 14713 'Protection Against Corrosion of Iron and Steel – Zinc and Aluminium Coatings – Guidelines' is available. Sections of this document deal with the important issue of design for galvanisers. Clear diagrams are included to assist designers and fabricators to achieve the optimum design of components, so that galvanising can be carried out simply, cost effectively and safely.

Venting is particularly important for enclosed spaces. During immersion in the zinc bath any liquid, e.g. from pickling, contained in enclosed areas may vaporise and the pressure may be sufficient to cause distortion or even bursting of the steel.

Distortion may also occur in the fabrication, particularly if steels of widely differing thickness are used. This arises mainly from uneven heating in the bath. Thinner materials below about 3 mm may need stiffeners to avoid distortion. At crevices and joints, acid may remain after pickling. When the fabrication is galvanised the acid may become entrapped by the zinc coating leading to attack on the steel interface under the coating.

The molten zinc must be able to reach all parts of the fabrication while it is in the bath. To ensure such flow, holes should be provided wherever necessary. It is always advisable to discuss the design with the galvaniser to ensure that adequate coating of the steel is achieved.

Dross is formed during the galvanising process from iron–zinc alloys, which, being heavier than zinc, sink to the bottom of the bath and can cause 'pimples' on the surface of the zinc coating. Zinc oxide forms on the surface of the molten zinc and if it is not skimmed away from the steelwork before immersion and during withdrawal it also can lead to poor appearance.

Galvanised coatings often have a bright spangled appearance but this is not an essential requirement for a good protective coating. Steels with high silicon content often produce a dull grey coating. This is likely to be thicker than the brighter coatings and provides sound protection where appearance is not important.

The general roughening of the galvanised surface may be produced in a number of ways, including over-pickling and excessive immersion times. Although not pleasing in appearance, such coatings may provide perfectly sound protection.

7.1.2   Sprayed coatings

When metal spraying, or thermal spraying, as it is sometimes known, is used for corrosion protection, the metals most commonly used are zinc, aluminium or their alloys. An important variation to this range is an aluminium/magnesium (95/5) alloy specified by Norwegian offshore operators.[2]

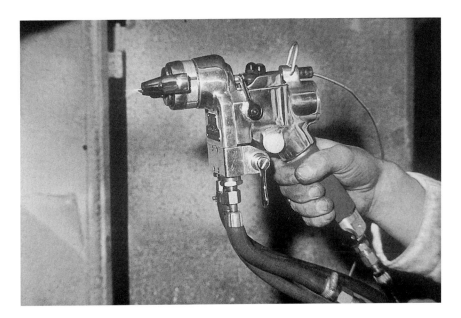

Figure 7.1   Metal spray gun, gas type.

Metal coatings applied by spraying should not be confused with paints such as zinc-rich or inorganic zinc silicate. Although these may have a high percentage of zinc in the final coating they also contain a binder. Sprayed metal coatings are purely metallic and are sprayed directly from a gun (see Figure 7.1) onto the steel surface, so no binders or solvents are involved.

The essential features of the process are as follows. Metal in the form of a powder or wire is fed through a nozzle with a stream of air or gas and is melted by a suitable gas–oxygen mixture. The melted particles are then sprayed onto the steel surface by compressed air or gas. The particles of metal impact onto the surface and a coating consisting of a large number of these particles is formed. Such coatings are porous because they are made up of many discrete particles which are not melted together. There is no metallurgical bonding at the steel surface as occurs with hot-dipping processes. There are also no alloy layers; the coating is virtually the same composition as that of the wire or powder used for the process. Wire is now more commonly employed than powder for corrosion-resistant coatings.

Another method of application is by arc spraying (see Figure 7.2). This process is based on the formation of an electric arc immediately before the jets of compressed air. The metal is broken into a fine spray of particles by the arc and is projected onto the steel surface. The process operates at a

Figure 7.2   Metal spray gun, arc type.

much higher temperature than does the more commonly used gas–oxygen system for melting the metal. Consequently, some incipient welding may occur between the sprayed particles and the steel surface. Whether for this reason alone or in combination with others, improved adhesion is obtained by arc spraying. This process is sometimes used for aluminium but not for zinc coatings.

Since the adhesion of sprayed coatings does not rely upon a metallurgical bond (cf. hot-dipping), some form of mechanical keying is required. Therefore, steel must be blast-cleaned with angular grit to provide a suitably roughened surface of high cleanliness. To ensure sufficient adhesion for aluminium, the steel must be cleaned to the highest standard, i.e. Sa3 (International scale) or the equivalent. A similar degree of cleanliness is advantageous for sprayed zinc coatings, although adhesion may be satisfactory if cleaning is to a slightly lower standard, i.e. Sa2$\frac{1}{2}$.

BS EN ISO 14713 'Protection Against Corrosion of Iron and Steel in Structures – Zinc and Aluminium Coatings – Guidelines' gives information on the design requirements of components to be thermally sprayed.

Other thermal spray processes use plasma spray and high-velocity oxygen fuel (HVOF) equipment with powdered metals mainly to produce engineering coatings, which are designed for modification of surfaces or repair of mechanical parts.

7.1.3   Coatings produced by diffusion

Although diffusion occurs in the hot-dipping process, some coatings are produced purely by diffusion processes in which the coating metal, usually in the form of a powder, is reacted with steel at a temperature below the melting point of the metal. There is diffusion of the coating metal into the steel and some reverse diffusion of iron from the steel. Many metals can be used to provide such coatings, but only zinc is used to any extent for corrosion resistance purposes. Aluminium and chromium diffused coatings are also produced on steel but not primarily for protection against corrosion. The term 'cementation' is sometimes used for the method when powders of low-melting-point metals such as zinc or aluminium are used.

The process for producing zinc diffusion coatings is called 'sherardising'. The coatings themselves are generally termed as being 'sherardised'. Sherardising is used primarily for small components such as nuts and bolts; it is not used for the main steelwork of the structures. In this process the components are first degreased and pickled. They are then packed in zinc powder or dust, usually with a diluent such as sand, contained in a steel drum. This is slowly rotated and the temperature is raised to just below the melting point of zinc, e.g. 400°C. The drum is rotated for some hours, depending on the size of the components, and the coating thickness required (see Figure 7.3).

Figure 7.3   Rotating drum for diffusion process.

The coating produced is an Fe–Zn alloy. An advantage of this method is that there is only a very slight change in the dimensions of the component after coating. This is particularly useful for threaded components such as nuts and bolts. Since the coating is an Fe–Zn alloy it will discolour on exposure to an atmospheric or aqueous environment, producing a 'rusty appearance'. This does not indicate corrosion of the steel, but is unattractive. Furthermore, sherardised coatings are usually only about $25\,\mu$m thick. So, for both appearance and corrosion performance, it is generally recommended that they be painted. Sherardising is the only diffusion coating process commonly employed for components used for constructional steelwork.

Aluminium coatings can be applied by a cementation process similar to sherardising. It is called 'calorising' and although such coatings can be used for fasteners for steelwork, they are much more commonly used to resist high-temperature conditions. 'Chromising' is the term used for the diffusion process used to produce chromium coatings. They can be produced from powders, from chromium halide vapours or from a combination of both. These are mainly used for oxidation resistance or for special purposes related to chemical environments, e.g. the prevention of intergranular corrosion of certain alloys.

Other metal coatings can be produced by diffusion processes but will not be considered here because they are not used for general corrosion-resistance purposes.

7.1.4 Electrodeposited coatings

Electrodeposition, or electroplating as it is more commonly known, is a process used to apply many different metals to both steel and non-ferrous metals. Many of the coatings produced are mainly decorative, e.g. silver plating. Chromium plating has been widely used for car trims and for some nuts and bolts normally used for interior purposes. However, for constructional steelwork, zinc is the metal most commonly applied by this method, usually to fasteners and other small components. The only other metal used to any great extent is cadmium, which is considered to provide better protection in some environments. However, cadmium-plated fasteners are not generally used for the bolting of heavy constructional steelwork.

The electroplating process is an electrochemical process with the same basic mechanism as that discussed for corrosion in Chapter 2. The metal to be coated, e.g. steel, is the cathode and the coating metal, e.g. zinc, is the anode of the cell. The electrolyte contains salts of the coating metal and other additions often of a complex nature. Proprietary solutions are often used for commercial electroplating. The steel must be thoroughly cleaned, usually by pickling, before it is electroplated. Components such as nuts and bolts may be plated in batches in a revolving barrel while steel wire and strip are usually treated by a continuous process. Generally, comparatively thin coatings are produced (about 2.5 μm) and these require additional protection for exposure outdoors.

There is also a form of chemical or electrolysis plating; this is a chemical process and is widely used for nickel 'plating' of comparatively expensive components to provide corrosion and wear resistance. There is no alloying between the steel and the coating in either of these methods.

7.1.5 Other application methods

Although cladding of one metal onto another, e.g. by rolling, is used to produce stainless coatings on carbon steel, this type of coating is used more for process plants than for constructional work. Cladding is used to protect steelwork in the splash zone of offshore structures, although usually only the hot riser pipes. The cladding is generally provided by wrapping 'Monel' (a Ni–Cu alloy) sheet round tubular members and then welding it into place. It is a comparatively expensive method, and, if damaged, there may be ingress of seawater behind the sheeting, leading to galvanic attack on the steel.

For other critical areas of ships, process plant and offshore platforms, highly resistant but expensive metals or alloys may be applied as weld overlays. In this method the alloy, e.g. Inconel 625, is welded to the steel and then treated to produce a smooth surface.[3] The method is normally used in limited areas, for example it has been used for trunnions on lock gates.

A summary of metal coatings used for constructional steelwork is given in Table 7.1.

7.2 Corrosion mechanism of metallic coatings

The corrosion protection afforded by metal coatings is basically that of an environmental barrier. Its effectiveness is determined mainly by the thickness of the coating and its ability to resist attack from the environment.

Table 7.1 Metal coatings for constructional steelwork

| Application method | Alloy layer | Metals generally applied | | General thickness (μm) |
		Structural steel	Components	
Hot-dipping	Yes	Zn	Zn, Al	75
Spraying	No	Zn, Al	Zn, Al	100
Electroplating	No	Nil	Zn, Cd	5
Diffusion	yes	Nil	Zn, Al[a]	20

a Generally only for heat-resistant purposes.

Zinc is amphoteric but stable in the pH range 6–10. In air a protective layer of zinc oxide forms on zinc and in dry conditions attack is negligible. The presence of moisture leads to the conversion of the oxide to zinc hydroxide which reacts with carbon dioxide in the air to form a basic zinc carbonate. This is insoluble, so in comparatively unpolluted atmospheres the corrosion is low. However, in the presence of industrial pollutants such as sulphur dioxide, acidic surface moisture dissolves the zinc. Although basic salts are reformed, over a period of time zinc corrosion progresses and there is a good correlation between the corrosion rate of zinc and the amount of pollution in the atmosphere. Chlorides have less effect than industrial pollutants but their presence leads to the corrosion of zinc.

Zinc coatings behave in the same way as zinc metal but both hot-dipped and diffused coatings have Fe–Zn alloy layers. These are at least as corrosion-resistant as zinc (some authorities consider that the corrosion resistance of the alloy is 30–40% greater than that of zinc[4]), but they behave differently, producing a brown, rust-like corrosion product.

Generally, there is a fairly linear relationship between the thickness of a metal coating and its protective life.[5] Metal coatings, unlike paints, corrode and it is the rate of the corrosion in a particular environment that will determine the loss of coating thickness and the period of its effectiveness as a barrier. Consequently, the lives of metal coatings are directly related to coating thickness and to the conditions of exposure (see Table 7.2). The method of application has an influence on the corrosion performance, in part because of the thickness of coating produced by the different methods and also because the nature of the coating varies with different forms of application. The most straightforward example is that of an electrode-posited coating, which is, in effect, a protective layer of non-ferrous metal with no metallurgical bonding to the steel substrate. Such coatings are usually thin and when exposed to an environment they corrode until all the coating metal has been converted to corrosion products, leaving the

Table 7.2 Corrosion rate data for various environments taken from BS EN ISO 14713:1999

Corrosivity	Corrosion risk	Zinc corrosion rate μm/yr
Exposed rural inland	Medium	0.1–0.7
Urban inland or urban coastal	High	0.7–2.0
Industrial with high humidity or high salinity coastal	Very high	2.0–4.0

steel substrate to be attacked by the environment. The period of protection will be determined by thickness of coating and the corrosion rate of the particular metal in the environment of exposure. In warm, dry interiors, the rate may be very low, so a comparatively long period of protection may be achieved. However, in most exterior or immersed environments, the life afforded by such a thin coating will be short and additional protection will be necessary. In fact, for most constructional purposes electrodeposited coatings of zinc or, where employed, cadmium, should be considered as temporary protection prior to painting.

Hot-dip coatings of zinc will initially behave in much the same way as electrodeposited coatings, because there will usually be a layer of nearly pure zinc at the surface. The thickness of this layer will vary. On continuously rolled sheet most of the coating will be zinc, whereas on many structural sections only a thin zinc layer will be present. Eventually this will corrode away leaving the Fe–Zn alloy layer exposed to the environment. Although the Fe–Zn alloy layer provides protection to the steel, the appearance is not necessarily acceptable for all structures. Therefore, for purely cosmetic purposes, some form of paint coating may be considered as a requirement. These circumstances do not arise under immersed conditions or where appearance is not important.

Diffused coatings, such as sherardised zinc, are composed virtually entirely of Fe–Zn alloy layers, so the brownish corrosion products are formed at an early stage. Furthermore, the coatings tend to be thin (12–25 μm), although thicker coatings can be obtained. Therefore, such coatings are usually applied to components such as nuts and bolts, where the change of dimensions in the threads is slight, and finally painted.

Both aluminium and zinc sprayed coatings are similar in that small globules of the molten metal strike the surface and are elongated as they solidify. Both coatings are porous but aluminium is usually somewhat more so than zinc, up to 10% although usually about 5%. Particles in the coating are surrounded by oxides but eventually there are enough discontinuities to allow sealing of the pores by the corrosion products.

Problems can arise with aluminium sprayed coatings exposed to fairly mild conditions because the steel substrate may rust slightly, producing some stain on the coating. Eventually the pores are blocked but the appearance may be affected by the rust staining. The influence of corrosion product formation may have effects on paints applied over sprayed coatings that have been exposed before painting. This is likely to be more of a problem with zinc than aluminium coatings. The presence of soluble zinc corrosion products under the paint film can lead to its premature failure by blistering and eventual flaking. In most circumstances, sealing of sprayed coatings rather than painting is recommended (see Section 7.3).

The metals commonly used to protect structural steel and the components used with it, zinc, aluminium and cadmium, are all anodic to steel.

This means that they will protect steel at damaged areas and edges of sheets. Furthermore, sacrificial protection will occur at small pores in the coating. This additional protection is clearly beneficial but is not the main reason for using such coatings: this is for overall protection of the steelwork. Sprayed aluminium coatings provide this form of galvanic protection when immersed in seawater, but it is by no means so effective with electrolytes of high resistivity, e.g. in river waters or in many atmospheric environments.

Where the protection to the steelwork is in the form of cladding or weld overlays, then over the general areas of coating the corrosion rate will be determined by that of the metal used for the particular form of protection. In the case of welded coatings, the performance will be similar to that of the cast rather than wrought form of the alloy, and these may differ. The main difficulty with cladding, where the protective metal is rolled onto the steel, arises at edges, and these must be given additional protection. With 'wrap round' cladding, e.g. of Monel on offshore platforms, any tear or other form of damage may allow water ingress behind the cladding. In the absence of suitable precautions such as applying a thick coat of paint to the steel, serious bimetallic attack on the steel may occur. Furthermore, of course, wrap round cladding can generally be applied only to tubular members.

7.3 Painting of metallic coatings

Diffused and electrodeposited coatings nearly always require additional protection, mainly because these methods tend to produce thin coatings. Both spraying and hot-dipping are capable of producing comparatively thick coatings and in many situations these will be sufficient to protect the steel for a considerable length of time, so additional coatings may not be required. Both hot-dip galvanised and sprayed coatings tend to produce a rather unattractive appearance after a period of exposure. This is more noticeable on large sections than on smaller items such as lamp posts. Some form of additional decorative coating may therefore be considered to be necessary. It should, however, be appreciated that future maintenance intervals will then be those of the metal plus paint system which is generally less than the bare metal coating, but longer than for a similar paint system applied directly to bare steel. The corrosion resistance of metal sprayed coatings can be enhanced by the application of specially formulated sealers that penetrate the pores and reduce the total area of exposed metal. Sealers not only extend the useful life of the metallic coating but provide a smoother finish and can give some colour as desired.

Sealers must have low viscosity to ensure penetration and must be suitable for the subsequent environment, e.g. aluminium pigmented silicone resin for aluminium metallic coating required for heat resistance. If further paint coatings are to be applied to a sealed metallic coating, the sealer must be compatible with these coatings. Sealers should also be applied as soon as possible after the metallic coating application, since the spray coating is extremely sensitive to condensation and moisture which, due to the porosity of the surface, is often not readily apparent. Powdery zinc oxide films can occur on unsealed zinc and oxides with a rusty appearance rapidly become visible on unsealed aluminium.

The sealer must be compatible with the sprayed metal. Some formulations which perform well on steel, will fail rapidly on zinc or aluminium, for example straight oil-based paints, such as alkyds, are not generally satisfactory. Typical single-pack sealers are solutions of vinyl chloride/acetate co-polymers. Two-pack epoxy or urethane sealers can be used providing that, if they are to be subsequently overcoated, the procedure necessary to ensure good adhesion is fully appreciated. It is also possible that the stress developing in such materials during curing may aggravate and reveal any inherent weakness in the adhesion of the metallic coating. Some specifications call for the application of poly vinyl-butyral etch primers but these materials are sensitive to moisture in the early stages of their cure and should not be left without further overcoating.

The painting of freshly galvanised surfaces provides problems of a rather different kind. Often adhesion between the paint and zinc coatings may be poor, leading to fairly rapid flaking of the paint. This arises from the nature of the freshly, galvanised surface, which is often shiny and too smooth for good adhesion of most coatings. A number of approaches have been proposed to deal with this difficulty and they can be summed up as follows:

(i) The use of sweep blasting, i.e. using low pressure (40 p.s.i.) blast-cleaning with a fine grade non-metallic abrasive, can be very effective, but only when undertaken with a high degree of expertise. It is used to best advantage on flat surfaces. Excessive blasting can remove all the zinc coating and expose bare steel, especially on edges and fasteners.

(ii) Application of a solution of T-wash, a non-proprietary formulation obtainable from many paint companies to a degreased surface. It is basically a solution of phosphoric acid in methylated spirits containing a small quantity of copper carbonate which reacts with the zinc surface to turn it black; the surface is then suitable for painting. If the surface does not turn black then the treatment has to be repeated. This method must be used with some caution because an acidic solution is involved. Furthermore, during application the T-wash reaction with the zinc produces a rather unpleasant smell. Care should be taken to avoid ponding or puddling of the T-wash and any excess should be removed by washing with water. Although widely and successfully used for many types of paint, it may not be a satisfactory preparation for two-pack epoxies and urethanes because the etching may not be sufficient to provide a good key.

(iii) After new galvanising has been exposed in the atmosphere for at least twelve months, a process in this instance called weathering, preparation of the surface for painting is much simpler. All the surface requires is using either abrasive pads or a stiff bristle brush to remove all loose material, corrosion products, etc. Care must be taken to ensure that the surface is not burnished or polished to become too smooth. Ideally this process should then be followed by a hot water detergent wash, then rinsing off with fresh clean water of drinking quality. T-wash should not be applied to weathered surfaces. In marine environments where chloride levels are high, weathering is not the preferred option.

(iv) Etch primers of a type suitable for applications to galvanised surfaces are also used. Because etch primers are fairly colourless and must be applied in very thin, wash-like films, it is often difficult to see if the surface has been fully covered. Also, major differences arise in the surface condition when single- or two-pack materials are used, the better quality of surface being achieved with the latter. The single-pack primers, although easier to use, have lower levels of reactive materials in their formulation, whereas the two-pack products carry higher levels which are activated as a result of mixing.

(v) Water-based proprietary products, claiming to aid adhesion on fresh and aged galvanising, being recoatable with either one- or two-pack materials, have recently become available.[6]

Generally, there are no problems in painting sherardised and electroplated coatings.

If the zinc coating has virtually corroded away leaving rusted and pitted steel, then it should be cleaned in the same way as other rusty steelwork.

Problems may arise if the zinc coating has been given a chromate treatment as a protection against wet storage stain. Such coatings are commonly applied to galvanised sheet steel but more rarely to sections. Advice should be sought from the paint manufacturers if galvanised steel has been chromated and requires to be painted.

7.4 Performance of metallic coatings

There is a considerable amount of data on the corrosion of zinc and zinc coatings. Where relative humidity is below 60% the corrosion rate of iron and steel is negligible, but above that level it can be 10–40 times greater than that of zinc, the higher ratios being in high chloride environments. Less data are available for aluminium coatings, although what has been published tends to show their high level of corrosion resistance. Many of the data on zinc coatings have been obtained from the exposure of zinc metal specimens, so actual corrosion rates have been obtained.

Although similar tests have been carried out on aluminium specimens, corrosion rates are not necessarily quoted as loss in micrometres per year because the metal tends to pit, unlike the more general corrosion of zinc (and carbon steel). Consequently, many of the test programmes concerned with aluminium coatings provide qualitative evidence of corrosion performance based on appearance. Nevertheless, this evidence is sufficient to show the excellent properties of aluminium metal coatings. In most aggressive environments they are more protective than zinc coatings of the same thickness; the main exception is under alkaline conditions.

As noted previously, coatings of aluminium and zinc alloys are produced in sheet form by continuous hot-dipping methods. Both the main groups of alloys 55 Al 45 Zn and 95 Zn 5 Al (with rare earth additions) are superior to zinc in corrosion resistance. Only limited data are available. In one series of tests the 55 Al–Zn coatings corroded about 2.2 μm/year in a marine environment, less than half that of the zinc in these particular tests. On the other hand, aluminium corroded at about half the rate of the alloy coating. However, the results are not directly comparable and the full results should be studied.[7]

7.4.1   Performance of zinc coatings

7.4.1.1   Atmospheric

For some years the corrosion rate of zinc was based on data supplied by the Zinc Development Association and published in the now redundant Code of Practice BS 5493 (1977). The new British, European and International Standard 14713 'Protection against corrosion of iron and steel in structures – Zinc and Aluminium – Guidelines' updates this information. It points out that, in recent years, environmental legislation to control pollution of the atmosphere has resulted in a substantial reduction in pollutants, especially sulphur (see Figure 7.4). For example, the standard states that 85 μm minimum of hot-dip galvanising, 100 μm minimum of sealed sprayed aluminium and 100 μm of sealed sprayed zinc will have a typical life-to-first-maintenance time in excess of twenty years. This is for exposures as aggressive as the interior of swimming pools, chemical plants and exterior exposure for industrial inland or urban coastal areas. It also states that a similar long life can be achieved in very high corrosion risk areas such as exterior industrial with high humidity or high salinity, by increasing the hot-dip galvanised thickness to a minimum of 115 μm and sealed or unsealed zinc or aluminium metal spray to a minimum of 150 μm thickness. Even when immersed in seawater, providing that it is in a temperate climate, the Standard claims that 150 μm minimum of sealed aluminium metal spray or 250 μm of sealed zinc metal spray will also last in excess of twenty years. These corrosion rates are based on 1990 and 1995 data and even lower rates of corrosion can be expected in the future as pollution continues to fall.

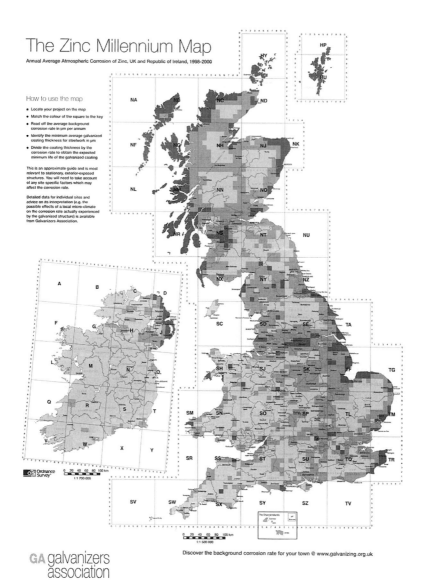

The Zinc Millennium Map

Annual Average Atmospheric Corrosion of Zinc, UK and Republic of Ireland, 1998–2000

How to use the map

- Locate your project on the map
- Match the colour of the square to the key
- Read off the average background corrosion rate in µm per annum
- Identify the minimum average galvanized coating thickness for steelwork in µm
- Divide the coating thickness by the corrosion rate to obtain the expected minimum life of the galvanized coating

This is an approximate guide and is most relevant to stationary, exterior-exposed structures. You will need to take account of any site specific factors which may affect the corrosion rate.

Detailed data for individual sites and advice on its interpretation (e.g. the possible effects of a local micro-climate on the corrosion rate actually experienced by the galvanized structure) is available from Galvanizers Association.

GA galvanizers association

Discover the background corrosion rate for your town @ www.galvanizing.org.uk

Zinc corrosion rates are represented by three categories indicated by the colour codes shown below.

Corrosion category	1	2	3
Average corrosion rate (µm/year)	0.5–1	1.5	2–2.5
Average life of 85 µm galvanized coating (years)	170–85	57	43–34

Figure 7.4 1999 MAFF Chart, typical UK corrosion rates for zinc.

Source: UK Galvanizers Association.

A publication on behalf of the UK Galvanisers Association[8] gives corrosion rates obtained from the exposure of zinc metal specimens at regular intervals throughout the UK. Based on several of these surveys, it could be estimated that a typical $85 \mu m$ hot-dip galvanised coating has an average lifetime of between 40 and 50 years. Since the corrosion resistance of zinc is roughly proportional to its thickness, other methods of application, such as electroplating or diffusion (sherardising) will give proportionally shorter lives.

7.4.1.2 Immersion in seawater

The corrosion rate in seawater is about $15-17 \mu m$ per year. A range of tests has shown corrosion rates from $10-25 \mu m$ per year. In the splash zone, higher rates of corrosion may occur, particularly on some tropical and surf beaches where inhibiting magnesium salts may not be present in the seawater. In most natural waters zinc performs well, but this is to some extent determined by the formation of protective scales, so corrosion will be higher in soft waters.

7.4.2 Performance of aluminium coatings

Tests carried out in an industrial environment in 1967 indicated that aluminium coatings gave even longer maintenance-free lives than those based on zinc. For constructional steelwork only sprayed coatings are of direct interest. Based on tests carried out in the industrial environment at Sheffield, aluminium coatings are usually more resistant than those of zinc. This is confirmed in the American Welding Society 19-year tests where, in many environments, aluminium was still protecting the steel over this period. In seawater immersion tests, sealed aluminium coatings of thickness $150 \mu m$ were still protecting the steel after 19 years, although there was evidence of corrosion of the aluminium. Zinc coatings did not protect the steel for this period when immersed in seawater. Even under alternate exposure to the atmosphere and seawater of splash-zone conditions, the sealed aluminium coatings over $80 \mu m$ in thickness performed well.

In tests carried out on hot-dip aluminium specimens,[7] again aluminium, and incidentally 55% Al–Zn coatings, performed much better than zinc coatings.

7.4.3 Performance of cadmium coatings

The corrosion rates for cadmium in different environments have been established in a number of test programmes.[9] Typical corrosion rates in micrometres per year are as follows: industrial, 10.2; urban, 2.3; and marine, 1.3.

7.5   Comparison between metallic and paint coatings

A number of points must be taken into account when considering metal coatings as the method of protecting steel. The obvious comparison is with paint coatings, but even where metal coatings have been selected, the choice of metal and, in the case of zinc, the method of application have to be considered. Compared with paint coatings there are advantages and disadvantages; these can be summarised as follows:

ADVANTAGES OF METAL COATINGS

(i)   Application is more straightforward than with paint systems and is more easily controlled.

(ii)   The life is more predictable and, usually, premature failures do not occur.

(iii) Handling is easier because there is no requirement for drying and metal coatings resist damage to a greater extent.

(iv)   If damaged there will be sacrificial protection by the coating metal. This is more certain with zinc than with aluminium coatings.

(v)   They have good abrasion resistance. Considered to be a factor of 10 or more than most conventional paint systems.

(vi)   Unlike paint coatings, they generally have a thick coating on edges.

(vii) Standards exist for most metal coatings and this makes specification more straightforward and ensures a higher level of certainty of performance.

A particular advantage of hot-dip zinc coatings is that faults in the coating are readily detected. This is not, of course, the case with paint coatings, nor are faults so readily detectable on sprayed metal coatings. On hot-dip and sherardised coatings there is a metallurgical bond with the steel, ensuring the highest level of adhesion.

DISADVANTAGES OF METAL COATINGS

(i)   If the metallic coating is thinner than the recommended thickness for a particular environment or if decoration is required, the metallic coating has to be painted and this is probably a more difficult operation than the application of paint to steel.

(ii)   Where welding is carried out after the metal coating has been applied or where severe damage to the coating occurs, it is difficult to treat such areas to provide the same standard of protection as the rest of the structure (Figure 7.5). (Similar problems also arise with paint coatings.)

(iii) With hot-dip galvanising there is a limit to the size of fabrication that can be treated and a limit on the availability of galvanising

Figure 7.5   Failure at weld on galvanised railings.

plant. Also the smaller the section the thinner the zinc coating, which inevitably leads to fixings, nuts and bolts, etc. having poorer protection than the main structure.

(iv) Without additional protection, metal coatings often develop an unattractive appearance.

ADVANTAGES OF PAINT COATINGS

(i) Painting facilities are widely available in shops and on-site.

(ii) A wide range of colours is available and paint coatings can provide a high cosmetic appearance, e.g. gloss.

(iii) A range of materials is available for different requirements. Unlike metallic coatings, paint coatings are available with good resistance to acidic conditions.

(iv) There are no limits on size or type of structure.

(v) Paint application is generally straightforward.

DISADVANTAGES OF PAINT COATINGS

(i) The main disadvantage of paint coatings arises from the application process. This provides endless opportunities for poor workmanship which can often not be detected after the next stage in the process has been completed. This can, however, be taken into account by adopting suitable quality control procedures.

(ii)   It is often difficult to predict the life of paint coatings. Even where standards or specifications for paint are available, they do not provide the same degree of certainty regarding the product as do standards for metal coatings.

7.6   Choice of type of metallic coating

For all practical purposes the choice of metal coating for structural steel sections lies between sprayed aluminium, sprayed zinc and hot-dip galvanised zinc.

The advantages of spraying over hot-dipping arise from the limitations in size of the galvanising bath and the ability to spray metal coatings on-site, although in practice this is rarely done. Hot-dipping is more easily controlled and, if steel is blast-cleaned, reasonably thick coatings can be applied. Problems can arise if freshly galvanised steel is to be painted, but these problems can be overcome. In most circumstances, hot-dip galvanising would probably be preferred to sprayed coatings if only because of the difficulties that may arise from painting sprayed coatings after some years of exposure. Sprayed aluminium coatings have been shown in tests to provide a high level of corrosion protection to steelwork, and although such coatings have been applied to structures quite successfully, there are many fewer coated with aluminium than with zinc. Nevertheless, it is somewhat surprising that so few structures are coated with aluminium in view of its corrosion-resistance. This may arise in part because the amount of aluminium used for coatings is a very small percentage of the total use of the metal; therefore, there is no great incentive on the part of aluminium producers to encourage its use.

This is quite a different situation from that concerning zinc, where the use of the metal for coatings is a significant percentage of the total production. Consequently, more data are available and a considerable amount of marketing is carried out. The standard of surface preparation probably has to be higher with sprayed aluminium than with sprayed zinc coatings to ensure satisfactory adhesion.

The choice between hot-dip galvanising and sherardising for components such as nuts and bolts is not clear-cut. The advantage of sherardising is the control of tolerances, particularly useful for threaded components. However, the method is more expensive than hot-dipping for equivalent coating thicknesses. Generally, fasteners are painted and the sherardising coating provides a suitable surface for paints.

Electrodeposition is only economic for very thin coatings. For exterior application, such coatings should be considered as providing temporary protection prior to painting.

7.7 Treatment of welded areas

Where metal coatings are applied after fabrication, welded areas can be cleaned and treated in the same way as the rest of the steelwork. However, where field welding is required, welded areas must be treated on-site. Although the most satisfactory way of dealing with welds would be grind them, then blast clean and metal spray with zinc, this is often not practicable and is rarely carried out. The usual way of dealing with welded areas is to clean them, often by wirebrushing and scraping, and then to apply a zinc-rich paint. The application of a suitable thickness of a zinc-rich paint to a blast-cleaned surface provides a reasonable measure of protection although, in many situations, less than that of the metal coating. However, application of the paint to a surface that has not been well prepared will result in early breakdown at the weld areas. This reduces the advantages of using metal coatings for long-term maintenance-free protection. Usually, the areas in the vicinity of the weld are not metal coated, or the coating is removed before welding.

To protect the weld areas of hot-dip galvanised steel, a fusible stick, similar to solder, can be applied to the area after suitable cleaning. This can be an effective method but is rarely used.

7.8 Wet storage stain

A white, porous, powdery deposit may appear on zinc surfaces. This is often called 'white-rusting' and occurs, usually on freshly galvanised surfaces, as a result of wet or damp storage conditions. This type of corrosion is most likely to occur with sheets that are not properly stored but can also manifest itself with components such as steel angles and occasionally even on sections.

In freely exposed atmospheric conditions, zinc coatings corrode to produce zinc oxide and zinc hydroxide, which is transformed by carbon dioxide in the air to basic zinc carbonate. Where, however, there is restricted access of air, and moisture or condensates are present on the surface, the basic carbonate may not form. Typical situations are the crevices formed by sheets closely packed together. The 'white-rust', basically a layer of zinc oxide and zinc hydroxide, is voluminous, up to 500 times the volume of the zinc from which it is produced. It does not adhere well to the zinc, so while any moisture is present, oxygen is available for the corrosion process to continue. This form of attack does not usually occur on zinc that has been weathered and so has a protective carbonate layer.

Although white-rusting often appears to be serious, there is usually only a slight attack on the zinc coating. This may be sufficient to cause problems with thin electroplated coatings but not generally with hot-dip coatings. Proper storage conditions will avoid white-rusting attack by ensuring that moisture does not come into contact with freshly galvanised surfaces; steel items should be arranged so that they are well ventilated with a good flow of air.

Sheets cause a particular problem because there is a tendency to stack them together; protection against storage stain may be provided by chromating the zinc surfaces after galvanising; such treatments often produce a yellow-green appearance on the zinc. Advice should be sought from paint manufacturers regarding the adhesion of their products to such treatments.

7.9   Fasteners

There are clear advantages to be gained from using metal-coated fasteners even for painted structures. Steelwork is often painted in the shop, with a final coat being applied on site after erection. If bolts are used then, if they are given just the final site coating, which often happens, the paint will fail on the fasteners after a comparatively short period. Additionally, if black or pickled steel bolts are used, they may well rust before the site coat is applied. The use of sherardised or hot-dip galvanised nuts and bolts will overcome such difficulties.

7.10   Health and safety matters

7.10.1   Hot-dip galvanising

It is necessary to be cautious about touching items from the galvanising bath because they can stay hot for a long period. During the galvanising process it is not uncommon for splashes of molten metal to be ejected from the bath at great speed. For example, this can be due to water vapour in a hollow section expanding rapidly.

7.10.2   Metal spraying

The quantity of dust evolved in the metal spraying depends upon the efficiency of deposition.[10] For example, the very high amperage used with arc spray produces large amounts of unstable energy where the wire is being melted and this results in a non-uniform spread of the metallic particles. The same thing happens if the flame in the gas spray method is not controlled correctly. In both instances the molten wire particles deposited on the surface will cool in the atmosphere and form dust. The dust creates a respiratory hazard for metal spray operators and respiratory protective equipment must be used. Full breathing apparatus enclosing the head is the only acceptable protection when arc spraying and full visor protection is also required when gas spraying.

A further hazard occurs with high concentrations of dust, which can be explosive and flammable. This is particularly so with aluminium and airborne concentrations. As little as $35\,mg/m^3$ have been known to result in explosion and fire when subjected to a source of ignition.[10] The actual concentrations at which the mixture may become explosive will vary and depend upon particle size. The size from a spraying operation can vary from sub-micron, which is respiratable level dusts, up to $50\,\mu m$. As a general rule the larger the particle size the greater the concentration required to become explosive.

The dust hazard may well only become apparent when the spraying operation is over and the dust has settled. Also the dust hazard can be aggravated by the very operation of trying to clean it up. Extraction is the method normally used to maintain dust levels below the explosive limits and the possibility of dust build-up in the extraction ducts is something that must be taken into consideration.

Ultraviolet light from the arc spray process can damage the eyes in a very short time without the operator being aware of the problem immediately. Dark glasses, or visors, similar to those used for welding, should be worn by the operator and anyone else in the vicinity.

Both arc and gas spray create a considerable amount of noise. It varies with the number and types of equipment being used and the confinement of the working area, but for example two operators working in the same area using arc spray have been recorded at 115 dB and with gas spray at 100 dB.

The metal spraying process releases a quantity of ozone, which is toxic, but generally not considered harmful to the operator in this instance. However, there is a possibility of absorption of fumes through the skin or even skin penetration by particles must be considered a real risk. Therefore all exposed extremities of the operator's skin should be adequately protected.

Overexposure to fumes from both arc and gas spray zinc can cause flu-like symptoms, along with irritation of the eyes from the arc, skin rash due to penetration by metallic particles and sunburn symptoms due to secondary radiation.

During metal spraying, the metallic dusts could cause short circuits with electrical equipment and therefore appropriate earthing and circuit breaking measures should be ensured.

References

1. Smith, W. J., *Corrosion Management* (Jan./Feb. 2001).
2. Hoff, I., *Protective Coatings Europe* (Dec. 2000) 13–16.
3. Powell, G. A. and Davis, R. V., *Proceedings, UK National Corrosion Congress,* London, Institution of Corrosion Science and Technology, Birmingham, 1982.
4. Kucera, V. and Mattsson, E., Bulletin 78, Korrosioninstitutet, Stockholm, 1976.
5. Iron and Steel Institute, Sixth Report of the Corrosion Committee, Special Report No. 66, London, 1959.
6. Peemans, R., *Protective Coatings Europe* (Workshop 11 March 1999).
7. Townsend, H. E. and Zoccola, J. O., *Materials Performance,* **18** (1979) 10.
8. Smith, W. J. and Baron, D. M., *Corrosion Management* (March/April 1999) 16–19.
9. Clarke, S. G. and Longhurst, E. E., *Proceedings, First International Congress on Metallic Corrosion.* Butterworths, London, 1962.
10. Fitzsimons, B., *Protective Coatings Europe* (Sept. 1996) 34–8.

Chapter 8

Writing effective specifications

A painting specification has five main objectives:

  (i)   As part of the contract to state the means by which the required life of the protective system is to be achieved ('means' includes materials, surface preparation, application, storage, handling, erection and inspection at all stages).
 (ii)   To serve as a basis for accurate pricing and tendering.
(iii)   To be a complete reference document for suppliers of materials, contractors, sub-contractors and all other parties to the contract.
 (iv)   To provide guidelines and authority for the painting inspector.
  (v)   To provide the basis from which any subsequent disputes, failure investigations and possibly arbitration can be resolved.

Writing effective specifications is an important factor in achieving a successful painting or repainting contract. The specification could be considered as 'the rules of the game'. If these are unclear to the players, the spectators or the referee, the game becomes an argument between the players and the quality of the game is lowered.

The specification should say what it means and mean what it says. There can never be any disadvantage in preparing a sound specification. It will probably be criticised by those involved in implementing it. Contractors often consider the painting specification to be too long, too inflexible, not based on common sense and with convoluted clauses and pseudo-scientific mystique. Although the idea of leaving it all to their knowledge and experience to 'do a sound workmanlike job as usual' is not realistic, there is often some justification in their complaint.

The expertise and experience of the contractor should not be disregarded and it may be advisable to make changes to the specification on their advice. However, such changes should be made only after careful consideration. They should not be made purely to suit the convenience of the contractor if there is any possibility that this will result in a reduction in paint performance, particularly when the contractor is off-site or has no further contractual responsibility.

The drafting of a specification requires technical expertise and should be prepared by somebody who understands the technology involved in coating steel. The wide use of computers has led to the adoption of standard specifications with inserts to meet the specific project. Too often such specifications contain clauses that are not relevant, do not allow for any advances in technology and often with conflicting requirements as, for example, with the paint manufacturer's paint data sheets. The final specification must always be studied carefully to ensure that it does provide relevant requirements. It must provide clear guidance for every stage of the process. It should be unambiguous, practicable and achievable. There is often vagueness in specifications, which arises from the specifier having some idea what is required, but not knowing how it can be achieved.

A typical example is writing statements such as:

'Remove all surface contaminants likely to impair the subsequent performance.'

The specifier may sleep soundly at night believing that all contingencies have been covered whereas, with the best will in the world, the requirement is unenforceable. Whereas to state in the specification exactly what contaminants, and their unacceptable limits, at least gives the contractor the opportunity to agree to and make allowance for, any extra work that may be required.

Without specific instructions the contractor will inevitably and understandably take the easiest and cheapest path. This may not cause problems with the coating system immediately but it is probable that its life will be reduced, leading to increased whole-life maintenance costs.

Another common inclusion in specifications is the words 'or similar'. Unless further defined, this is too vague to be of use. The requirement must be defined as exactly as possible. A good painting specification should be a help not a hindrance to all parties, including the painting contractor, who should then know exactly, not only what is required, but that it can and will be enforced.

8.1   Scope of the specification

Ideally the painting specification should contain the following:

(a) Scope.
(b) Documents.
(c) Pre-job conference.
(d) Surface preparation.

(e) Materials.
(f) Control of coating materials and testing.
(g) Coating application.
(h) Workmanship.
(i) Areas of special treatment.
(j) Handling and transport.
(k) Remedial work.
(l) Inspection.
(m) Safety.

Each of these items is now considered in detail.

8.1.1   Scope

It is useful to define the scope of the work and to provide general information that will assist the contractor to carry out the work satisfactorily. Aspects that can usefully be included are:

(a) Type of structure.
(b) Location of structure.
(c) Location of coating operations.
(d) General environments where coating is to be carried out.
(e) Areas of special attention.
(f) Problems of access and timing of operations, particularly for maintenance painting.

8.1.2   Documents

This should include all relevant Standards and Codes of Practice.

In the past, at least British Standards were definitive documents and provided a set of requirements that were contractually binding, whereas a Code of Practice, such as the former BS 5493:1977 'Code of Practice for protective coating of iron and steel structures against corrosion' set out all the options and advised rather than specified. Currently with International, European and identical British Standards, the distinction is less clear. An example that illustrates this is BS 7079-D2-2000 (ISO 8504-2-2000) 'Preparation of steel substrates before application of paints and related products. Part D2 Abrasive blast-cleaning'. This gives a Table of abrasives and their properties, different blast-cleaning methods and their effectiveness and limitations and procedures to be used before, during and after the cleaning process. All very useful information but not definitive, the specifier must make the selection to suit requirements. This can be illustrated with the following example.

Within the Standard there is a Note to the effect that: 'It may be advantageous to remove heavy, firmly adherent rust and scale by hand or power tools before blast-cleaning.' In practice, if there is heavy rust or scale, there is no question that this is the quickest way to get the job done. However, most blast-cleaning operators object strongly to putting down the blast nozzle and using hand or power tools, even assuming that they are available.

So, for any particular job and the conditions that exist on that job, the specifier must not specify that all work shall be carried out according to BS 7079 Part D2, but must state specifically which method is required from that Standard. Then the contractor knows exactly what is expected and can quote a realistic price for the work.

8.1.3   Pre-job conference

This is an American idea that is not normally included in UK specifications, but one that can be very valuable. It ensures that all interested parties, such as the project team, paint supplier, painting contractor, inspector and safety engineer, meet before the work starts to study the specification and paint manufacturer's data sheets and resolve conflicts, misunderstandings, ambiguities, etc. at an early stage. This has often proved valuable in practice. It is important to note that this is a meeting after the contract has been let but before the work starts.

It is not a pre-contract meeting, which has a different motive altogether.

8.1.4   Surface preparation

The method of surface preparation and standards required should be defined to obtain both what is possible and what is necessary. Ideally for maintenance painting this information is obtained from a pre-contract survey and feasibility trial (see Chapter 11). Specifications for blast-cleaning or hand or manual cleaning are straightforward with the Standards now available. In practice Standards for visual cleanliness seldom match well with the cleaned surfaces; the type of steel and the type of abrasive influences the appearance. Consequently, it is not unusual to specify special test panels, which are blast-cleaned to an agreed standard and are then used as a reference for subsequent work.

Where rusting of the steel surface has occurred, and particularly if there is significant pitting and the subsequent exposure is to wet conditions, i.e. condensation, rainfall, or immersion, then if long-term maintenance-free life is required, it will be necessary to specify a cleaning method capable of removing soluble salts from the surface. It is also important to specify, for example, ultra-high pressure water-jetting (see Chapter 3), and the means of monitoring that the visually cleaned surfaces are also adequately chemically clean (see Chapter 9), and the acceptable limits for any contamination.

8.1.5 Materials

Because of the difficulty in specifying paint composition in detail, there are very few meaningful paint standards. Therefore, in specifications they have to be defined by choice of manufacturer or generic type. However, generic description may provide only a limited indication of the composition and performance characteristics of a particular paint. If several proprietary materials are specified it is important to ensure that they have similar properties. This is best determined by careful study and comparison of the respective data sheets.

If any differences are noted, the paint suppliers involved will undoubtedly provide an explanation.

8.1.6 Control of coating materials and samples

The specification should cover the requirements for storage of paints and other coating materials. Where appropriate, any requirements regarding temperatures for storage should be included. It should also provide clear instructions regarding the facilities for storage and the requirements for receipt and issue of paints to ensure that early deliveries are not retained beyond their shelf lives, while later deliveries are not issued immediately.

Many large organisations, for example the UK Highways Agency, carry out routine sampling and testing of paint materials, both as delivered and as being used by the applicators. It is claimed that this has proved to be worthwhile and in many instances resulted in an improvement in the paint manufacturer's own quality control but more importantly with less problems with the actual paint application.

Where paint testing is to be carried out, the specification must state clearly whether this is to be done and reported upon before the painting commences, or whether samples are to be retained for subsequent testing. The number of samples required and whether they are to be in the form of unopened containers, or whether samples are to be taken during paint application, must also be made clear. The responsibility of the contractor or whoever may be responsible for collecting and dealing with any samples should be specified and, where appropriate, the method to be used for submitting them for test and the name of the test laboratory that will carry out the tests should also be included.

The tests will usually be specified separately unless they are to be carried out on-site, in which case the details will be given. Site tests will usually be of a comparatively simple nature, generally carried out by the inspector. If the test results are required before any painting is commenced, this should be stated.

The preparation of paints for use is an important factor in ensuring sound application and appropriate requirements must be included in the specification. These might include mixing instructions, which are particularly important for heavily pigmented and two-pack materials. The specification may call for all paint manufacturers' data sheets, but it is usually advisable to draw particular attention to critical points. These may include the restrictions on using certain two-pack materials in low-temperature conditions and the strict use of thinners and solvents in accordance with the data sheets. Any relaxation of these requirements should be made only by agreement with the client or his representative, and this should be stated.

8.1.7 Coating application

Application has a significant influence on the quality of coating produced, but is one of the more difficult parts of the process to control. Often the specification covers little more than the dry film thickness requirement. There are a number of factors involved in paint application that should be considered for inclusion in a specification. The manufacturer's data sheets should be followed implicitly, except where the client agrees to changes. In particular, attention must be paid to the following:

   (i) Drying time.
  (ii) Minimum and maximum interval before overcoating for the appropriate ambient temperature, where applicable.
 (iii) Use of only the recommended thinner.
 (iv) Only the recommended volume of thinner must be used for the particular application method chosen.
  (v) Correct nozzle orifice to be used as recommended by coating manufacturer for the particular method of spray application.
 (vi) Nozzle pressures as recommended.
 (vii) Mixing instructions for paints to be followed.
(viii) Pot and shelf lives to be noted for the appropriate ambient temperature, and paints to be disposed of if they have reached such lives before application.

Specifications should stipulate the requirements for the proper cleaning of equipment and the necessity for checking that it is operating properly. The working conditions will influence the quality of the coating and the requirements will vary depending on whether the painting is carried out in the shop or on-site, but they should be clearly specified.

Ideally, the facilities in the shop should be thoroughly checked before any painting is undertaken. However, where this has not been done, certain requirements should be stipulated:

(i)   The minimum and, where appropriate, maximum temperatures for paint application.

(ii) Prohibition of heaters that pollute the shop with exhaust products. Only indirect heating should be allowed where the ambient temperature has to be raised, and the heating should not impinge directly onto the painted surface.

(iii) Coatings should not be applied under conditions where condensation or moisture on the surface has occurred or is likely to occur. This can be specified by reference to the relative humidity, dew point and steel temperatures (see Section 9.5.3.4).

(iv) Lighting must be adequate; a figure of 500 lumens falling on the surface is considered to be satisfactory for most operations, but this will depend on factors such as the type of shop and the colour of the coating. In some cases an increased amount of light may be required.

(v)   Blast-cleaning facilities must be sited so that dust does not blow onto the areas where coating application is being carried out. Where appropriate, tests can be specified. Dust should preferably be removed by vacuum methods rather than by sweeping it from one surface to another.

(vi) Ventilation must be at a level to ensure proper safety standards, both regarding threshold level values and explosive limits. The latter, which is normally higher than the former, can be reached if operators are using a mask supplied with fresh air and therefore do not notice or concern themselves about the toxicity of the threshold level values.

On-site, additional restrictions may be necessary on the conditions acceptable for paint application and these should be stipulated. These will generally cover the suspension of operations during periods of rain, snow, etc., or where condensation has occurred or is likely to do so. Where relative humidity checks are specified, there should be a requirement that they be carried out close to the steelwork in question. Often condensation will occur on ground-facing or sheltered steelwork, even though freely exposed steelwork may be dry. Relative humidity and steel temperature measurements taken some distance from areas liable to condensation may not reveal the potential problems. Where shelter, e.g. of suitable sheeting, is specified so that site work can continue under adverse conditions, it will still be necessary to check for changes in relative humidity and this should be included in the painting requirements.

It is always necessary to ensure that surfaces to be painted are free of dust, grease, etc., and methods of checking these requirements are considered in Section 9.5.1.3. However, when working on-site additional clauses may be required to ensure that all foreign matter, e.g. cement dust, is removed before applying paint. In marine situations, account must be taken of chloride contamination even when the surface appears to be dry. The specification should call for checks that chloride contamination is absent or is at an acceptably low level for the coatings being applied.

8.1.8   Workmanship

The work shall be carried out strictly in accordance with the specification and paint maker's current instructions and work performed by skilled workman in a safe and defined workmanlike manner. The obvious problem is that the terms 'skilled' and 'workmanlike' are vague and difficult to define. The specifier may be able to insist on some proof that the operators are capable of carrying out the work to a satisfactory standard. One of the advantages of employing independent paint inspectors is that if they are suitably qualified they should be experienced in paint application themselves and will soon detect those that are not.

8.1.9   Treatments of special areas

The above discussion has been concerned with the cleaning and treatments of bulk steelwork but, in practice, steelwork is fabricated. This may involve both the welding and bolting together of sections. Furthermore, fabrication involves cutting and drilling steelwork. These operations and the overall design of the structural elements will have an influence on coating requirements. The influence of design on corrosion and its control are considered in Chapter 10 and it is recommended that designs should be audited to reduce the likelihood of problems arising because of unsatisfactory features such as water-traps. So far as the specification is concerned, it is important to check beforehand that areas that will be difficult to coat, either initially or during maintenance, are suitably protected; if necessary by extra coats of paint. Additionally, the stage at which protective coatings should be applied in relation to fabrication must be taken into account and reflected in the specification. If areas of the steelwork are to be fully or partially enclosed, then they would be coated before such work is completed. Painting is similar to welding in that access for the operator is necessary if the work is to be properly carried out. Treatments in *welded areas* should be clearly specified and the following points should be covered:

(i)   Where a suitable blast primer of the 'weld-through' type is specified, then no special treatment of areas to be welded is required prior to the actual welding operation.

(ii) Where such primers have not been specified, then areas adjacent to the weld should be suitably masked or painted with the full system, which must be removed before welding is carried out. The distance from the weld where areas are to be masked or where paint must be removed will depend upon the size and thickness of the steel sections, but a strip of 50 mm width is considered to be satisfactory for most purposes.

(iii) Although grinding of the weld to a smooth, rounded profile may not be required from the standpoint of structural reliability, it may be advantageous to ensure sound performance of the coatings applied to the weld area. This should be specified if required.

(iv) The surface preparation requirements after welding should be stipulated: mechanical cleaning to remove weld spatter, slag, etc., followed by blast-cleaning will provide the best surface for paint application, even where the welds have not been ground flush. Nevertheless, whatever choice is decided upon, the methods should be specified. If alkaline deposits are likely to be produced by the welding process, then if the weld is to be hand cleaned, e.g. by wire-brushing, suitable washing with clean water should be specified. Additionally, a requirement for a check on pH, using suitable test papers, should be included. This will ensure that alkaline deposits are removed.

Welding areas are often weak points with regard to the performance of coatings applied to them, and proper treatments should be specified. Coatings in the vicinity of welding may well be affected by the heat produced in the process, which may be conducted to areas some distance from the weld. This may result in the disintegration of coatings and, with some materials, the production of corrosive residues, e.g. acids. Such problems should be avoided by suitable specifications.

Fasteners, such as nuts and bolts, present problems of a different nature and the specification should be drafted to ensure that they are coated to a standard equivalent to that of the rest of the steelwork. Where steelwork is coated in the shop and bolted on-site, only the finishing coat may be applied on-site, unless the specification is written so as to ensure that the fasteners receive an adequate degree of protection.

Where manual cleaning is used to prepare the surface of the main steelwork, then black bolts may be similarly treated; but where higher-quality cleaning is specified, e.g. blast-cleaning, either the black bolts should be blast-cleaned *in situ* or be replaced by pickled or zinc-coated bolts.

High-strength friction-grip (HSFG) bolts should be treated in the same way as above, but pickling should not be specified unless suitable precautions are taken to avoid hydrogen embrittlement. HSFG bolts depend upon a suitable level of friction between the faying surfaces to be bolted. Consequently, the specification should ensure that the surfaces are blast-cleaned and masked, or coated with a material that will provide suitable slip coefficients, e.g. zinc silicate or sprayed metal. Where masking is used, methods of removing the adhesive should be specified. At bolted joints it is advised to specify immediate painting round the joint to prevent ingress of moisture and, where appropriate, application of a suitable mastic. For non-HSFG-bolted joints, it will be advantageous to paint the contact surfaces before bolting, and if practicable to bolt up while the paint coatings are still wet. The specification should take into account that various services, pipes, etc., will be fitted to the main steelwork. Where this is part of the overall design then suitable precautions can be taken by stipulating the remedial measures to be taken where steel is drilled and the coatings to be applied to brackets, etc., that will be fitted. However, this work is often carried out in an 'ad hoc' manner or designs are changed. It may be difficult in some circumstances to deal with the problem in the specifications for coating the new work because those fixing the various services are unlikely to receive a copy. In such cases, these matters can only be dealt with at the hand-over inspection or at the first maintenance repaint.

8.1.10   Handling, transport and storage

A good deal of damage may be inflicted on coatings during handling and transport. It is therefore advisable to specify the fundamental requirements necessary to reduce such damage to a minimum, and to ensure that where remedial work is necessary it is carried out correctly so that the coating at damaged areas is equivalent to that originally applied. The selection of the original coating and the decision whether to apply most of the coating in the shop will take into account the probability of damage and the difficulties likely to be encountered in repairing coatings. (Notes on remedial work are provided in Section 8.1.11.)

The following points should be borne in mind when preparing this section of the specification:

(i)   Lifting devices should be of suitable material to prevent mechanical damage, e.g. nylon slings and rubber protected chains (Figure 8.1).

Figure 8.1   Crane hook with soft renewable contact surface to minimise coating damage.

(ii)   Where appropriate, special lugs should be welded to large sections to facilitate lifting.

(iii) Attention should be paid to the packaging of components and small items.

(iv)   Coated steelwork should be packed to avoid contact between surfaces. Various forms of wrapping and packing should be considered, depending on the size of the steelwork.

(v)   All steelwork should be stacked correctly and raised from the ground on suitable supports, e.g. timber. The height of the stack should be limited to prevent undue pressure on the coatings on the steelwork at ground level.

(vi)   Where steelwork is protected by covers, e.g. tarpaulins, these should be arranged to allow ingress of air to provide ventilation.

(vii) Sections should be marked and stacked to allow for removal with minimum interference of the stack.

(viii) Steelwork to be transported as deck cargo must be coated to provide protection during the journey and for periods spent in storage on dock sides, etc.

(ix)   Steelwork should be stored so as to prevent 'ponding', i.e. the collection of water in pools on horizontal surfaces (Figure 8.2).

Figure 8.2   Incorrect storage gives an aggressive start to the life of primed steelwork.

Considerable damage and deterioration of coatings can occur during the period between shop painting and final erection. The specification should provide clear recommendations for handling. In some situations transport and storage is the critical part of the whole coating operation and may determine the type of coating to be used, and the relative parts of the system to be applied on-site and in the shop. Coating systems for shop application should be chosen so that they can be repaired effectively on-site. For example, if the blast-cleaned steelwork is coated with a two-pack epoxy primer and a two-pack epoxy micaceous iron oxide paint which, for ease of maintenance, is to be top-coated with a chlorinated rubber system, then all chlorinated rubber paint should be applied on-site. Otherwise, if damage down to bare metal occurs during transit, it is almost impossible to repair the system to the specification and to the same standard. For example, top coats dabbed onto spots of damage may look satisfactory but will eventually prove weak links in the overall protection. Damage to a paint system requires that all of the damaged coats are repaired to the correct thickness and with adequate drying times between coats. This is likely to prove unpopular with those concerned with the construction programme.

8.1.11   Remedial work

The specification should cover the following points:

(i)   All remedial repair and painting should be carried out as soon as is practicable and before further coats of paint are applied.

(ii)   Where damage does not extend to the steel surface, the paintwork should be washed to remove contamination. The original paint system should then be applied to the overall film thickness specified. All paintwork should be chamfered to ensure a smooth coating. Where appropriate, e.g. for two-pack epoxies, the surface should be lightly abraded to ensure good adhesion (see Section 13.6.20).

(iii)   Where damage extends to the steel surface, the surface should be cleaned by the original method specified to the required standard. An area of approximately 25 mm from the damaged area should be cleaned and the adhesion checked. The cleaned area should then be coated to provide the specified thickness. Paint coatings should be chamfered to provide a smooth surface. Where appropriate, surrounding paintwork should be abraded to ensure sound adhesion.

Other points to be covered include:

(i)   Removal of all contamination on any sound paintwork to be re-coated to ensure good intercoat adhesion.

(ii)   Use of mastic to fill small gaps.

(iii)   Removal of surface dust and debris before painting.

8.1.12   Inspection and quality control

The specification is the inspector's guideline to the coating work. Inspection is covered in detail in Chapter 9. Its purpose is to ensure compliance with the specification and should be independent of the coating operational work. The inspector acts as the client's representative and his main concern is to ensure that the technical specifications are properly followed.

Inspection requirements may be included in the individual clauses or in a single section. The specification should clearly indicate the inspector's responsibilities and authority. It is also advisable to indicate the type of instrument or equipment that will be used to measure various requirements, e.g. coating thickness. Problems can arise when the contractor and inspector use different instruments and methods.

Proper records are essential and this requirement from the inspector should be specified. The records will deal primarily with such matters as relative humidity, temperature and thickness readings related to the inspection work, but observations regarding the progress of the work may be useful. These records are for the client and should be so specified.

Paint testing is not usually carried out by the inspector but the specification may call for simple tests on viscosity or density. The methods should be clearly specified. Inspectors may be responsible for the collection and despatch of samples for further paint testing. The methods and timing of sampling and the information required on the samples should be specified.

Paint testing will usually be carried out in properly equipped laboratories. The types of test and the required information should be specified.

8.1.13 Safety

The specification must draw attention to safety requirements. In particular, nothing must be specified that breaks statutory safety regulations or that can be construed as likely to cause a hazard or danger to personnel.

Proper safety equipment should be specified for all stages of the work. Even though it may be the contractor's primary responsibility, the specifier should draw attention to the safety requirements expected. These should include:

(i)   Proper ventilation and precautions against explosive hazards in paint shops and confined spaces.
(ii) Suitable precautions to ensure that operators avoid skin contact with hazardous solvents or paints.
(iii) Adequate first-aid equipment available.
(iv) Proper protective clothing to be worn where appropriate.

8.1.14 Other aspects of specifications

Most corrosion control specifications for structural steel relate to protective coatings. Other methods may also be included, either as separate specifications or sometimes as part of the coating specification. These include the use of inhibitors and silica gel to control humidity in enclosed spaces, cathodic protection procedures and the use of special components, e.g. Monel bolts.

8.2 International standards

An International Standard ISO 12944 'Corrosion Protection of Steel Structures by Protective Paint Systems' was published in 1998. The standard is also recognised as a European Standard and therefore puts obligation on members of the European Union for its use. The Standard is in eight parts and Part 8 is titled 'Development of specifications for new work and maintenance'. This document gives guidance on the preparation of project specifications, paint system specifications, paint work specifications and inspection and testing specifications. Various annexes deal with particular aspects, such as planning of the work, reference areas and inspection and also provides models and forms intended to facilitate the work.

Chapter 9

Quality control of coating operations

9.1 Introduction

Quality control is applied to most industrial processes and is a standard and relatively easy function for paint applied in a continuous operation, for example coating of cars, domestic appliances, etc. Quality control of the coating of structural steel, however, encounters some unique factors.

Most paint coatings are, at least superficially, remarkably tolerant of variations in application conditions and procedures. However, it is generally impossible to tell from the appearance of the coating whether it has been applied over a suitable surface or has formed the correct polymer to give optimum performance. There are, as yet, no positive tests that can be applied to a paint film *in situ* that will provide this assurance. Many workers are investigating electrical measurements of paint films, such as AC impedance testing, but such tests are, as yet, not fully correlated with actual long-term performance.

One problem is the manner in which coatings fail. Loss of adhesion, visible corrosion, blistering, etc., which occur within a year or two of application are obvious faults. It is in everyone's interest to avoid such failures, particularly since adequate repair work is nearly always more expensive and troublesome than the initial work and almost inevitably to a lower standard. However, less obvious failures are those where there are shorter times than necessary between maintenance and with the requirement for extensive and expensive surface preparation.

It must be added that often there are many people in the coatings chain who are not interested in very long-term durability. However, some authorities faced with a likely maintenance programme that was beyond their resources and likely to get worse, addressed the problem some years ago and are now reaping the rewards.

The UK Highways Agency have now greatly increased the maintenance repainting intervals on motorway bridges. According to those responsible for arranging or undertaking the maintenance work, the majority of existing structures are now undergoing major maintenance at intervals in

excess of 20 years.[1] This has been achieved by comprehensive specifications, testing and monitoring of paint materials, particular concern with
surface preparation and removal of water-soluble contaminants and,
last but not least, independent painting inspection of the entire coating
operation.

In the early 1980s British Petroleum adopted QA/QC procedures that
included full-time independent painting inspection and now consider that
the higher standards of workmanship achieved have shown considerable
economic benefit.[2]

One large chemical plant in Germany has also found that intervals
between maintenance can be extended considerably although they operate
with a fairly high ratio of one inspector to 15–25 painters.[3]

Another large manufacturer of chemicals in the United States has estimated that the use of professional painting inspection can, over 10 years,
save $941 000 for every $1 000 000 dollars of the original cost of painting.

As with all quality control activities, they must be carried out indepen-
dently of those involved in production, i.e. the operators carrying out
surface preparation and coating application. These operators should, of
course, be properly supervised and carry out their work to meet the contract requirements. However, they may have priorities other than those
provided by the technical specification, not least the meeting of deadlines
and payment on the basis of the amount of work done in a given time.
There is, therefore, often a genuine conflict between quality and production; something not unique to coating processes. The user has to decide
whether the additional cost of inspection, for example 5% on painting
costs, is worthwhile. Usually there is no guarantee of performance of the
coating system in relation to the inspection, so the judgement as to its use
must be based on factors such as the experience of the user with respect to
inspection carried out on previous projects; the data concerning the performance of coatings that have been subjected to inspection compared
with those that have not; and the importance of the project itself.

The situation is summed up by S. T. Thompson,[4] who states

...it depends upon four cornerstones: good specifications, quality
materials, qualified contractors, and effective inspection. Failure in any
of these areas, like the proverbial weak link in the chain, will result in
decreased performances and increased costs – often substantial.

Generally, any inspection is preferable to none, although there are exceptions to this; for example, where inspection reaches a level of incom-
petence that leads to either unnecessary delays, possibly entailing
considerable additional costs, or a failure to apply the correct technical
procedures. In some cases it could be said with some truth that 'poor
inspection is worse than no inspection'. Most users who regularly specify

inspection for their projects will be aware of these problems, but for those with less experience it may be difficult to decide on the benefits or, indeed, sometimes the nature of the inspection. In the following sections, the various aspects of inspection will be considered.

9.2   Inspection requirements

Strictly speaking, inspection is carried out to ensure that the requirements of the coating specification are met by the contractor. Sometimes other duties are added and this may well be advantageous, but the prime responsibility of the inspector is clear: to ensure that the work is carried out in accordance with the specification. Although inspection may be carried out by any group independent of those involved in production, including personnel from the user's own organisation, it is often done by independent firms that have been specially formed for this purpose.

Sometimes such firms are invited to prepare the coating specification and even to provide a measure of supervision of the coating operations. This may well prove to be beneficial but does not alter the basic requirements already described. Inspection is a quality control measure and should not be confused with other important requirements involved in coating procedures. The coating specification is the essential basis for providing a sound protective system. This is considered in Chapter 8, but it must be said that some specifications are not of a standard likely to provide the highest quality of coating performance. Although the inspector may comment to this effect, he has no authority to alter the specification. Consequently, if his advice is not taken he may well find himself in a position where he carries out his duties conscientiously ensuring that an inadequate specification is correctly applied which may eventually lead to a coating of poor durability. It must be appreciated that in such a situation the fault lies with the specification, not the inspector; where inspection is to be used, the requirements should, of course, be included as part of the specification.

The user must be quite clear in his inspection requirements and must not expect more than is intended. If a consultancy–management operation is required then this should be decided at the outset. No matter how competent the inspector may be, he cannot be expected to act as a substitute for a sound specification, good management and proper planning. The basic requirements for a good inspector are considered in the next section and the user will be well advised to ensure that these criteria are met by those called upon to carry out quality control work on his behalf.

The nature and importance of the requirements for quality control are being increasingly appreciated, so the approach used should also be considered in a more critical manner.

In some ways the term 'coating inspection' is a misnomer because it may imply that it is confined to an assessment of the final coating, whereas it covers the whole coating process. The term 'quality control' is probably more correct.

9.3   The approach to quality control

The aim of quality control is to ensure that coatings attain their full potential of performance. Inspection is one of the steps involved but, to gain full advantage from it, other aspects must also be considered. The main stages in obtaining the optimum performance from coatings can be summarised as follows:

(i)   A full consideration of the coating requirements and the selection of a system suitable for the particular conditions.

(ii)   Appraisal of the design of the structure relative to coating application and performance.

(iii)   The choice of coating processes applicable to the fabrication techniques involved, or sometimes the choice of fabrication to suit the coating.

(iv)   Preparation of a concise unambiguous coating specification.

(v)   Tendering and acceptance of the requirements by the contractor.

(vi)   Checking, where appropriate, on the quality of the materials specified and supplied.

(vii)   Inspection at all stages of the coating process, including handling, storage and erection.

For maintenance painting, as opposed to new work, there may be additional requirements, as discussed in Chapter 11, e.g. a survey of the paintwork and tests to determine the feasibility of the maintenance procedures.

Clearly, inspection alone will not overcome problems arising from poor decisions in (i)–(vi) above, although in some situations the organisation responsible for inspection may be involved in these other areas. At the very least, such matters should be discussed so that the inspector is absolutely clear regarding the requirements and responsibilities for his part of the work. He cannot reasonably be expected to interpret unclear or ambiguous specifications and in fact it may be inappropriate for him to do so. Any doubts or problems should be cleared up by discussions among the interested parties before the work begins.

In the authors' experience, the majority of problems that arise over inspection, particularly in relation to the standards required, could have been avoided by adequate discussion at a pre-job conference. Unfortunately, the cost of a pre-job conference, which can be considerable, is seldom allowed for by those concerned, and yet apart from clarifying the specification it can also settle important issues such as what action can or cannot be taken by the inspector, what the specific safety requirements are, and how and to whom the inspector should report problems. Problems will arise that cannot always be foreseen, but there is no excuse for work being held up while, for example, arguments take place regarding whether the coating thickness in the specification was intended to be minimum, maximum or average.

Costs are obviously attributable to all the stages listed above and an opinion must be taken by the engineer of the amount considered reasonable for the project. For minor works, particularly where hand cleaning of the steelwork is to be used, limitation on cost may be necessary. However, for important projects, e.g. submarine pipelines, tank linings, bridges, etc., no long-term advantage will accrue from carrying out any stage with less than fully competent personnel with suitable expertise for the roles they undertake. Before considering inspection in detail it is worth discussing some of the related aspects which, while not falling within the strict definition of inspection as expressed earlier, nevertheless will be of benefit to the inspection work.

Although many large organisations have specialists capable of selecting coating systems and preparing coating specifications, others do not carry this expertise in-house. Advice can be sought from the coatings suppliers and generally this will be of a high standard. However, it will clearly be biased towards their own products. This may not be particularly important with some paint products that are made to a similar formulation and standard by a range of manufacturers. It becomes more crucial, however, with specialist coatings. Often the products from different manufacturers are not the same, even though their descriptions may be covered by similar generic terminology. In these cases, independent advice may be preferred. Again, by the nature of tendering, contractors will often offer a product which is unfamiliar to the engineer and independent assessments may be advantageous. The engineer may then take advice from a protective coatings consultant or from the organisation involved with the inspection. Provided the organisation has sufficient expertise, the following may be considered as appropriate areas for advice and discussion.

(i) Carrying out an audit of the design in relation to its susceptibility to corrosion and the problems of coating application. Offering advice on changes to improve the situation, taking into account cost and the overall effects of such changes.

(ii) Advising on the selection of suitable protective systems and discussing the requirements with coatings suppliers, taking into account future maintenance requirements and overall costs.

(iii) Advising on the preparation of a suitable coatings specification and the levels and requirements of inspection to be incorporated.

(iv) Discussions and advice on the contractor's quotations, with particular reference to the technical merits of any alternative protective systems proposed.

(v) Carrying out an audit of the contractor's premises where coating application is to take place. Assessing the probability that the coating will be applied to provide the requirements of the specification; typically, that blast-cleaning areas are positioned so that dust and abrasives are contained in that area and are not carried over to the facilities for coating application. Checking of all equipment and plant and ensuring that there are proper facilities for handling coated steelwork.

(vi) Discussing the programming of the coatings work, the equipment and manpower requirements and the methods to be used for handling, storing and transporting steelwork.

(vii) Discussing the storage and handling of steelwork on-site and other points concerned with the coatings, e.g. touch-up of damaged areas and cleaning and coating of welds.

All this goes beyond inspection but is considered here because many so-called 'inspection organisations' operate beyond the strict interpretation of the quality control requirements. Undoubtedly, the above approach will prove beneficial to many engineers.

9.4   Requirements for an inspector

The term 'inspector' here is taken to cover the person actually carrying out the work, but on larger projects a team of inspectors may be involved and the organisation responsible for the inspection may provide different inspectors for specific parts of the work. To simplify the discussion this is also covered by the term 'inspector'. Basically, the inspector must have the required technical competence for the work but, in view of the nature of the work he has to carry out, there are also personality requirements. Often the inspector is working on his own and because he is involved only with quality there is a basis here for potential conflict with those attempting to complete the work on time, i.e. the contractors. He must, therefore, be adept in human relationships, capable of communicating in a clear and concise manner and be capable of resisting pressure to accept standards below those required. He should have a good knowledge of the coatings industry and, most important, should be properly trained for the role he has to undertake (see Section 9.4.1).

As in all jobs, some inspectors, through experience, knowledge and personality, are able to achieve more than others. However, such qualities are not universal and, providing the inspector is technically competent and understands his responsibilities, he will be capable of carrying out his work satisfactorily. It is implicit that all inspectors, by the nature of their work, must be trustworthy and conscientious.

9.4.1   Training and certification of inspectors

One problem since the early days of coating inspection has been the lack of recognised training programmes and systems for the qualification of coating inspectors. Anybody can describe themselves as a painting inspector. Ex-painters see the job as steadier, more prestigious and less arduous than actual painting. On the other hand, they do have a considerable advantage in that they are familiar with the problems, can communicate with the operatives and can even assist the less experienced. They also know all of the short cuts and can be effective in detecting them on the basis of 'poacher turned gamekeeper'. The disadvantage is that, in ignorance, unless properly trained, they can perpetuate bad practices and also their experience of the wide range of protective coatings now available may be limited.

British Gas were the first to recognise this problem, and some years ago set up the Approval Scheme (ERS) for Paint Inspectors. This took one day and consisted of written and practical examinations and an interview. The UK Institution of Corrosion (I.Corr) later introduced a more comprehensive assessment scheme that took place over a one-week period. The initial scheme eventually lapsed because inspection companies did not receive recognition or financial return for their outlay.

None of these programmes involved any training, only examinations and peer review.

To remedy this lack of an adequate programme in which coating inspectors could be trained before certification, and to provide a programme for uniform assessment of experienced inspectors throughout the world, the National Association of Corrosion Engineers (NACE) in America, in conjunction with I.Corr, developed a Coating Inspector and Certification Programme. NACE declared[5] that the aims of the scheme were:

1   To provide professional and independent recognition to coating inspectors if their skills, knowledge and experience are to a sufficiently high standard.
2   To build confidence in the persons recognised under the scheme and receive from them an Attestation that they will apply themselves diligently and responsibly to their work, behave in an ethical manner and only profess competence in those areas in which they are qualified by knowledge and experience.
3   To provide the individual with a sense of achievement, since the qualification represents advancement in the chosen field.
4   To provide an opportunity for training coating inspectors with a wide range of backgrounds, including those with no previous work experience.

The Attestation referred to above is a document that all applicants are required to sign. There are five clauses to the effect that the inspector must recognise and acknowledge safety requirements, conservation of resources, cooperation with other trades and that the quality of the inspector's work and personal conduct will reflect on coating inspection as a whole. Failure to comply with the requirements of the Attestation can result in a disciplinary action by NACE. This includes withdrawal of the Certification which, in the USA in particular, where it is difficult to obtain inspection work without it, results in such obligations being taken very seriously.

As the NACE International coating inspector certification scheme expanded in North America, increasingly the programme included more and more American National legislation and procedures. Although the agreement with I.Corr at the outset of the scheme was that the NACE International scheme would be recognised on a world-wide basis, these changes in the programme have resulted in a number of similar schemes being established in other countries during the 1990s. Australia and Norway have developed schemes similar to the original NACE/I.Corr programme but have included national requirements within the framework. The lack of uptake of the NACE International scheme in the UK and Europe has resulted in I.Corr developing a similar scheme to the NACE International programme. It consists of 3 separate week-long periods of classroom and shop training sessions and both written practical and oral examinations given at the completion of each week's programmes. An inspector can stop at any one of the levels and receive recognition and membership of the Institute of Corrosion. It is now being increasingly recognised by owners of structures who undertake major painting contracts that a level 2 or on the more important contracts, a level 3 I.Corr painting inspector is employed to ensure high standards are maintained. In parallel with the NACE International scheme, failure of a qualified painting inspector to maintain adequate standards means that certification and, in a serious case, membership of the Institute can be withdrawn. The final examination is taken at the end of level 3, which would be taken after satisfactory completion of levels 1 and 2, and with a further 2 years' relevant work experience, and this standard is of a high level. A number of candidates fail at this point but the standards imposed by I.Corr have resulted in a significant improvement and standard of coating inspection since its introduction in 1995. The need for such a qualification has resulted in significant numbers of new and existing inspectors gaining this qualification and in 2001, approaching one thousand candidates had been successful in obtaining certificates in one of the three levels.

9.5   Methods of inspection of paint coatings

A good painting specification should define the inspector's duties and give the necessary authority to demand the specified requirements and limits. The inspector should never allow deviation from the specification without written authority from the appropriate body. The inspector should never insist on a higher standard than specified. If the contractor fails to meet the specification to any significant degree and the inspector has made reasonable efforts to gain compliance, the matter should be brought to the attention of the appropriate authority as soon as possible. The inspector should never direct or even appear to direct the work of a contractor's employees. If the quality control inspector takes on the role of supervisor there is a danger of relieving the contractor of all contractual obligations.

The methods used to inspect each phase of the coating operation will be considered below, with notes on the more common types of instruments and aids employed by the inspector.

9.5.1   Surface preparation

There are three aspects that the inspector may be called upon to examine: visual cleanliness, surface profile and freedom from salts, e.g. ferrous sulphate ($FeSO_4$). Of these, cleanliness is most commonly specified. Requirements relating to the state of the steel surface, i.e. the amount and type of rusting, before blast-cleaning is carried out may also be specified and this is often an inspection requirement.

9.5.1.1   Steel surface before blast-cleaning

The specification may cover requirements for the steel surface before blast-cleaning. Various clauses may be used to cover this aspect, e.g. new steel with virtually intact millscale. The original Swedish Standard covered four grades of steel surface designated A, B, C and D, ranging from new steel with millscale to fairly badly pitted and rusted steel. ISO 8501-1:1988 which replaces the Swedish Standard SIS 05 59 00-1967, includes the same photographs. The inspection in such cases will be of a visual nature, using either the Standard or some other suitably specified method.

Defects other than corrosion, e.g. laminations and shelling, will also be examined visually but often this is more usefully carried out after blast-cleaning.

9.5.1.2   Visual cleanliness

Current Standards for cleanliness of substrates prior to coating rely entirely on visual assessment of freedom from rust and millscale. The assessment is made with the aid of one of two methods. The first and probably the most widely used is by comparison with photographs of surfaces in varying degrees of visual cleanliness. This is the method used in the International Standard on surface preparation, which came into force in October 1989. The second method is to use a written description which includes a percentage of visual contamination allowed for each grade. This method is used in the NACE Standards.

In general there is no problem in identifying the highest standard, i.e. Sa3, white metal, etc., by either method; the problem arises with the lower standards that allow some residues left on the surface, such as Sa$2\frac{1}{2}$, NACE 2, SSPC, SP10, etc.

The difficulty with the Swedish Standard photographs, which incidentally are the same as those now in the International Standard and with the same prefix (Sa3, etc.), is that it is often difficult to match the photographs exactly with actual blast-cleaned surfaces. One reason is that the type of abrasive used affects the colour of the surface owing to embedded particles. The Swedish Standard photographs were produced from panels prepared by sand-blasting. Sand gives a whiter, brighter finish than for example, copper slag. Another reason is that the depth and shape of the surface profile affects the amount of shadow on the surface: deep, sharp-edged pits give more shadow, shallow depressions give less. Also, the Swedish Standard photographs are produced from specific rusted surfaces as represented by the Rust Grades A, B, C and D. Surfaces that have rusted to a greater or different extent will have a different appearance after blast-cleaning. Furthermore it is not easy to tell from photographs the exact nature of the residues remaining on the surface. The standards allow slight stains in the form of spots or stripes but not particles of scale in spots or stripes and it is difficult to distinguish these using a photograph. Finally, it is difficult to obtain consistent reproduction of colour photographs to the standard required, incidentally, a much more stringent standard that is normally necessary for colour illustrations. Comparison of different editions of the Swedish Standard shows significant differences between the reprints.

The International Standard ISO 8501-1:1988 has tried to overcome these objections as follows:

(i) It mentions in the text that appearances may be different with different abrasives and therefore provides a supplement which shows representative photographic examples of the change of appearance imparted to steel when it is blast-cleaned with different abrasives.

(ii) The ISO Standard differs from the Swedish Standard in that the Standard is represented by verbal descriptions and the photographs are representative photographic examples only. This distinction is more for the situation where there is a legal dispute rather than for practical use.

(iii) Considerable care is now being taken over monitoring the colour reproduction of each reprint. A special working party has to approve each colour print run.

In North America the situation is complicated; there are two large, prestigious but separate organisations that have over the years produced Standards for surface preparation. The National Association of Corrosion Engineers (NACE) and the Steel Structures Painting Council (SSPC) both have Standards consisting of written descriptions that are similar but not identical. For example, for second quality, i.e. NACE 2 and SSPC SP 10:63, the former states that 95% of the surface shall be free of residues, the latter says 95% of each square inch, and so on. NACE further supplemented the descriptions with visual comparators consisting of steel panels encased in clear plastic that illustrated the grades of cleaning when starting with one steel. One is for air-blasting with sand abrasive,[6] and the other for centrifugally blasting with steel grit.[7] SSPC visual standards to supplement the written specification were based, with some exceptions, on the Swedish Standard.

More recently there has been some cooperation between NACE and SSPC to produce common standards. NACE have also produced a visual comparator of plastic-encased steel to illustrate the surface preparation grades on new steel air-blasted with slag abrasive.

It is to be hoped that with a few years' experience of all the new standards there will be a distillation of the best and that this will become accepted world-wide, particularly since the intentions are the same and only the method of expression is different.

It should be noted, however, that in Europe the majority of painting specifications call for second-quality blast-cleaning, i.e. Sa2½, whereas in North America there appears to be a general tendency to call for lower standards, which are evidently considered as acceptable. For example, there is wide use of commercial blast (equivalent to Sa2) or brush-off blast (equivalent to Sa1). More recently they have produced a standard known as Industrial Blast. Basically this allows for tightly adhered millscale, rust or old paint remaining on no more than one-third of the surface providing that it is defined as dispersed. The definition of tightly adherent is defined as impossible to remove by lifting with a dull putty knife.

Hand- and power-tool cleaning also have their visual standards. In ISO 8501-1:1988 there are two grades, St2 (thorough) and St3 (very thorough cleaning). The photographs are identical to those in the Swedish Standard.

ISO 8501-1:1988 also includes photographs taken from a German DIN Standard and these show the rust grades A, B, C and D after flame cleaning. These are given the prefix Fl.

The problem, basically, with specifications for surface cleanliness arises from the need to make assessments rather than measurements.

Experienced inspectors are able to assess surfaces by eye with reasonable accuracy but the method is subjective and may lead to disagreement. This tends to occur rather more with the use of certain types of abrasives on previously rusted steel and with Sa2½ or equivalent grades. The problem is commonly overcome by preparing test panels to an acceptable standard prior to blast-cleaning the main steelwork. The standard is agreed upon by all parties and used for all assessments. Sometimes an area of the steelwork itself is blast-cleaned to the required standard and used as a reference.

9.5.1.3   Surface contaminants

Apart from the removal of millscale and rust, the absence of other contaminants, e.g. oil, grease and dust, may be specified.

Oil. In some continuous coating operations in factories, a combustion method is used to detect the presence of oil, but this does not appear to be practical for use in the field.

The water-break test, i.e. observation of the behaviour of a droplet of water on the surface, or the Fettrot test can be used. The Fettrot test consists of applying one drop of 0.1% solution of Fettrot BB dye in ethanol to the surface. On horizontal surfaces free from oil the drop spreads out rapidly and a circular residue remains. On vertical surfaces free from oil, the drop runs away quickly leaving an oval residue. On horizontal surfaces not free from oil the drop remains in its original size until it evaporates leaving a sharply pointed residue. On vertical surfaces it leaves a long trace on the surface. If the dye Fettrot BB is not available, the test will work equally well with either a 1% solution of Crystal Violet in ethanol or a 1% solution of fluorescein in ethanol.

The UV test for oil can be useful in some situations. The procedure involves exposing the surface to ultraviolet light and some, but not all, oils will fluoresce. The test has particular value for the maintenance painting of the internals of oil tanks provided the oil that is likely to be present does fluoresce under UV irradiation.

Soluble iron corrosion products. Apart from the cleanliness as indicated by standards (e.g. Sa2½), the steel surface, even after blast-cleaning, may be contaminated with salts produced by the corrosion process (see Chapter 3 for an explanation). These are commonly ferrous sulphate ($FeSO_4$) and various chlorides of iron, e.g. $FeCl_2$. There are at present no national or international standards available for specifying the amount acceptable for various conditions. Consequently, specifications may call for the absence of such salts, or a limit may be specified, based on previous experience.

If inspectors are called upon to determine the presence of such salts, there are test methods available. Generally, these salts, often termed 'soluble iron corrosion products', can be detected in the pits on the steel surface by using a magnifying glass of about ×15 magnification. The visual appearance is not, however, always a reliable method of identifying the presence of such salts. The salts are usually colourless and are present at the bottom of pits. Over a period of time they react with the steel, producing a darkening of the surface, so one possible method of checking their presence is to moisten the steel. This is effective but is not sufficiently rapid as an inspection method. It should be noted that the rapidity with which a newly cleaned surface re-rusts is also an indication of the presence of soluble salts, and there have been suggestions that this might be quantified as a test method.

A test method based on the use of filter papers impregnated with potassium ferricyanide was rejected by the ISO and BS committees dealing with surface preparation, on the grounds that its sensitivity was such that the results were misleading. A more suitable test method, originally devised in the CEGB Paint Testing Laboratory in 1977, is based on 2,2-bipyridene test strips and has been published as ISO 8502/1. This involves swabbing the test area with distilled or deionised water and testing the sample with proprietary test strips specific for ferrous ions (Figures 9.1 and 9.2). The colour change on the test strips is a semi-quantitative measure of the ferrous salts present. This is a quick, cheap method of test. Its weakness is that the method of sampling is crude and probably only extracts about 25% of the ferrous salts present.[8]

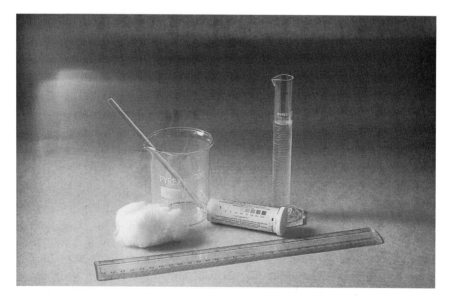

Figure 9.1   Apparatus for Merckoquant test for soluble ferrous salts.

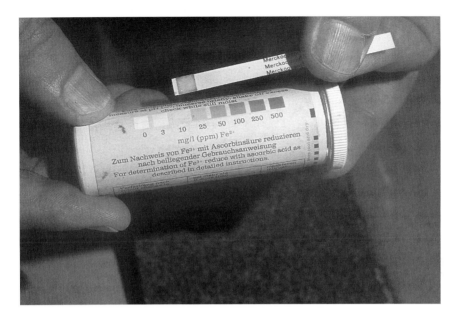

Figure 9.2   Indicator strips for ferrous salts.

The use of the Bresle sample patch (ISO 8502/6) is more accurate. (See under 'Sampling' later in this section.)

Moisture. Instruments have been devised to measure surface moisture but the problems arise in calibrating them to provide a useful assessment for specification and inspection purposes. Usually, visual examination is employed. A crude method is to sprinkle a little talcum powder on the surface. If the surface is dry and free from contaminants then the powder can be readily removed by gentle blowing. The method is most usefully employed on smooth paint surfaces but can sometimes be used on blast-cleaned steel.

Some paints are supplied for application to damp surfaces. Problems can arise in determining the amount of moisture that is acceptable and methods of assessing it. In such situations, advice should be sought from the paint supplier.

Chloride. In a marine environment the major source of soluble contaminants is seawater. Detection is necessary on both the bare steel and painted surfaces to be overcoated. ISO 8502/1 is specifically for ferrous salts and will not detect sodium chloride. ISO 8502/2 is a laboratory method for the determination of chloride. ISO 8502/10 is a field method using a relatively simple titrimetric determination of chloride. There are also proprietary test kits for chloride determination available.

Most paint suppliers to the marine industry prefer to use a conductivity method. This has the advantage of measuring all dissolved salts, i.e. it will measure both ferrous chloride from corrosion and chlorides from sea-water. The disadvantage of the method is that it is very sensitive and needs great care in handling. For example, sweaty hands will add to the apparent contamination and the original sample water needs to be of a very high standard of purity. Also the method does not distinguish between the ions causing the rise in conductivity and not all ions have equal corrosive power. ISO 8502/9 is a field method for the conductometric determination of water-soluble salts.

Dust. ISO 8502/3 is a method of determining both the density and size of dust.

The test is a simple one consisting of pressing a pressure-sensitive tape onto the surface. To standardise the pressure applied to the tape the ISO Standard gives the option of using a specially designed, spring-loaded roller to apply the tape. The tape with the dust adhering to it is then placed on a white background and compared with pictorial ratings for both quantity and particle size. In practice, this is difficult to use without contaminating the top surface of the tape and is less effective than human fingers in pressing the tape into a pitted surface; it therefore seems to be an unnecessary refinement.

As yet, there are no recommendations for acceptable levels of dust. This is a necessary requirement since it is unlikely that any surface in an industrial atmosphere will be dust free. The acceptable level will depend upon the method of paint application, brushing being the most tolerant, and the type of paint being used.

Sampling. One of the problems with measurement of surface contamination is the difficulty of sampling from blast-cleaned and possibly pitted surfaces. Swabbing the surface is a crude technique that is neither effective nor consistent.

As far back as 1959, Mayne[9] suggested the use of a limpet cell, but this is difficult and cumbersome on a rough surface. A sampling method was developed in Sweden by Bresle and is now ISO 8502/6:2000. This consists of a flexible cell consisting of a self-adhesive plastic patch, about 1.5 mm thick and with 40-mm diameter punched hole in the centre. The hole is covered by a thin latex film to provide a sample area. Water is injected into and withdrawn from the sample patch by means of a hypodermic syringe (see Figure 9.3).

The volume of the sample is small but by leaving the syringe needle in position it is possible to inject and extract the water several times to provide

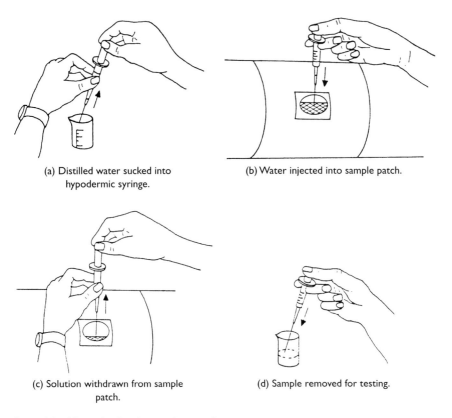

(a) Distilled water sucked into hypodermic syringe.

(b) Water injected into sample patch.

(c) Solution withdrawn from sample patch.

(d) Sample removed for testing.

Figure 9.3   Using the Bresle sampling patch.

some agitation on the surface and increase recovery. This procedure also prevents leakage and enables the patch to be used in any position. The disadvantages of the Bresle patch are that they are relatively expensive, can only be used once and, with a diameter of 40 mm, sample only a relatively small area. If used to detect soluble iron corrosion products the result obtained may well depend on the number and depth of corrosion pits within the sample area. Other similar sampling devices are available from different manufacturers but, as yet, they are not included in an ISO Standard.

Conductivity. Conductivity measurements of sample wash liquids from a blast-cleaned surface may well prove to be the easiest and most reliable method of detecting water-soluble contamination. Small conductivity meters (about the size of a pen torch) are now available and only require a small sample of water. This would be particularly useful to use in conjunction with the Bresle 'patch' since this would reduce the possibility of contamination, for example from sweaty hands, which would affect the

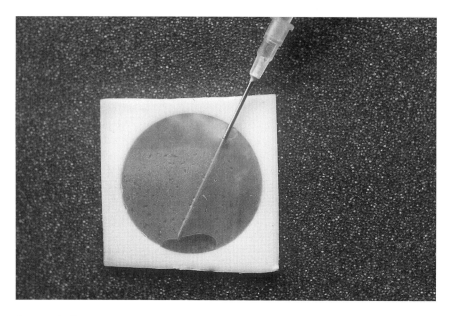

Figure 9.4   Bresle sample patch in use.

reading. Several models of larger conductivity measurement cells are available and can be used.

Shell Research UK have developed a conductimetric apparatus including a limpet cell that can measure conductivity of steel surfaces. The method was primarily designed to check the deterioration of zinc silicate primers but can also be used to detect soluble salts. A Japanese firm, Toa Electronics has also designed a sampling apparatus for measuring liquids on surfaces.

A salt contamination meter, known as SCM 400, has been developed in the UK. Its method of use is to apply a wet filter paper to the surface to be tested, remove, and measure the conductivity of the filter paper in a special apparatus.

The advantage of conductivity measurement is that it includes all forms of water-soluble contamination, i.e. sodium chloride as well as ferrous chloride. The disadvantage is that it does not distinguish between the ions causing the rise in conductivity.

It has been argued that conductivity measurement used to detect salts will be influenced by the pH of the solution and are, therefore, unreliable. Work by Igetoft[10] concludes that an unusually high conductivity measurement of wash water from a blast-cleaned surface will always indicate the presence of ionic species regardless of the pH of the solution and that this is enough to show that the surface is not suitable for painting.

9.5.1.4   Surface profile

There is no direct relationship between the surface profile and surface cleanliness and specifying to the ISO Standard does not imply any profile requirements. The surface profile is the term used to indicate the undulating nature of blast-cleaned surfaces, i.e. the roughness. Over the years a number of devices have been developed to measure surface profile but, for field conditions, three methods are generally used to measure or assess the distance between the summit of any peak on a blast-cleaned surface and the bottom of an immediately adjacent trough:

(i)   A specially converted dial gauge for direct measurement.
(ii)   The production of a suitable replica of the surface, which is measured, usually by a dial gauge.
(iii) Specially prepared panels used to compare the roughness of the surface being inspected and so assess the profile.

All three are used but (iii), the use of a special comparator, has been standardised by the appropriate ISO committee and has some advantages that will be considered later.

Dial gauge method. This was the first method developed for use in the field. Dial gauges are specially manufactured for measuring surface profile and their operation is straightforward. The foot of the gauge is a flat anvil through which protrudes a pointed probe. When the anvil is placed on a perfectly flat surface the base and the pointer are level and the gauge reads zero. Placed on a blast-cleaned surface, the probe drops into a valley and the gauge reads the depth. Obviously, the probe does not always fall into the lowest point and several readings must be taken to produce an average profile measurement.

The method is quick and easy to use. It has been a popular method in the UK but hardly used in the USA.

Replica method. The method above can be used fairly easily on large, flat, horizontal areas but problems may arise on undersides and areas with difficult access. In such situations replicas may be produced and then measured with an instrument such as a Tallysurf or by optical methods. The measurements would be carried out in a laboratory, not usually by the inspector. However, there is a replica method that can be used directly by inspectors in the field.

The replicas are produced from a commercial material called Testex Press-o-Film Tape using a specially adapted dial gauge. The tape is made up of two separate layers. One is an incompressible Mylar backing of thickness $50 \mu m$; the other is a compressible material of virtually nil elasticity. By placing the tape with the compressible layer on the steel surface and rubbing the backing with a blunt instrument, a replica of the surface is produced. The tape is then measured with the special dial gauge. The maximum profile can then be calculated by subtracting the thickness of the non-compressible backing, i.e. 50 μm, from the dial reading.

This method only provides a measure of the *maximum* profile; it does not indicate an average. However, it has a number of advantages. It is comparatively easy and quick to use, it can be employed on curved surfaces and in positions with difficult access. Furthermore, the tape can be maintained as a record of the profile reading. On the other hand, the use of tapes adds to the overall cost. Two grades of tape are available: (a) Coarse for profiles 20–50 μm, and (b) X-Coarse for profiles from 40–120 μm. The tapes are of American origin and the specially adapted dial gauge is calibrated in mils (10^{-3} in.), not in microns. The tapes are usually supplied in a dispenser containing a roll of 50 test tapes. The method is particularly favoured in North America, where it is a NACE standard recommended practice.[11] It is also likely to become an ISO Test Method.

Profile comparators. Special steel coupons prepared to provide a range of different surface roughnesses have been available for many years, e.g. 'Rugo test' specimens. These have not been specifically prepared for assessing blast-cleaned surfaces but can sometimes be used for this purpose, generally using a tactile assessment. More recently, however, specially produced gauges have been available to compare average profiles produced by blast-cleaning. The Keane–Tator surface profile comparators are commonly used in America but to a lesser extent in Europe. The comparator consists of a disc with a central circular hole and five projections. Each of the projections or segments is prepared to a different average profile depth. The disc is placed on the blast-cleaned surface to be assessed. A battery-operated ×5 illuminated magnifier is placed over the central hole; the surface to be examined can then be compared with the various segments and a match of the profile made. This may be a direct match, say at 50 μm, or may be between the 50 and 75 μm segments, and so regarded as 60 or 65 μm. Being of American origin, these comparators are in practice calibrated in mils. Three different discs are produced to represent surfaces blast-cleaned by sand, metallic grit and steel shot. The discs are a high-purity nickel electroformed copy of a master, which has been checked for profile height with a microscope. Another set of comparators, again produced in the United States, is the 'Clemtex Coupons'. These are stainless-steel specimens treated to provide a range of profiles.

A Work Group of the ISO Committee on Surface Preparation carried out extensive trials on all methods of measuring surface profile. They decided that the precision and accuracy of all of the methods was poor and this was probably due to the considerable variation in roughness that occurs on a blast-cleaned surface. In addition, each method tended to give a different result. They considered, therefore, that to give an answer in 'numbers', itself implied an unjustifiable accuracy that was unnecessary for its purposes. In addition, none of the methods was likely to measure rogue peaks. They therefore decided to produce a special surface profile comparator.

ISO Standards 8503 Parts 1 to 4 and the identical BS 7079 Parts C1 to C4, describe the specification for the comparators, the method of use and the two reference methods to calibrate the comparators, namely by focusing microscope or stylus instrument.

The comparators consist of a square with four segments and a hole in the centre (see Figure 9.5). Segment 1 is the finest and each segment increases in roughness. The blast-cleaned surface is compared with each segment and if between 1 and 2 it is called fine, between 2 and 3 it is called medium and between 3 and 4 it is called coarse. There is one comparator for grit-blasted surfaces and one for shot-blasted surfaces. Unfortunately, because shot cannot produce the same roughness as grit, the terms coarse, medium and fine represent different levels between the two comparators. In the opinion of the authors, it would have been more logical to keep the same levels of roughness between both comparators and to leave the coarse segments blank for the comparator for shot blasting.

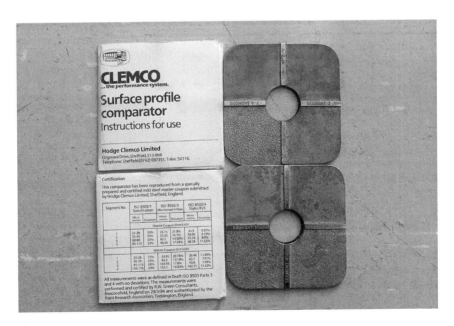

Figure 9.5   ISO surface profile comparators.

9.5.1.5   Weld areas

Weld areas are vulnerable and likely to be the first point of failure of a painted structure. Their treatment is often neglected in specifications. References to visual cleanliness, such as Sa2$\frac{1}{2}$ or to surface profile are not applicable.

Often the welding inspector's requirements are different from those of the painting inspector. Most of the chemical contamination will arise from welding rods and small portions of the surface should be wetted and tested with pH papers. Most welding rods give an alkaline deposit which must be removed before painting, particularly for oleo-resinous paints which could be saponified. The major concern with welds, however, is the roughness: sharp edges, undercutting, pinholes, etc., which cannot be coated properly and must be eliminated.

The International Standard 'Preparation Grades of Previously Coated Steel Substrates after Localized Removal of Coating', ISO 8501 Part 2 includes photographs taken from a German DIN Standard,[12] which shows examples of treatment of welds. For this important subject the ISO Committee is producing a standard to illustrate types of surface defects. This standard will use diagrams of the defects and these give a much clearer picture of each defect than a photograph. Unfortunately, the extent that such defects need not be treated is the subject of a difference of opinion between those concerned with paint performance and those concerned with the integrity of, for example, the welds. There is also a commercial implication if all specifiers selected the highest standard of treatment for all steelwork regardless of its final use. Both sides have a good case and a compromise has been reached whereby there will be two or three grades of different degrees of remedying surface defects. It remains to be seen when the standard is published, whether specifiers can be persuaded to opt for the lower grades in any circumstances.

NACE has addressed the situation by producing a weld replica in plastic which illustrates the varying degrees of surface finish on welds prior to coating. The weld comparator includes full seam welds, butt welds, lap welds and skip or tack welds[13] (see Figure 9.6).

9.5.1.6   Blast-cleaning operations and equipment

Many specifications rely on the inspection of the final blast-cleaned surface as the criterion for acceptance. This is a reasonable approach based on the view that the contractor will have to operate his equipment satisfactorily in order to achieve these requirements. However, if the contractor's equipment or the operators using it are not particularly efficient, problems and delay will arise; to avoid this, inspectors may be required to carry out certain checks. This will be concerned mainly with assessing the quality of

Figure 9.6   NACE weld comparator.

the equipment and its probable effectiveness in achieving the requirements of the specification.

This section is concerned with blast-cleaning, but hand tools are
often used, particularly for maintenance work, and they should also be
checked, e.g. for wear (in the case of wire-brushes and scrapers). Warped
and blunt tools are much less effective and lead to slower cleaning,
so where appropriate suitable action should be taken. Power-driven
tools should also be inspected and, where necessary, the attachments
such as wire-brushes, grinding discs, etc., should be replaced. The hoses,
cables and other equipment involved should also be inspected to
ensure that they are capable of continuous operation and can reach all
parts of the structure to be cleaned. Delays due to poor cleaning equipment can largely be avoided through careful inspection before the work
commences.

There are two main types of blast-cleaning equipment: one using an air
compressor and the other relying on rotating wheels to provide the
centrifugal force necessary to provide the abrasives with a suitable velocity
for cleaning.

Compressed air blast-cleaning machine. The following general checks
should be made on the compressor and ancillary equipment:

(a) Correct capacity for the work in hand and ability to maintain the required air pressure and volume: this can be checked from the manufacturer's data sheets.

(b) Compressed air should be free of contaminants and moisture. This can be checked by blowing the air onto a piece of filter paper or white cloth of suitable size. If there is no discoloration or sign of moisture, then the air is clean. On the other hand, a trace of oil or moisture would indicate that the various filters and traps are not functioning properly. Serious contamination of the test paper or cloth may indicate a requirement for further inspection of the air source. Where the specification calls for a complete absence of moisture in the air, then a change of equipment or the use of a special drying installation may be called for. The risk of water condensation carrying through the system increases with increasing relative humidity. When the humidity is high, sometimes the water-separating devices cannot cope and the inspector must advise the supervisor.

(c) The blast-cleaning machine should be checked overall, including all the valves and separators.

(d) The hoses should be of suitable length so that they can be properly operated but they should not be unnecessarily long, as this will reduce the pressure available at the nozzle. All couplings should be sound and tight. The pressure at the nozzle can be checked with a special hypodermic needle gauge which is inserted in the hose close to the nozzle during the operation of the equipment, i.e. with abrasives being used. The small hole which is made seals, and does not affect the operation of the hose. This check is easily carried out and will ensure that the machine is operating at a suitable pressure (550–690 kPa; 80–100 psi) at the nozzle. The pressure gauge on the equipment will indicate the pressure before the air has been forced through the hoses. The hoses, particularly 'whip ends', the small pieces of smaller-bore hose used at the nozzle end by the blast-cleaner to ease handling, will result in some pressure drop. Blast-cleaning may be completely ineffectual if the pressure at the nozzle is too low. The nozzle is also important and should be chosen for the specific work in hand. Nozzles should be checked for wear and damage; special gauges are available for checking the nozzle diameter. If badly worn, the nozzle should be replaced.

Centrifugal blast-cleaning machines. Only limited checks can be carried out on this type of equipment, which is in a self-contained unit, without separate compressors, hoses, etc.

(a) Check that the sections to be cleaned can be properly treated. The arrangement and number of wheels used in the plant will influence the area cleaned. Adjustments can be made to the equipment but the presence of re-entrant angles and complexity of shape may result in 'hidden' areas, which will have to be blast-cleaned by hand later.

(b) The degree of cleaning is determined to a considerable extent by the speed of the sections through the machine and, if necessary, this can be adjusted.

(c) The abrasives are recirculated in this type of machine and correct screening is required to ensure that the specified profile is obtained.

Water jetting. The specification requirements for water jetting have to take into account that the standards used for dry blasting, e.g. the Swedish Standard, are not directly applicable to this method of cleaning. Furthermore, there is a likelihood of flash rusting on-site from the moistening of the surface by the water used for cleaning.

Nowadays, many enlightened paint manufacturers are prepared to accept a degree of oxidation of blast-cleaned surfaces. To be acceptable, such oxidation should not be powdery and should be ginger in colour and shown to be free of soluble salts. Such 'gingering', as it now tends to be called, is due to the differential aeration corrosion cells formed as clean water evaporates from the surface; dark brown rust, particularly in spots, would be an indication of the presence of soluble salts and would be unacceptable. Various marine paint manufacturers have produced photographic standards for the visual appearance of suitably cleaned water jetted surfaces and also the amount of re-rusting that is acceptable. Those considered the most suitable are incorporated in the new ISO Standard.

9.5.1.7   Abrasives

The degree of inspection will be determined by the specification requirements and may cover either or both of the following:

(a) Check on type of abrasive, e.g. steel shot.

(b) Size of abrasives. This can be checked by suitable screening with standard sieves, e.g. to BS 410 or ASTM D 451-63.

The sample tested must be representative and should not be taken from the top of the bag where there is likely to be a preponderance of larger particles. The bag selected for sampling should be representative of the batch and the contents should be thoroughly mixed prior to the actual taking of the sample. The test is carried out on the following lines:

(i)   A correctly prepared sample of 200 g of abrasive is weighed.

(ii)   The mesh sizes for sieves required for the particular abrasive are selected by reference to a suitable standard, e.g. BS 2451.

(iii) The sieves are arranged in order (they fit together) with the largest mesh size at the top and the receiver under the smallest mesh size.

(iv)   The weighed sample is poured in and vibrated or shaken for 5 min.

(v)   The contents of each sieve are weighed and the weight is expressed as a percentage of the original sample.

(vi)   The results are expressed in accordance with the particular standard being used.

If the BS 2451 method is used, the results would be expressed as follows.

Taking G24 abrasive as an example: total sample passes 1.00 mm mesh; at least 70% is retained by 0.710 mm mesh; at most 15% passes 0.600 mm mesh; and none passes 0.355 mm mesh.

All sieves must be cleaned prior to the screening and this is best achieved with a hand brush. Abrasive should not be pushed from the sieve with a pointed metal object as this is likely to damage the mesh and provide incorrect results. The above test can be carried out reasonably quickly but it is sometimes sufficient to screen an unweighed sample of about 400 g through the largest and smallest sieves only, by shaking for 5 min. If all the abrasive has passed through the top sieve and has been retained by the bottom sieve, this would be considered as satisfactory.

(c) Checks for contamination of the abrasive may be called for, particularly the presence of dust and oil. Dust can be checked using the sample prepared for the screening test on sieves. About 100 g is placed in a clean container and water is poured in so that the abrasive is just covered. After stirring to ensure wetting, any dust from the sample will be visible on the surface of the water.

The presence of oil can be checked by placing another sample of 100 g in a suitably sized glass beaker. Clean solvent, e.g. xylene, is poured so as to just cover the abrasive; after stirring, some of the liquid is poured carefully onto a clean glass plate. The solvent is allowed to evaporate and the presence of oil and grease will be detectable on the glass, as a smear. As solvents are flammable, suitable safety precautions are necessary when carrying out the test.

(d) There may be a requirement for a minimum soluble chloride content with some abrasives. This can be obtained only by standard analysis in a laboratory.

For a quick check on contaminants, a small sample of the abrasive can be stirred in distilled or demineralised water in a glass container. Spot checks can then be made for chloride, ferrous ions, etc., or tested in general for soluble salts with a conductivity test.

There are ISO Standards for abrasives (see Chapter 3, Section 3.2.3.5). These are mainly concerned with laboratory determinations on new abrasives. Probably the only involvement for a quality control inspector is to ensure that samples are obtained correctly.

9.5.2 Testing of liquid paints

Increasingly, and particularly for important projects, paint users are checking the quality of paint materials prior to work starting and during the work. Most of the tests must be carried out in a laboratory (see Chapter 16), but the painting inspector is frequently required to obtain samples and also to carry out some simple tests on site.

Sampling is generally by selecting unopened tins at random from stock. If samples are required from large containers, it is important to ensure that sampling is carried out correctly, for example to ISO 1512:1991 Paint and Varnishes, 'Samples of products in liquid or paste form' and ISO 1513 'Examination and preparation of samples for testing'. Special deliveries of single tins of paint arranged by the contractor are generally not acceptable as a representative sample.

On-site, a painting inspector may be required to check paint supplies for type, colour, suitability for the chosen method of application, condition, gravity and viscosity. These latter methods are described in Chapter 16.

As the work proceeds, and particularly with brush application, the inspector may take samples from painters' 'kettles' for gravity checks. These will indicate whether unauthorised thinning has been performed. Results should be within 10% of the paint manufacturers' declared figures but in practice, if thinning has occurred, it is likely to be so gross as to be obvious. The tests are generally only required to give a scientific basis for the complaint against the applicator. Fortunately, although this was a common fault in the past, it is less likely with spray application since excessive thinning gives the operator little advantage.

9.5.3 Coating application

The quality of the coating application has an important influence on the durability of coatings and sound inspection techniques will play an important role in ensuring that application is carried out to the required high standard. Inspection should cover not only the physical application of the coating but other aspects which also affect the quality of the final coating. These include checks on the ambient conditions at site or in the shop to make sure that they satisfactorily meet the specification requirements and also checking of paints, including storage and mixing. The application of metal coatings also requires inspection and the inspection methods will be determined by the method of application and the particular metal being coated.

Good housekeeping has an important influence on the quality of work and the inspector should endeavour to ensure that as high a standard as possible is maintained and should seek the cooperation of the contractor's supervisor in this respect.

9.5.3.1   Storage and preparation of paint

The following checks are necessary to ensure that the paint is correctly stored and prepared to use:

(a) Paint delivered to the site corresponds to the specification requirements, including type, e.g. primer, and colour.

(b) All paints are correct for the method of application to be used, e.g. airless spray.

(c) Condition of storage: a properly prepared store that does not suffer from extremes of temperature should be used. When not in use it should be locked and the key held by a responsible person.

(d) Batch numbers should be recorded and paint should be withdrawn from store in the correct sequence. Withdrawals should be properly recorded.

(e) Sufficient paint should be available in store, either on-site or at a central depot, for completion of the work.

(f) The inspector should be present when paint is issued at the start of the work period.

(g) Single-pack paints must be thoroughly stirred. This particularly applies to paints containing heavy pigments such as micaceous iron oxide or zinc dust.

(h) Two-pack materials must be mixed strictly in accordance with the paint manufacturer's data sheets. Furthermore, other requirements such as induction periods and pot life must be strictly adhered to. Material must not be used after the expiry of its pot life; it must be discarded. Temperature has an effect on pot life. If the temperature is markedly higher than that quoted on the paint data sheets, the pot life may be decreased and advice should be sought if any doubts exist regarding this aspect.

(i) Additions to paints must be strictly in accordance with the manufacturer's recommendations. Too much thinner will result in reduced film thickness, too little may cause dry spray, pinholes or poor appearance. The wrong type of thinner, i.e. one other than that given in the manufacturer's data sheet, may cause coagulation of the paint. In some cases, this can be quite spectacular as the coating gels in the spray lines.

(j) If test results on paints are required before painting commences, it must be checked that they are available and have been submitted to the client.

9.5.3.2   Paint application equipment

All equipment must be in good condition if sound paint coatings are to be achieved. Brushes and rollers should be of the correct size and shape and must not have worn to an extent where coating application will be affected. All spray equipment should be checked to ensure that it is in sound working condition and that reserve equipment is available if required.

The information in the manufacturer's paint data sheets should be followed regarding pressure, tips, etc. Where appropriate, hoses, filters and compressors should be examined to ensure that they are in sound condition and suitable for the work in hand. All the equipment must be clean and should be thoroughly checked before operations are started.

The condition, cleanliness and suitability for the work to be carried out are essential requirements for paint application equipment and must be thoroughly assessed and checked beforehand. The lack of suitable equipment or its malfunctioning can lead to serious delays or poor application of the paint, both of which will add to the overall costs.

9.5.3.3 General conditions in the shop and on the site

The suitability of any workplace for proper paint application should be assessed prior to placing a contract. However, this is often not done and the inspector should then satisfy himself that the conditions in the shop are suitable for the work in hand. The various requirements can be summarised as follows:

(a) Maximum and minimum temperatures and their control. Suitability of equipment for eating, e.g. combustible products should not be produced inside the shop.

(b) Ventilation: sufficient to maintain low concentration of fumes and vapours; solvent vapours must be kept below the TLV (threshold limit value) or Occupational Exposure Standard.

(c) Lighting: sufficient intensity to allow for adequate painting and inspection.

(d) Positioning of various pieces of equipment, e.g. blast-cleaning areas must be properly separated from painting areas to avoid contamination with dust, abrasives, etc.

(e) Suitable areas for storage of painted steel and adequate equipment for handling the painted product.

(f) Availability of proper protective equipment for operators.

(g) Sound health and safety procedures.

(h) At site, as opposed to inside a shop, there can be no direct control of the ambient conditions. However, where protective sheeting has been specified, e.g. to allow for the continuation of work during adverse weather conditions, the adequacy of such measures should be checked.

(i) There must be adequate access to all surfaces to be painted but a situation frequently overlooked is when parts of the scaffolding (albeit only relatively small areas) are fixed too close to the work surface.

The storage conditions for painted steelwork coming to the site should be suitable, with proper foundations for the stacked steelwork, ensuring that sections are not resting in mud or on gravel which can damage the coating. The sections should be stored to avoid ponding, i.e. collection of water on horizontal areas. Suitable slings, etc., should be available for handling.

9.5.3.4 Measurement of ambient conditions

Specifications may require paint application to be carried out under certain ambient conditions and the inspector will need to measure and assess these requirements. The conditions generally taken into account are those appertaining to air temperature, steel temperature, relative humidity and dew point.

Air temperature and relative humidity can be measured automatically using recording hygrographs and thermographs, which record the information on charts to provide daily or weekly records. These instruments may be operated by clockwork or electrically. Generally, however, for inspection work, various forms of hygrometer and thermometer are used. Hygrometers, or psychrometers as they are also known, are instruments for measuring the relative humidity indirectly. The readings obtained with these instruments provide wet and dry bulb temperatures and these are converted to relative humidities by reference to suitable tables. The most commonly used instrument of this type is the whirling hygrometer or sling psychrometer, as it is called in some parts of the world. Relative humidity and dew point are important requirements for satisfactory paint application and their influence has been considered in Chapter 5. As both are quoted in specifications, the inspector must have suitable and convenient methods for measuring them. ISO 8502/4 is a guide to the estimation of the probability of condensation on a surface to be painted.

Whirling hygrometer. The whirling hygrometer (see Figure 9.7) contains two identical thermometers, one of which is covered with a small piece of fabric or wick which is saturated with water; this is called the 'wet bulb', the other being called the 'dry bulb'. The dry bulb records the ambient temperature and the wet bulb records the effects of water evaporating from the wick. The rate of evaporation is influenced by the relative humidity which indicates the amount of moisture in the air. The lower the humidity, the faster is the evaporation rate. The two temperatures are then compared with standard tables which provide a figure for relative humidity.

Certain precautions are necessary with the operation of these hygrometers. They should be examined before use and continuity of the mercury columns in the thermometer checked. The fabric covering the wet bulb should be clean, secure at both ends and wet. The small container in the instrument should be filled with distilled water.

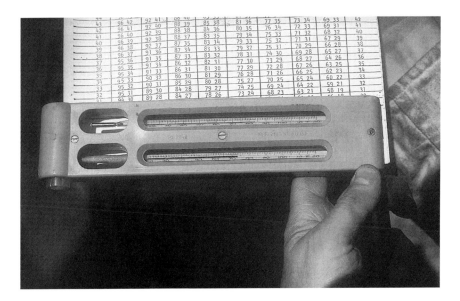

Figure 9.7   Whirling hygrometer.

The following sequence is used for measuring the relative humidity with this type of hygrometer:

(i) With the hygrometer prepared for the measurements, it is rotated or whirled at about 180 rpm, or slightly faster, for 20 seconds.

(ii) In a still atmosphere, the operator should walk slowly forward during the whirling operation to avoid any effects from his body. The operation is best carried out away from direct sunlight.

(iii) The temperatures on both thermometers are noted, the wet bulb first, immediately after the completion of 20 seconds of whirling.

(iv) The whirling is repeated for another 20 seconds at the same speed as before.

(v) Both temperatures are read.

(vi) The procedure is continued until the temperature reading on each thermometer is constant for two successive operations.

(vii) The two temperatures are recorded.

(viii) Suitable meteorological tables, supplied by national authorities in most countries, are consulted. The relative humidity and dew point can then be read from the tables.

The whirling hygrometer may not operate satisfactorily if the air temperature is below freezing point (0°C) and other methods such as direct-reading instruments may be required. These are, however, expensive and are not usually considered to be a necessary part of an inspector's equipment. Digital instruments, which give instant readings of the dry and wet bulb temperatures (see Figure 9.8), and electrically operated instruments which incorporate a fan to draw air across the wet bulb are available. These instruments can give substantially different readings from the whirling hygrometer and this can be the cause of dispute between inspector and contractor. The probable reason for this is that the static apparatus samples a smaller and more localised quantity of air. Some of the electronic types can be calibrated with standard humidity cells but this merely establishes the precision of the instrument not the accuracy of the determination. Obviously, the whirling hygrometer is more dependent on its correct use by the operator. However, if the standard procedure is followed and the operator continues to take measurements until the readings stabilise, then this apparatus, however cumbersome and old-fashioned it looks, is a standard reference method.[14]

9.5.3.5   Measurement of steel temperature

Specifications commonly require that steel temperatures should not be less than 3°C above the dew point, to avoid the possibility of moisture condensing on the surface during painting operations. It is, therefore,

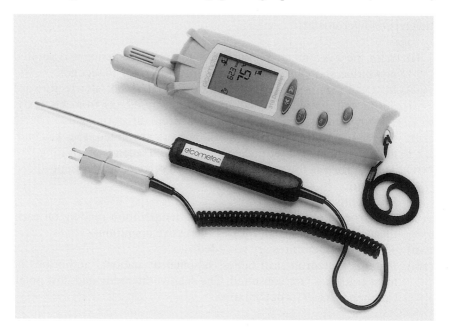

Figure 9.8   Electronic relative humidity and dew point gauge.
Source: Elcometer Instruments Ltd.

necessary to have a simple means of measuring the steel temperature. This is usually measured with a contact thermometer which has magnets attached for fixing to the steel. The thermometer contains a small bimetallic couple which acts in a way similar to a thermocouple and records the temperature on a dial. These instruments are normally cheap but some models can be very inaccurate. Each instrument should be checked before use and, since there is generally no method of calibration, discarded if more than 10% out.

Digital temperature gauges with separate probes are available which provide a direct reading of the surface temperature in a matter of 20 seconds compared with a stabilisation time of as much as 30 min with the dial type, and with greater accuracy. However, these instruments are not necessarily intrinsically safe and may not be permitted for use in hazardous areas.

9.5.3.6   Measurement of paint film thickness

A range of instruments is available for the measurement of dry film thickness. Provided they are regularly and correctly calibrated and used properly, they will provide reasonably accurate measurements of thickness. Other qualitative methods are used such as adjusting the opacity of the paint so that the underlying coat is just obscured when the correct coating thickness has been applied. Again, strong contrasts between successive coats enable both the painter and the inspector to check on the continuity of coatings and to ensure that all coats in a system are in fact applied.

Apart from checking dry film thickness, tests can be done to determine the thickness of the wet film and this has the advantage that adjustments can be made immediately. It is often assumed that provided the minimum thickness of coating is attained, then additional thickness is advantageous to the client. This is not always correct, however, and in some cases a maximum thickness may be specified. Clearly, problems arise if the dry film is greater than this.

Wet film thickness measurement. For paint applied to steel surfaces, this method is used mainly as a guide to the painter and the inspector; dry film thicknesses are usually specified. Wet film thicknesses are most commonly measured with a small comb gauge. This has a number of projections similar to a comb. The two at the end are the same length and those in between progressively vary in height (see Figure 5.6). The gauge is pressed into the wet film perpendicular to the surface with the two end pieces in contact with the steel surface. It is then removed and examined. Some of the teeth will have been wetted by the paint whereas others will have remained proud of the surface of the paint coating, so will not have been wetted. Each of the teeth or graduated steps is designated by a thickness in micrometres (or mils) and the thickness is taken as being the average of the highest step covered and the lowest one not covered, e.g. if the $50 \mu m$ step is wetted and the $75 \mu m$ step is not, then the thickness is considered as between those two measurements. A range of gauges is available, so if none of the teeth is wetted, another gauge should be used. Gauges are available in stainless steel and these must be cleaned after use. Disposable plastic gauges are also used.

Another type of wet film thickness gauge consists of a wheel with two outer rings of equal diameter with a graduated eccentric centre ring. The gauge is held between a finger and thumb and rolled over the painted surface. The inner ring is then examined and the thickness is indicated by the graduation mark where the paint no longer wets the ring.

The wet film thickness is only useful if it can be related to the dry film thickness. The ratio is as follows: dry film thickness = wet film thickness × % solids (vol) ÷ 100.

Dry film thickness measurement. The specification may call for a minimum, average or maximum thickness, but the specifier should take into account the inevitable variations in film thickness that occur in practice and provide a realistic requirement for film thickness. Otherwise, unnecessary problems are likely to arise during inspection.

Many instruments have been developed for the measurement of dry film thickness. Only those commonly used in high-quality inspection will be considered here. The instruments can be used for measuring the thickness of metal coatings on steel, with certain limitations, as well as paint coatings.

The non-destructive method most commonly used depends upon measuring the magnetic flux between the instrument probe and the ferrous substrate. The weaker the flux, the bigger the gap and therefore the thicker the apparent coating. The instruments are divided into three broad types: (a) magnetic pull-off, (b) fixed probe, and (c) electronic induction. All of these instruments, to varying degrees of accuracy, operate on the principle of the coating acting as an air gap, and this gives rise to errors. Rust, millscale, dirt, vacuoles, etc., will add to the apparent coating thickness. Measurements on very soft paint films will give low results. This can be counteracted by making measurements over a calibration foil of known thickness. This spreads the load of the probe and prevents indentation. Obviously the thickness of the foil is then subtracted from subsequent readings. There are other factors which may influence readings. There can be appreciable differences in the magnetic properties of steels of different composition. Differences between most low-carbon steels are probably insignificant but may be greater in higher alloy steels. It is always advisable that the gauge should be calibrated on the same type of steel to which the coating is to be applied.

Magnetic gauges are also sensitive to geometrical discontinuities, such as holes, corners or edges. The sensitivity to edge effects and discontinuities varies from gauge to gauge. Measurements closer than 25 mm from the discontinuity may not be valid unless the gauge has been calibrated for that location. Some of the electronic gauges may be sensitive to the presence of another mass of steel close to the body of the gauge. This effect may extend as much as 75 mm from an inside angle.

Magnetic gauge readings will be affected by the curvature of the surface. This may be overcome by calibrating the gauge on a similarly curved but uncoated surface. All probes of these instruments must be held perpendicular to the surface to provide valid measurements. Strong magnetic fields, as from welding equipment, can interfere with the operation of electronic gauges. Residual magnetism in a steel substrate can affect the result. Most magnetic gauges operate satisfactorily between 4°C and 49°C. However, if such temperature extremes are met in the field, the gauge should be checked with at least one thickness reference standard after both the standard and the gauge are brought to the same ambient temperature. Since tolerance to the above effects can vary considerably between makes of instrument, the manufacturer's instructions should be followed carefully.

The most difficult problem is the interpretation of results over a blast-cleaned surface. A rough surface alone, without any paint, will give a reading on the instrument of anything from 15 to 50 μm, depending on the roughness.

Since with thin films, in particular, it is necessary to know the film thickness above the peaks of the profile, allowance must be made for this in the instrument reading. Opinions differ as to the best way this should be accommodated and in fact different makers of the same types of instrument give different advice. The most usual methods are as follows:

(i)   Place a shim of known thickness, appropriate to the film to be measured, on the unpainted blast-cleaned surface and adjust the instrument to read the shim thickness. All subsequent measurements on that surface can then be read directly.

(ii) Place an appropriate shim on a polished flat steel plate and calibrate the instrument. Then take a number of measurements directly from the blast-cleaned surface. Obtain an average figure which can be used as a factor to be subtracted from all future results on that surface.

(iii) Assess the roughness of the blast profile by use of the ISO Comparator (see Chapter 3, Section 3.1.4) and allow a correction factor depending upon the Comparator grade of roughness, for example minus 10 μm for a 'fine' profile, minus 25 μm for a 'medium' profile and minus 40 μm for a 'coarse' profile.

Of the three methods, the third is recommended for most situations. The problem with the first method is that shims are likely to deform under the pressure of the instrument probe. Also, the roughness of any blast-cleaned surface is very irregular and the calibration could be made over an area that was unrepresentative.

The ease and accuracy with which instruments can be adjusted varies with type and make, and manufacturer's instructions should be followed. The various types will be discussed separately.

There are relevant ISO Standards as follows: ISO 2178 'Non-magnetic Coatings on Magnetic Substrates–Measurement of Coating Thickness – Magnetic Method' describes the limitations of dry film thickness gauges. ISO 2360 'Non-conductive Coatings on Non-magnetic Basic Metals' is concerned with the eddy current method and ISO 2808 'Paints and Varnishes. Determination of Film Thickness' describes several dry film thickness measurement methods. SSPC (The Steel Structures Painting Council, USA) have also produced a comprehensive specification PA2 'Measurement of Dry Coating Thickness with Magnetic Gauges'. This includes recommendations for the number of measurements necessary for conformance to a thickness specification. This is important because it is impossible to apply coatings to a strictly uniform thickness, particularly if they are applied as thick films. Paint manufacturers' data sheets often give a recommended film thickness but without specifying whether that is a minimum or average value. Most engineers interpret it as a minimum, but this does raise the average significantly, which may not always be desirable. Film thicknesses can be quoted as 'nominal'. The definition is that over any square metre the average of the readings taken should equal or exceed the nominal thickness and in no case should any readings be less than 75% of the nominal thickness. The disadvantage of this definition is that it does not allow for even one rogue result from the instrument and does not define a maximum thickness permissible.

Concerning frequency of measurement, the SSPC specification states that for structures not exceeding $300\,\text{ft}^2$, five measurements should be taken over an area of $100\,\text{ft}^2$ spaced evenly over the area, each reading to be the average of three readings taken close together. The average of the five readings should be not less than the specified film thickness and any one reading (an average of three determinations at one spot) should not be less than 80% of the specified thickness. For larger structures the number of sampling areas is proportionally increased.

The SSPC specification also describes the different types of magnetic instrument available and suggests their method of calibration. Some of their recommendations differ from those given by the instrument manufacturers and it is understood that the subject is under review. ISO 12944 has similar recommendations (see Table 9.1).

Table 9.1   Measurement and acceptance criteria of the dry film thickness and for the frequency of measurement

Area/length of inspection area/ difficult area (m² or metres)	Minimum number of measurements
up to 1	4
1–3	10
3–10	15
10–30	20
30–100	30
above 100	add 10 for every additional 100 m²

Magnetic pull-off. There are a number of instruments in this category and they are known by various names according to the country of supply. However, they can be sub-divided into a number of groups, the simplest in operation being the Pencil Pull-off or Tinsley Gauge. This is the size of a small pencil, with a clip attached for placing in a pocket. The principle of operation is straightforward, with a magnet attached to a spring held in a housing and a reference mark that moves along a scale. When the force exerted by the spring balances the magnetic attraction, the magnet pulls from the paint film and a small marker can be read from the scale. The instrument must be used perpendicularly to the painted surface under test. These instruments are cheap and easily carried about, but the operating principle makes accurate measurement difficult to achieve.

Another instrument of this type works in a somewhat more complex way but again relies upon the tension in a spring. A hair spring is connected to a lever at the fulcrum and fitted to the instrument housing, the outside of which carries a marked dial fixed to the lever. The instrument lever has a small magnet at one end which protrudes through the casing and is counterbalanced at the other end. The instrument is placed on the coating and the dial is rotated so that the spring tension is increased to a position where the magnetic force balances the spring tension. The magnetic force is, of course, related to the effect of the coating thickness, which acts as an air gap. Just beyond the balancing point the small magnet on the lever springs from the surface and the dial reading, in microns or mils, provides a direct reading of the coating thickness. Errors that can arise with this instrument include:

(i)   Use on overhead surfaces, which may produce inaccurate readings.
(ii)   The effect of vibration in the area of testing may cause loss of magnetic attraction between the magnet in the instrument and the steel surface. This can result in readings that are too high.
(iii)   If paint films are soft or tacky, the magnetic probe of the instrument

may press into the film and produce a low reading of the true film thickness.

(iv)   If the probe is not regularly cleaned, metallic abrasives may accumulate at the magnetic tip and lead to incorrect readings.

Fixed probes

(i) *Twin fixed probe.* These were one of the original types of paint film thickness instruments. They rely on permanent magnets within the instrument casing. Nearby strong magnetic fields, as for example from welding equipment or nearby power lines, will interfere with the reading. It is recommended in all cases that two readings are taken, the second with the probes turned by 180°.

(ii) *Single fixed probe.* This instrument functions mechanically on a self-balancing magnetic principle. The fixed jewelled tip probe is placed on the coating and a digital reading is obtained from a revolving mechanical device.

Electronic type. Electronic gauges use either electromagnetic induction or Hall-effect probes to measure non-magnetic coatings on a ferrous substrate[15] (see Figures 9.9 and 9.10). The maximum coating thickness that can be measured with both types depends on their size and design. Typical thickness ranges are: 0–250 μm for small probes, 0–1500 μm for standard size probes and right angled configurations, 0–5000 μm (0–5 mm) for probes used with

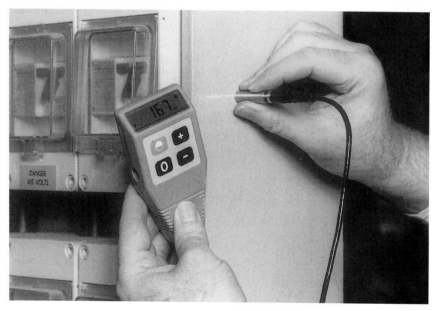

Figure 9.9   Basic electronic thickness gauge.

Source: Elcometer Instruments Ltd.

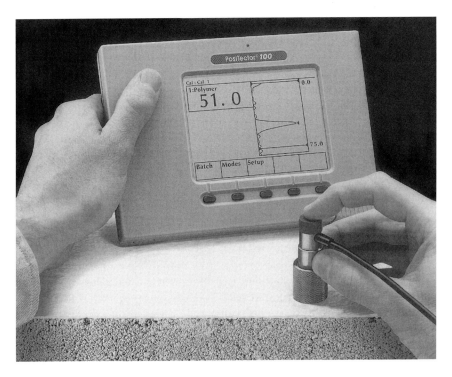

Figure 9.10   PosiTector 100 – a non-destructive gauge measuring coating thickness on various substrates, using ultrasonic technology.

Source: DFT Instruments Ltd/De Felsco Inc.

fire-resistant coatings, 0–$13000\,\mu m$ (0–$13\,mm$) for large probes used for foamed coatings. To measure non-conductive coatings on non-ferrous metals, an eddy current gauge is used. Electronic gauges vary with manufacturer, some or all of the following features could be included:

(a) Digital display of thickness readings.
(b) Memory for calibration settings.
(c) Integral probes for one-handed operation or separate probes for areas of difficult access.
(d) Statistical calculations of mean value to characterise a set of readings, standard deviation to assess the spread of a set of readings, highest and lowest readings to assess the range of values, and number of readings to determine the validity of the sample.
(e) Batch storage of readings to keep different sections of work separate.
(f) Data transfer to a printer or computer for record-keeping and further analysis.
(g) Date and time marking of readings.

Other instruments and methods for assessing dry film thickness. The instruments considered above are, in various forms, those most commonly used for measuring the dry film coating thickness. They are all non-destructive and this is an advantage. However, there are also instruments that require a small cut in the paint film and could be called 'destructive'. This would be correct but gives the wrong impression because, with most coatings, the small area of damage can be repaired without difficulty. However, it is not a routine measurement and should not be undertaken unless specified. Otherwise the contractor is liable to claim compensation for the repair of damage, however small. The method is mainly used when disputes arise or on non-metallic substrates.

The principle of operation is based on viewing a small cut or a small conical drill hole made in the coating at an angle, using a small microscope held perpendicular to the surface of the coating. A graticule is incorporated into the eyepiece and this allows measurement of the coating thickness. Cutting devices with different angles are available and simple conversions can be made to determine the value of the graticule graduations. This information is supplied with the instruments. The conical hole is made by a small hand-rotated drill which must be operated to remove the coating to the steel base. The other type relies upon a knife-type cut and, on brittle materials, it may be difficult to provide a sharp cut; the drill-type may be preferred. The advantages of these instruments lie in their ability to provide a measure of the individual coatings in a system. This cannot be achieved with magnetic instruments which provide a measurement for total thickness only, although the individual thickness of coatings can be measured during application.

Coating thickness can also be calculated from the amount of paint applied to a known area of steelwork. This method is not generally used for inspection purposes but is employed in some maintenance and general testing work.

9.5.3.7   Detection of discontinuities in coatings

Obviously, there is a tendency for applicators to miss areas that are out of sight or have difficult access. An important qualification for a painting inspector is ability and willingness to climb and work at heights. Observation by binoculars from the ground has limited value. With fairly close access, the use of a small angled mirror attached to a telescopic handle is useful for getting into awkward areas, undersides of pipes, etc. Some consider that the mere visible possession of such an instrument by the inspector has a worthwhile psychological effect on the painters.

Other discontinuities can be difficult to detect visually, for example porosity due to dry spray, or minute slivers of steel penetrating from the substrate and pinholing owing to incorrect atomisation or release of bubbles from a porous substrate.

Even in multi-coat systems, discontinuities such as pinholes and pores may occur in the coating. The term 'holiday' is widely used to describe such discontinuities. A few pores per square metre may be acceptable in some situations, but for immersed conditions, particularly for critical requirements such as tank linings or pipe coatings, a pore-free coating or one with a specified limitation on the number of discontinuities may be required. Instruments are available for checking the presence of these discontinuities. They are based on the insulating character of the coating so that an electrical circuit will be completed only when there is a discontinuity, i.e. no coating. The two common forms of test are based on the use of a sponge at low voltages or a sparking technique at high voltages.

Wet sponge test. The principle is straightforward. A sponge fitted to a metal plate is moistened with tap water and connected by a cable through a voltage source to the steel to which the coating is applied (see Figure 9.11). The wet sponge should be moved over the surface at about 30 cm/s using a double pass over each area and applying sufficient pressure to maintain a wet surface. If a discontinuity is detected, the corner of the sponge should be used to determine the exact location.

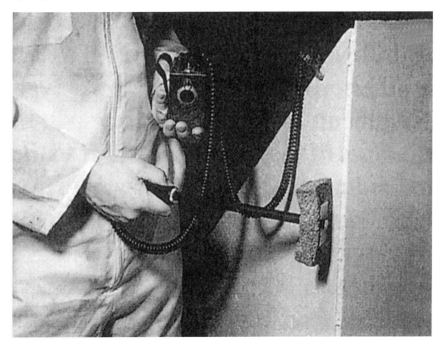

Figure 9.11   Low-voltage wet sponge pinhole detector.

The coating acts as an insulator so that current does not flow unless there is a discontinuity such as a pinhole present. In this case, the circuit is completed and a device in the circuit, e.g. a lighted bulb or an audible signal, operates to indicate the presence of a discontinuity in the film.

The voltages vary with different types and makes of instrument, but they generally range from 9 V to 90 V, the higher voltage being used for thicker coats.

Although simple in operation, certain precautions are necessary. The sponge must be sufficiently wet to allow proper operation but excess moisture on the surface or in the atmosphere may give rise to 'tracking', i.e. the water will conduct current to a pinhole some distance from the sponge. The coating to be tested should also be dry to eliminate tracking. The use of a weakened detergent solution, e.g. 'Teepol', may be advantageous, particularly for coatings of thicknesses above $250\,\mu m$.

The advantage of the method is that it cannot harm the coating. The disadvantage is that it is only effective at film thicknesses less than $500\,\mu m$.

High-voltage test. Although the principle is the same as for the low-voltage test, no sponge is used; instead a wire brush or other conductive electrode is used (see Figure 9.12). These instruments are generally either DC or DC pulsed type. A direct-current apparatus provides a continuous voltage; a

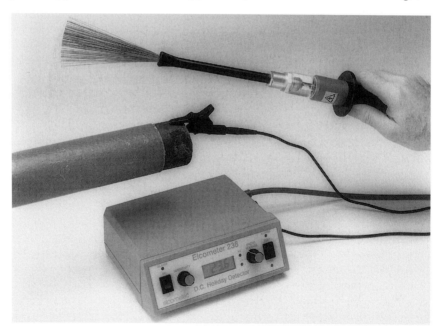

Figure 9.12 Holiday detector, 'spark tester'.

Source: Elcometer Instruments Ltd.

pulsed type discharges a cycling high-voltage pulse. Both types can accomplish the desired result but the pulsed type is considered more suitable for the thickest films. The DC voltage is variable up to about 30 kV, depending on the type of instrument. The circuit is similar to that used with the low-voltage system, with one end of a cable fixed to the steel structure and the other end to the electrode. This is passed over the surface of the coating at approximately 30 cm/s using a single pass. At a discontinuity a spark jumps, completing the circuit.

At the high voltages used, these instruments are capable of burning holes in the coating, so they must be used with discretion and only at the voltage recommended by the coating manufacturer and as stated in the specification. In situations where no voltage setting is specified, then a reasonable procedure to follow is: to use the ratio of 100 V per 25 μm of the nominal coating thickness, since most coatings have quite high dielectric breakdown strengths with 500 V to well over 2000 V per 25 μm being quite common. Even materials with relatively low dielectric breakdown strength, such as glass-fibre reinforced polyesters, would require 43 kV to break down a 3 mm coating. Reducing this by one-third gives 14 kV as a test voltage, which would present no danger of destruction to properly coated areas.

In the absence of any knowledge of the coatings or the instrument voltage outputs, the best method is to make a very small hole in the coating and starting with the apparatus at its lowest voltage, increase until the test hole is consistently detected.

Many applicators are strongly opposed to this test, with some justification if it is incorrectly carried out. However, much of the suspected coating breakdown is due to incorrect use of the apparatus, non-uniform thickness, voids, vacuoles or dirt inclusions in the coating. Retained solvents can also give erroneous indication.

Re-testing should not result in more indications of discontinuities than previously unless the speed of testing or test voltage has been changed or the coating has been in service or stressed so that its dielectric strength is lower.

Devices using AC current are less common but have an application for the testing of coatings that are conductive, for example coal-tar epoxies or rubber linings that contain carbon black pigment. The apparatus is not directly connected to the substrate but emits a blue corona which when passed over the coating surface will cause a spark to jump from the tip of the probe to the discontinuity. Surface contaminants and moisture can also cause the apparatus to spark. Since there is a higher risk of a severe electric shock from this apparatus, it requires greater care in handling.

9.5.3.8   Adhesion

Testing of coating adhesion is another destructive test that the inspector may be called upon to carry out in the event of a dispute. There are two types of adhesion test that can be carried out in the field. These cause small areas of damage to the coating and should only be carried out by agreement with all concerned. One method involves cutting a pattern through the film with a knife and assessing the amount of detachment of the film that this causes. For example, there is the Cross-Cut Test as described in ISO 2409 and ASTM D 3359. This requires a matrix of cuts to produce even squares. The cuts can be made individually or by using a multi-blade cutter. The spacing of the cuts and the similar ones at right angles, and hence the size of the squares formed, is determined according to the thickness of the coating. Adhesive tape is applied to the cuts and pulled away sharply. The amount of detachment is rated according to the scales illustrated in the standards.

The other cutting method, called the X-Cut – or St Andrew's Cross – is simpler, but requires an even more subjective assessment and simply involves making two knife cuts, each about 40 mm long and with the smaller angle between the cuts of between 30° and 45°. Again adhesion tape is applied and then pulled away and the adhesion rated according to the amount of paint film detached. Probing the cut edge with a knife can give erroneous results because the amount of detachment can depend upon the cohesive strength of the film.

The other type of adhesion test method uses a pull-off principle. It involves sticking a test dolly to the coating with a suitable adhesive and measuring the force required to detach the dolly and the coating, or part of the coating, from the substrate. There are several proprietary makes of this type of adhesion tester on the market using different methods to exert the force, for example: compressed spring assemblies, hydraulic pressure (see Figure 9.13) and pneumatic pressure. They are normally calibrated to display the force per unit area, taking into account the area of the face of the dolly that was stuck to the coating. This type of method, that could possibly be considered non-destructive, involves using a tensile test method to apply force to the dolly, but in this case the dollies are made to break at a certain force. It therefore becomes a go/no-go test.

It is then possible with a shear force to remove the dolly without damaging the coating. Each of the test instruments are affected by different factors, but some are common, for example:

(a) Low results will be obtained unless the coating is adequately cured before testing. With some two-pack epoxies, this may take several weeks, or even months before optimum strength is reached.

(b) The thickness of the substrate affects the result, for example, adhesive

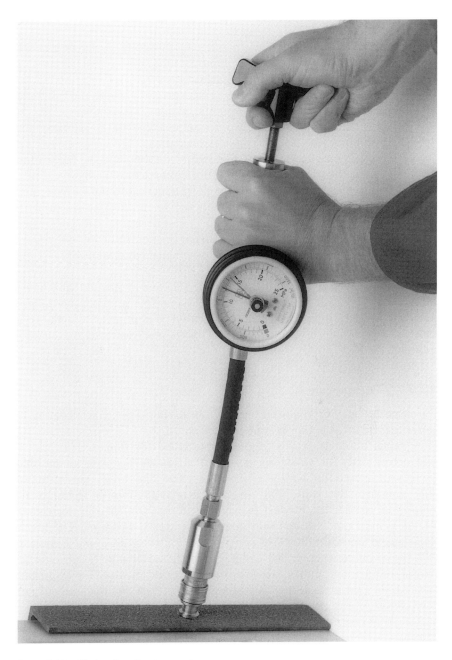

Figure 9.13 Hydraulic adhesion testing gauge.

Source: Elcometer Instruments Ltd.

rates for the same material can be nearly 60% higher on a plate 12 mm thick, than on a plate 5 mm thick.

(c) The force must be applied smoothly to the dolly. Jerky or uneven application can cause premature failure.

(d) The force must be normal to the surface.

(e) The rate of the application is also critical and, in general, the full force should be applied quickly, for example, within 100 seconds.

(f) The dolly must not be applied to an uneven coating surface, for example, one with sags, runs and dry overspray.

(g) If the coating is glossy it should be lightly abraded and the face of the dolly should be abraded to remove all glossiness or old adhesive.

(h) The coating around all parts of the dolly adhering to the surface must be scored, otherwise the measurement will be of the cohesive strength of the film not the adhesion.

(i) Hydraulic instruments tend to pull from the centre of the dolly, providing results of a higher order and lower standard deviation.

If the dolly fails at the adhesive level, the test is invalid and must be repeated. If it fails within the coating system this is a cohesive failure and will usefully indicate a weakness within a multi-coat system. If it fails at the coating/substrate interface, this is a measure of the coating adhesion.

9.5.3.9   Tests for cure

An inspector may be required to carry out hardness tests on a coating in order to establish whether it is adequately cured.

A pencil hardness test is frequently specified for thin stoved finishes and occasionally for air-drying finishes. The method consists of pushing pencils of increasing hardness across a paint film until one is found that scratches the paint film. In the USA the opposite technique is employed, so that the hardest pencil is used first and then in decreasing hardness until one is found that does not scratch the coating. The result will depend upon the make of pencil used, the pressure applied and the interpretation of whether the coating is scratched or not. It can only be considered a rough indication. Fish[16] has reported in work for ISO standardisation that even using one manufacturer's pencils, the test did not give consistent results.

For thick coatings, in excess of 800 μm, there are hand-held portable hardness testers. The two most commonly used for testing the cure of elastomeric coatings are the Barcol hardness tester and the Durometer. Both of these instruments have models in a suitable range for testing soft plastics. The testers are held against the coating surface and a spring-loaded plunger is driven into the coating. When the resistance of the coating equals that of the spring then a reading can be taken from the dial indicator on the instrument. Results are normally given in hardness for the particular instrument. Hardness results from different makes of instrument generally cannot be directly correlated. Therefore, both instrument and hardness required must be specified.

Another test for cure of coatings is the solvent sensitivity test or solvent wipe test. It can be used on chemically cured coatings such as epoxies but is frequently used to ensure the complete hydrolysation of inorganic zinc primers. There are several slightly different methods of carrying out the test, so the method required should be detailed in the specification. One method is to saturate a rag in solvent. The choice of solvent can make a difference to the result and should be checked with the paint manufacturer if not specified. The saturated rag is then wiped across the surface of the coating for a specified number of times. If the coating is not properly cured, it will stain the rag quite readily. If only a trace, or none, of the coating comes off on the rag, this will at least indicate that the surface of the coating is cured.

9.6   Inspection of metal coatings

There are fewer stages in the application of metal coatings so inspection is usually more straightforward than with paint coatings. There are four methods of applying metal coatings to constructional steelwork, or the components used in conjunction with it:

(i)   diffusion, e.g. sherardising of zinc.
(ii)   electroplating.
(iii) hot-dipping.
(iv)   spraying.

(i) and (ii) are usually factory-applied coatings and inspection is not normally carried out during the processing. The thickness of electroplated components on steel can be checked using the magnetic-type instruments employed for paints. Diffusion coatings contain a series of iron-based alloys, so usually the thickness cannot be determined with any accuracy using such instruments.

Hot-dip galvanised coatings also contain alloy layers, so again problems may arise when measuring the thickness with magnetic instruments. In practice, these instruments are used for determining the coating thickness of hot-dipped zinc coatings. This may be acceptable as a control check where a series of steel sections of similar size is being galvanised, but in the event of disputes other methods must be used. These are the determination of the weight of zinc per unit area, employing chemical methods, or measurements of a cross-section of the coated steel using a microscope. Both methods are outside the scope of normal inspection.

The thickness of sprayed coatings of zinc or aluminium can be measured with standard magnetic instruments.

9.6.1   Hot-dip galvanising

Sometimes the inspector may be called upon to check the process itself, but this is rare. The specification requirements are usually based on the applied coating, in particular its appearance.

The following defects would probably be grounds for rejection:

Spots bare of galvanising coating.
Flux inclusions (stale flux burnt on during dipping).
Ash inclusions (ash burnt on during dipping).
Black spots (including flux particles from flux dusting).
Warpage and distortion.
Damaged surfaces.

The following defects may be acceptable unless they are present in gross amounts or are specifically excluded by prior agreement:

General roughness.
Pinholes (entrapped particles).
Lumpiness and runs.
Dull grey coating.
Bulky white deposit.
Blisters (entrapped hydrogen drive-off during pickling).

The specification may stipulate certain requirements in the coating and these must be taken into account in the inspection procedures. Generally, however, apart from the defects noted above, the zinc coating should be continuous and reasonably smooth.

On larger steel sections, minor imperfections may be accepted, because removal of defects, e.g. by grinding, may be more harmful than leaving them, provided they are of a minor nature. Small bare spots may be repaired with a zinc-rich paint or, less commonly, using metal spray or a special solder.

9.6.2   Sprayed metal coatings

Blast-cleaning is essential for sprayed metal coating application and should be at least $Sa2\frac{1}{2}$ and possible Sa3 (or equivalent) for aluminium.

The cleanliness of the surface is checked as for cleaning prior to paint application. The thickness of the coating can be measured but, apart from this, inspection is mainly visual.[17,18] The coating must have a reasonably uniform texture and be free from powdery deposit or coarse or loosely adherent particles, protrusions and lumps. Adhesion checks may be called for in the specification; these will be carried out in accordance with the specified standard, e.g. BS 2569. Two useful guides to the inspection of sprayed metal coatings have been produced.[17,18]

9.7   Inspection instruments

The inspection requirements should be clearly specified and will involve a range of examinations and measurements. Special instruments have been developed to measure or assess many of the coating properties.

Because of the cost of such instrumentation, frequently the dry film thickness gauge is the only instrument possessed by some paint inspectors. The following are considered by many authorities to be the *minimum* requirements:

– Carrying case (see Figure 9.14).
– Mirror with a telescopic handle.
– Dry film thickness gauge or gauges for the appropriate range with spare batteries where appropriate.
– Calibration shims or thickness standards.
– Wet film thickness gauges of the appropriate range.
– Gauge and hypodermic needles to measure the air pressure close to blasting nozzle.
– Comparators to assess the profile.
– Visual standards to measure the cleanliness of blast-cleaned surfaces.
– Tape for dust measurement.
– Cloth test for oil in air supply.
– Test kits for the sampling and testing of soluble salt contamination of surface.
– pH papers.
– Penknife and scraping tools.
– Illuminated magnifier.
– Wet and dry bulb hygrometer, plus spare thermometers.
– Surface thermometer.
– Tables or calculator for determination of relative humidity and dew point.
– Method of detection of abrasive contamination.
– Device for marking defects.
– Appropriate National Standards and Specifications.
– Appropriate paint manufacturers' data sheets.

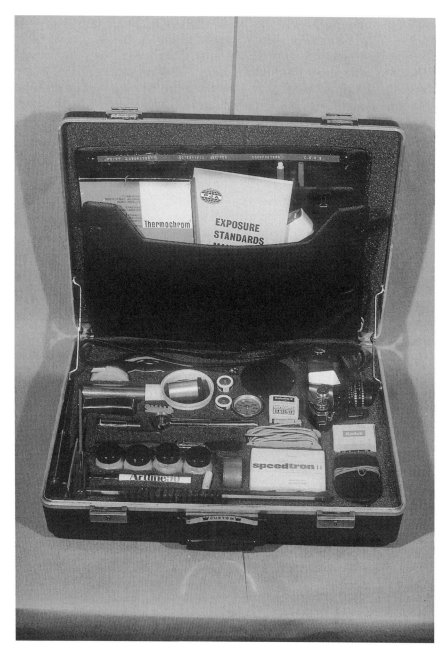

Figure 9.14 Case of instruments used by a coatings inspector.

- Log book, report forms and squared paper for diagrams.
- Project specification.

In addition, the following more specialised equipment which may not be necessary for each job should be available if required:

- Holiday detector of appropriate range and size.
- Set of sieves for checking abrasive sizing.
- Balance or scales.
- Eddy current thickness gauge for non-ferrous substrates.
- Density measuring cup.
- Viscosity flow cup and stop watch.
- Adhesion tester.
- Ultraviolet lamp (for oil detection).
- V-cut inspection gauge.
- Fibre-optic inspection instrumentation.
- Camera, plus close-up lens and films.

9.8   Reports and records

The inadequacy of records providing data on the performance of coating systems has had an adverse effect on the selection of systems suitable for specific situations. This lack of records often proves to be of concern when considering maintenance procedures because there may be no effective information on the original systems or even on the types of paint used for previous maintenance work. It is always necessary to have a detailed record of coating procedures, not least because it may provide essential information regarding the cause of premature failures or the problems concerned with particular types of coating application or performance. The inspector's records are not primarily maintained to provide material for a data bank but, if properly prepared and studied, they will be of considerable value to all those concerned with a project. Some of the essential requirements of a report are given below. The details will obviously vary with the nature of the work, the type of surface preparation specified and the type of coating selected. In some situations, additional records may be considered to be appropriate.

All records must provide dates of the operations concerned and must relate to specific structural items, with clear identification codes, otherwise it may prove impossible to use the report in any meaningful way. Furthermore, matters such as progress, safety precautions and general comments on standards of workmanship will prove useful.

(i)   SURFACE PREPARATION
     Initial condition of steel and details of pre-cleaning inspection.

Type of surface preparation adopted.
Details of quality demanded.
Method of quality control agreed.
Name of instrument agreed and number.
Details of method of preparation.

(a) *Blast-cleaning*

Type of equipment and manufacturer, ventilation arrangement, lighting.

Type of abrasive used and grade.

Arrangements for separation of debris and topping-up abrasive and arrangements for removing dust from work.

Shop conditions and location relative to initial painting area.

Comments on programming.

Inspection arrangement.

General comments.

(b) *Hand cleaning*

Type of tools agreed.

Quality control arrangements agreed.

Comments on programming, including pre-weathering.

(c) *Incident reports*

Any event which affects progress of work significantly, i.e. delays between stages; supply problems; disputes; staff changes, etc.

Records of any re-working enforced.

(d) General observations, especially on labour recruitment and discipline and supervision provided.

(e) Summary of quality control measurements on profile heights, surface cleaning and daily reports on progress.

(ii) PROTECTIVE SYSTEM

Details of system; coat by coat, with film thickness required, including any protective metal coating.

Details of quality of finish required.

Details of mixing ratios for two-pack materials.

Manufacturer's thinning recommendations.

Method of quality control measurements agreed.

Name of instrument and number.

Details of any special feature in application to be observed, i.e. ambient temperature, humidity, re-coating times, paint temperature for hot spraying.

(iii) APPLICATION
  (a) Air spray: air pressure at gun. Nozzle combination details. Compressor details: viscosity of paint.
  (b) Airless spray: airline pressure, fluid tip (orifice diameter – spray angle); air supply details; pump capacity; viscosity of paint.
  (c) Roller: type of roller and arrangement for brush infilling.
  (d) Brush: type of brush.
  (e) Details of any intercoat treatment, e.g. abrading required.
  (f) Details of shop conditions: heating, lighting, humidity, ventilation, housekeeping, incident reports; any event which affects progress of work significantly, e.g. delays between coatings or between cleaning and initial coating on steel, supply problems, disputes, staff changes: records or any re-working enforced.
  (g) General observations, especially on labour recruitment, discipline and supervision provided. Comments on application characteristics of materials used and on any equipment employed.
  (h) Details of quality control measurements, film thickness, continuity, pinholing, adhesion, general standard of finish.

(iv) SITE WORK
  The following additional information should be recorded:
  (a) Daily weather, including records of changes during working period and recordings of RH and metal temperature.
  (b) Condition of steelwork on arrival at site.
  (c) Record of any deterioration during storage, with details of time of storage.
  (d) Record of damage suffered during construction.
  (e) Details of any remedial work carried out.

It should be appreciated that the record is produced for the client and should not be distributed without his agreement. It must be considered as a factual document but, where appropriate, comments on the contractor's attitude, competence of work force, etc., may be valuable to the client. However, such comments may not be capable of substantiation to the satisfaction of neutral observers and so should be considered as no more than useful comments, preferably provided in a separate document rather than in the report, which may be passed, with the client's agreement, to others concerned with the project. Valuable though reports are, engineers are usually busy and will not have time to study them thoroughly. Consequently, a summary of the main points on a periodic basis, the time interval depending on the project, will prove useful to the client.

9.9   Health and safety matters

Generally the painting inspector does not have the qualifications or authority to be a safety engineer or supervisor. There is, however, a duty of care that requires everybody to look after their own safety, follow all specific safety requirements and immediately report any unsafe conditions or practices to the appropriate authority.

The painting inspector must wear appropriate personal protective clothing and use the correct form of respiratory protection, should such be necessary, and generally comply with the Site Safety Policy. Working from heights will necessitate the use of safety harnesses, with such work only being carried out when other personnel are in the area and aware of the inspector's presence.

Working in any area which may be regarded as a confined space will require the correct form of entry permit to be in place, with a suitable harness and the presence of a safety operative, remote from the confined space, but in contact with the inspector.

Any tools used must be intrinsically safe and be used in a safe manner. No actions of the inspector – by word or deed – must prejudice the safety of himself or others.

References

1. Deacon, D. H., Iles, D. C. and Taylor, A. J., Durability of steel bridges: A survey of the performance of protective coatings. Tech. Report 241 Steel Construction Institute, 1998.
2. Palmer, F., Improving specifications for offshore and onshore facilities. *J. Protective Coatings and Linings*, **6**(12) (1989) 25–8.
3. Hauck, J. C., Painting practices at West German chemical plants; an American's observations. *J. Protective Coatings and Linings*, **6**(12) (1989) 56–61.
4. Thompson, S. P., Managing a maintenance painting program to reduce costs. *Corrosion 82*, Pre print 146, NACE, Houston, 1982.
5. *International Coating Inspector Training and Certification Program*, NACE, Houston, 1982.
6. *Visual Standard for Steel Surfaces of New Steel Cleaned with Sand Abrasive.* NACE TMO 170-70. NACE, Houston, 1970.
7. *Visual Standard for Surfaces of New Steel Centrifugally Blast Cleaned with Steel Grit and Shot.* NACE TM 0.175-75. NACE, Houston.
8. Garrabrant, R. and Biesinger, J., *J. Protective Coatings and Linings* (Jan. 1999) 32–40.
9. Mayne, J. E. O., The problem of painting rusty steel. *J. Appl. Chem.*, **9** (1959) 673–80.
10. Igetoft, L. I., *Determination of Salts on Blast Cleaned Surfaces.* Swedish Corrosion Institute Report 61.044-27.
11. *Field Measurement of Surface Profile of Abrasive Blast Cleaned Steel Surfaces Using Replica Tape*, NACE RPO 287-87. NACE, Houston, 1987.
12. *Korrosionsschutz von Stahlbauten durch Beschichtungen und Überzüge*, DIN Standard 55928, Teil 4. Deutsches Institut für Normung, 1978.

13. *Visual Comparator for Surface Finishing of Welds*, NACE Standard RPO 178-89. NACE, Houston, 1989.
14. *Whirling Hygrometers*, BS 2482 : 1975, 1981. British Standards Institution.
15. Fletcher, J., *J. Protective Coatings and Linings* (Jan. 2000) 30–6.
16. Fish, R. A., An investigation of the pencil hardness test. *Paint Technology*, **34**(6) 14–17.
17. *Inspection of Sprayed Aluminium Coatings.* The Association of Metal Sprayers, Walsall.
18. *Inspection of Zinc Sprayed Coatings.* Zinc Development Association, London.

Chapter 10

Designing for corrosion control

The design of structures and buildings is usually considered in the context of shape, appearance and structural integrity. However, there are other aspects of design which are less obvious but can nevertheless have an important influence on the overall economics of a structure. The protection against corrosion both initially and over its lifetime accounts for a not inconsiderable part of the overall cost of a structure. Therefore, any aspects of the design that increase costs should be of concern. The purpose of this chapter is to indicate areas where the design itself may adversely affect the performance of coatings or increase the probability of corrosion.

In many structures the detrimental influence of certain design features may cause problems related to maintenance. However, the additional breakdown of coatings and consequent rusting is often assumed to be an inevitable situation that occurs with all steelwork. The possibility that it could be avoided by attention to the detailed design is often not taken into account. This is to some extent understandable because the designer has problems with other aspects of design and fabrication and even if aware of problems from corrosion tends to assume that these will be satisfactorily dealt with when the structure has been painted.

The adage 'corrosion control begins at the drawing board' is correct but presupposes that at this stage there is an appreciation of design in relation to its effect on corrosion. It is interesting to note that more attention appears to be paid to these matters for offshore structures compared with those built inland. This no doubt arises from the very aggressive situations that offshore structures have to withstand and the necessity to reduce repainting to a minimum because of the difficulties involved.

Although coating breakdown and steel corrosion are likely to occur over a longer timespan on structures exposed to less aggressive conditions, they still occur, so action should be taken to avoid any situation that unnecessarily causes such problems.

ISO 12944 Part 3 'Design considerations' provides information on the basic requirements for the design of steel structures. It includes diagrams showing examples of good and bad design features, such as water traps, mixture of dissimilar metals and the problems of accessibility for applying, inspecting and maintaining paint systems.

10.1   Environmental conditions

The nature of the conditions to which structures are exposed has a marked effect on the selection and performance of protective coatings and also on the overall effect of design. Generally, the more aggressive the conditions, the greater the influence of adverse design features on the performance of coatings. For example, bimetallic corrosion (see below) may not be appreciable when steel structures are exposed to comparatively clean inland conditions, but the same alloys coupled to steel under immersed conditions in seawater may cause serious problems. Other features, e.g. crevices, are also likely to cause additional problems in more aggressive environments, particularly under immersed conditions, where maintenance and remedial action will be much more difficult to carry out. It follows, therefore, that increased attention should be paid to all aspects of design where structures are exposed to severe environments. However, the effect of environment is often in terms of time to failure rather than the absence or occurrence of corrosion.

10.2   Materials

In plant construction, the materials used are of major importance. The selection of suitable alloys for pipework, heat exchangers, pumps, etc., is an essential part of the design requirement. For the types of structure covered in this particular book, the materials are usually fairly standard steel sections and slight variations in composition are unlikely to affect the overall performance of protective systems. Nevertheless, there are certain aspects of material selection that may influence corrosion.

  (i) Fasteners are often the weakest part of a structure with regard to protection. This arises from a number of causes such as the greater likelihood of damage to the coatings because fasteners act as projections. More important, however, is the stage at which they are protected. Often the main structural sections are painted in the works before transportation to site prior to erection. The sections are then bolted and often the fasteners receive inadequate protection; sometimes no more than the final on-site coat applied to the structure. Zinc-coated bolts will, in many situations, raise the standard of protection to the equivalent of that of the main structure painted in the works. For some situations, fasteners produced from alloys such as stainless steel or 'Monel' may be preferred to carbon steel. Such alloy fasteners are advantageous where the nuts and bolts require to be removed during the life of the structure. Carbon steel bolts tend to rust at the threads, which makes removal very difficult. Attention must, however, be paid to problems of bimetallic corrosion when using alloy fasteners and where appropriate they should be coated.

(ii) Galvanised wall ties used in the exterior brickwork cavities of buildings have, in some cases, corroded at an unacceptable rate and in many situations stainless steel is now preferred. This is a design problem because the cavities are not necessarily dry, particularly where the ties are fixed to the outer wall. Bricks are porous and may allow the ingress of moisture, particularly in situations of driving rain.

(iii) The employment of alloys or alloy-clad material may be preferred to the use of conventional coatings in some critical areas of a structure. Typical examples are the use of a 'Monel' cladding on the risers of offshore oil production platforms and 'Incalloy' welded overlays on parts of marine structures. Clearly, attention must be paid to galvanic effects in such situations and the different alloys must be insulated from each other. Sometimes carbon steel clad with titanium or stainless steel may be used; in this case sound sealing of the edges is essential.

(iv) Stainless steels may be used for certain parts of structures or buildings. When they are freely exposed to air, very little corrosion will occur. However, when they are immersed in seawater the presence of crevices or overlaps can lead to corrosion and the same precautions should be used at crevices as for carbon steel (Section 10.5). The reason for the corrosion is that the protective (passive) oxide film breaks down and because of lack of oxygen in the crevice is not repaired, so that the stainless steel acts like carbon steel. Deposits and fouling on the surface of stainless steel immersed in seawater can also cause pitting.

10.3   Bimetallic corrosion

The corrosion arising from connecting two alloys, one of which is more noble (or cathodic) than the other, leads to bimetallic corrosion of the less noble (or anodic) alloy. The relative nobility of metals is shown in various 'galvanic series', the most common of which is that produced for immersion in seawater (see Table 10.1). The extent of the corrosion depends upon a number of factors:

(i) The presence of an electrolyte and its conductivity: bimetallic corrosion does not arise under dry indoor conditions. It may occur under damp conditions in air, especially in areas that remain moist for prolonged periods, e.g. crevices. However, it is likely to be most severe under immersed conditions, particularly in high-conductivity seawater.

(ii) The relative positions of the alloys in the galvanic series: generally, the effects are likely to be greater when alloys are well separated and corrosion will be less when they are close together. Some alloys dependent on

Table 10.1   Simplified galvanic series of alloys and metal coatings used for structures[a]

Cathodic (protected)	Titanium
	Austenitic stainless steels (passive)
	Nickel alloys, e.g. 'Monel'
	Copper alloys
	Lead, tin
	Austenitic stainless steels (active)
	Carbon steel, cast iron
	Cadmium
	Aluminium
	Zinc
Anodic (corroded)	Magnesium

a Based on tests in seawater.

passive films for their corrosion resistance appear in two positions in the series, depending on whether they are 'passive' or 'active', i.e. whether or not the protective surface film has broken down. Stainless steels are usually far more noble than carbon steels, so will increase the corrosion of the carbon steel to a considerable extent, but when 'active', i.e. corroding themselves, they are a good deal less noble.

(iii) The relative size of the two alloys has a significant influence on the rate of corrosion. For example, a larger area of bare carbon steel in contact with a small area of stainless steel will not be severely attacked overall, but where the stainless steel is much greater in area than the carbon steel, considerable corrosion of the latter may occur. This situation may occur when stainless steel or other alloy fasteners, e.g. 'Monel', are used in conjunction with painted steel structures. If there are any small pores or pinholes in the coating, the small area of bare carbon steel will provide a very large cathode/anode ratio with the more noble bolts, and severe pitting of the carbon steel may occur. This is, of course, likely to be most severe under immersed conditions.

(iv) The formation of corrosion products on the less noble metal or other polarisation effects may lead to a reduction in corrosion or even to its being stifled. On the other hand, in some situations, e.g. in coastal marine atmospheres, the corrosion products may be hygroscopic, leading to prolonged periods of 'wetness' and increased corrosion attack.

There are a number of methods of combating bimetallic corrosion but wherever practicable it should be avoided by not mixing different alloys, especially in aggressive environments. However, where this is not practicable, the following approaches should be considered:

(i)   Insulate materials to prevent electrical contact: this may be achieved in some situations by the use of suitable plastic gaskets or

washers. In other cases the painting of both alloys may prove satisfactory.
(ii)   Where practicable, ensure that the more noble alloy is smaller in
area than the less noble one.
(iii) Ensure good drainage from bimetallic joints and avoid such joints in
areas of potential water entrapment and at crevices.
(iv)   Allow for adequate access for inspection of any bimetallic joints
that may be considered as critical parts of the structure. For
example, where facia cladding is fixed to a building, bimetallic joints
may be 'hidden' for 50 years or more and it is not possible to predict
with any certainty how they will react. Moisture may reach the
joints from leaks or deterioration of materials in other parts of the
building. Even the use of high-duty coatings will not guarantee protection over such prolonged periods.
(v)   Bimetallic joints should not be 'hidden' in concrete or other porous
materials without a sound protective coating.

10.4   Access for inspection and maintenance

Adequate access for inspection and maintenance of the protective system
is usually an essential requirement of any design. Sometimes the design is
such as to preclude reasonable access. This situation should be avoided but
where this proves to be impracticable suitable action must be taken to
ensure that undetected corrosion does not result in the loss of structural
integrity.

It is unlikely that any conventional organic coating can be relied upon
for periods over 20 years without maintenance. Three possible approaches
may be considered:

(i)   to build in what is often termed 'a corrosion allowance';
(ii)   to use a more corrosion-resistant alloy;
(iii) to employ a thick coating of concrete.

Each of these possibilities carries inherent problems. A corrosion allowance
is simply the addition of sufficient steel to the section to allow for the anticipated loss by corrosion. The allowance is usually conservatively calculated
so, in most situations, will satisfy the requirements for the environmental
conditions. This method is used for much sheet piling but generally any
increase in corrosion over that allowed for is not critical. Again, it is used for
some areas of bridges, e.g. at the abutments. However, when using this
approach, the possibilities of enhanced local corrosion or changes in the
initial environment must be taken into account. Often the addition of a
coating to the steel even where a corrosion allowance has been added is
worth considering, because it will add a safety margin of some years.

The use of an alloy that is more corrosion resistant than steel is some-times a sound approach, although in most situations the additional cost proves to be unacceptable. However, the characteristics of the alloy must be fully appreciated, especially where it has to be embedded in some other material because, while an alloy may overall be very corrosion resistant, it may be attacked by alkaline materials or at crevices formed during construction.

The use of concrete to protect the steel is widely practised with water pipes where, because of the compressive nature of the coating, it generally performs well. However, the thickness must be sufficient in atmospheric situations to allow for the carbonation layer of the concrete. Furthermore, if chlorides diffuse through the concrete layer and reach the steel surface, corrosion may occur despite the alkaline nature of concrete. The possibil-ity of monitoring the corrosion of steel in inaccessible areas is always worth considering because this does allow time to decide on a suitable course of action if problems arise.

The above discussion has been concerned with areas where access and inspection are generally physically impossible, but there are also situations where access is possible but very difficult. Alternatively, access may be possible for visual inspection but not for remedial work. In some ways such conditions cause more problems because often insufficient attention is paid to ensure that the steelwork is satisfactorily protected in such situ-ations. Access in this sense can have two aspects. The problems may be concerned with the cost and the difficulties of erecting scaffolding. Often this is inevitable, although on bridges and other structures suitable perma-nent methods of access are included in the design. However, the access problem may also arise from the difficulties of inspecting and later reach-ing the steel surface to allow for adequate repainting. These situations can be avoided, sometimes with comparatively small changes in design.

Figure 10.1 illustrates the situation where the design unnecessarily

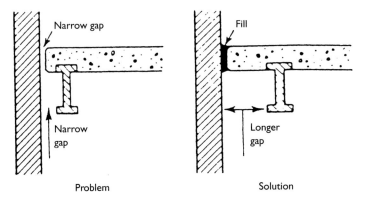

Figure 10.1   Insufficient access for maintenance.

precludes the possibilities of sound cleaning and repainting. Sometimes access is not impossible but because of the array of service pipes and other equipment leads to considerable difficulties, particularly with the cleaning of the steelwork before repainting.

The problems that can arise are illustrated in Figure 10.1. The relative positions of different structural elements may be arranged in such a way that access for maintenance painting is virtually impossible or even if there is some access, the cleaning and painting of the steel is likely to be inadequate. Often comparatively minor changes in design will overcome the problem.

Access problems of a somewhat different kind can occur where an array of pipes and other equipment leads to difficulties in cleaning and adequate painting. In such cases, it may be worth considering the 'boxing-in' of such areas. The inside of such 'boxes' where the services are located should, where practicable, be treated to remove moisture and some form of inspection cover should also be incorporated. Generally, the boxing-in of inaccessible areas is not recommended without some form of access for inspection because if, for example, there is leakage into the box, considerable corrosion may occur. The addition of services after construction of the main elements often poses difficulties, particularly if pipes are fitted too close to other steelwork so that it is virtually impossible to reach some parts for maintenance work. Often there is no reason why pipes should be fitted flush, although additional costs may be incurred in using extended brackets to allow for access. Alternatively, it may be possible to provide brackets that are easily removed to allow for painting of the whole pipe.

The use of galvanised steel for pipes, pipe hangers and fasteners is recommended in such areas. A particular problem arises with 'back-to-back' angles. The space between such angles is too small to allow proper cleaning and painting once they are in place. Consequently, rusting eventually occurs. If the angles are exposed to corrosive fumes inside a building, a serious situation can arise in a comparatively short time. The problem can be avoided by changing the design, e.g. using T-bars. Where the problem becomes apparent during maintenance, the gap can be cleaned to the best practicable level and mastic used to seal it. Another, but more costly way of dealing with the situation is to weld in a spacer plate.

Where access of any type is likely to be a problem, the design should be examined carefully to see whether an alternative approach can be adopted. In particular, it may be possible to re-design parts of the structure so that critical elements are fully accessible.

10.5   Crevices

A crevice is a very small gap between two steel surfaces which allow access of air and moisture, often through capillary action, but does not provide sufficient space to allow for cleaning and repainting during service. Typical crevice situations are shown in Figure 10.2. The presence of a crevice or overlap does not necessarily provide a corrosion problem. For example, two heavy steel plates bolted together may produce a crevice between the contact faces. However, although rusting may occur within this gap it may simply be stifled as the space between the plates is filled with rust. On the other hand, a similar situation with thin steel sheets bolted together could well result in buckling of the joint and shearing of the bolts. It is not possible to provide strict guidelines on situations which are safe but, generally, small gaps in bolted heavy steel sections will withstand the pressure set up by the formation of the rust.

As the formation of rust results from the reaction of the steel to produce an oxide or hydroxide, the corrosion product will have a considerably greater volume than the steel from which it is produced (Figure 10.3), because $2Fe \rightarrow Fe_2O_3.H_2O$. Rusts can produce high stresses within a

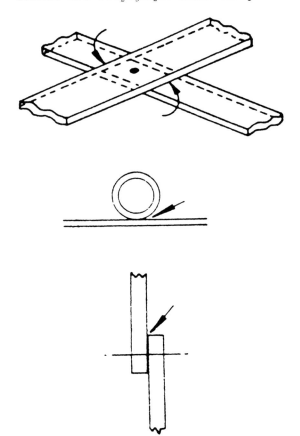

Figure 10.2   Typical crevice situations.

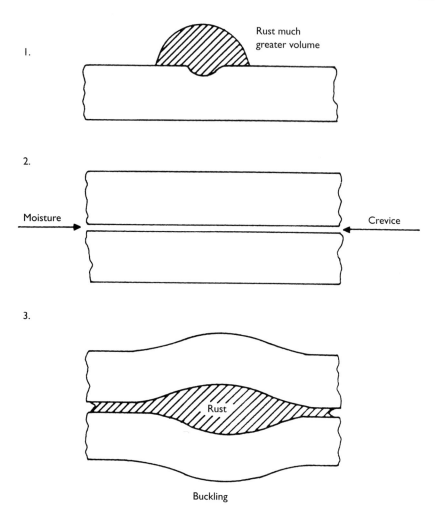

1.

Rust much
greater volume

2.

Moisture

Crevice

3.

Rust

Buckling

Figure 10.3   Rust formation in crevices.

crevice and Figure 10.4 shows the typical buckling effect on steel railings at the coast. This problem with the formation of corrosion products in a confined space can arise with alloys other than steel and the effect is sometimes descriptively called 'oxide jacking'. It is the same basic process that results in the spalling of concrete from reinforcements that are corroding. Crevices are often a more serious problem with alloys such as stainless steel because the passive film may break down at a crevice, particularly if chlorides are present. This can lead to severe pitting and may preclude the use of the alloy for many situations. Typically, crevices arising from gaskets at joints may produce serious problems.

Figure 10.4   The effect of crevice corrosion.

The difficulty with crevices arises because moisture and air gain access through the gap and so allow corrosion to proceed. However, because of the size of the gap it is not possible to clean or repaint it. Where practicable, crevices should be avoided or at least sealed. Sometimes welding in preference to use of bolted joints may solve the problem. Welding of the entrance of the crevice or filling with mastic are other approaches, although mastics may harden and contract or crack over a period of time, leading to the formation of a further crevice. Sometimes the solution is to increase the gap so that crevice conditions no longer operate. For example, it is preferable to allow for a gap between bricks or masonry and steel rather than to have close contact where rust exerts direct pressure on the brittle building materials, causing cracking. There is clearly an advantage in painting the contacting faces before they are bolted together in a structure. It is preferable to bolt the sections together while the paint is still wet to make sure that any gap will be filled with the paint.

10.6 Ground-level corrosion

The conditions at ground level are often more severe than on other parts of a structure. This arises from splashing that may occur from rain or the movement of vehicles on roads, and from the collection of water where insufficient attention has been paid to the detailing at the point where the steel enters the ground. Problems occur with most structures, but can be severe on lamp posts, railings and similar steel furniture which may be fabricated from comparatively thin material. The effects of salts exacerbate the situation in coastal areas. It is advisable to provide additional protection at ground level. This may be achieved by applying thicker coats of paint, but the use of properly designed concrete plinths is preferable (see Figures 10.5 and 10.6).

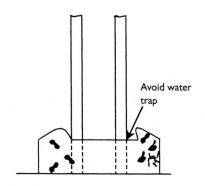

Avoid water
trap

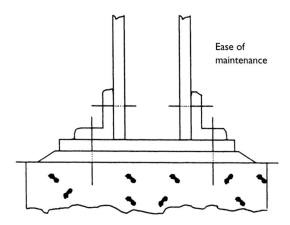

Ease of
maintenance

Figure 10.5   Design of column base.

Figure 10.6   Concrete plinth to protect steel at ground level.

10.7 Entrapment of moisture and condensation

The entrapment of moisture and dirt in elements of a structure can lead to enhanced corrosion or the reduction in life of coatings. Condensation also may lead to similar effects. The problem with moisture arises from the prolonged periods of contact that may occur with an electrolyte which, because it is generally well aerated, can attack the steel fairly rapidly at defects in the coating. Furthermore, where coatings are chosen to resist atmospheric influences they are seldom so protective when in more or less permanent contact with aqueous solutions. It should also be borne in mind that other design features such as bimetallic couples and crevices are likely to be more active under prolonged damp conditions.

Figure 10.7 illustrates typical situations where moisture may be entrapped and indicates possible methods of overcoming the problems.

Drainage holes often act as a simple solution provided they are properly sited. Clearly, they must be positioned at the places where water is likely

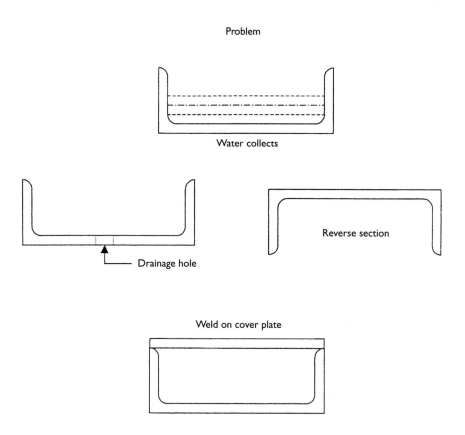

Figure 10.7 Various solutions to the problem of water entrapment.

to collect, generally at the lowest points. This can be checked fairly easily but, from observations on some structures, drainage holes are not always placed in the most appropriate positions. They must be of sufficient diameter to ensure that they do not become rapidly blocked with leaves and other debris. The use of suitable drainpipes will ensure that the water is not allowed to run over other parts of the structure. Ledges should, if possible, be avoided as they act to collect dirt and moisture. Often hygroscopic particles in the dirt collected will cause a permanently wet poultice of corrosive matter to remain in contact with the steel, again causing breakdown of coatings followed by corrosion. Condensation produced by warm humid air impinging on a colder surface can equally be a problem. In interior situations it can often be avoided by suitable ventilation, provided the surfaces are reasonably insulated from the exterior environment. Difficulties can arise with sheet steel roofs where the temperature of the metal surface varies from that of the ambient conditions inside the building.

Condensation inevitably occurs on the undersides of sections on many structures, e.g. on bridges over rivers. The extent will depend upon the actual conditions and size of the structure. The solution is usually based on the selection of suitable coatings to resist the conditions. Anti-condensation coatings are widely employed in highly humid conditions such as those experienced in parts of breweries.

10.8   Geometry and shape

The geometry and shape of structural elements influences corrosion in a number of ways. Smooth, simple surfaces are the easiest to clean and paint and so are most likely to provide the best resistance to corrosion. Most design problems arise basically from the protrusions, changes of plane, sharp angles and edges and general lack of clean lines which are inevitable with the fabrication and erection requirements of complex structures. Although such lack of homogeneity cannot be completely avoided, steps can be taken to reduce its overall influence to a minimum. For example, it is easier to coat tubular rather than angular sections. Clearly, for a variety of reasons, round sections cannot always be used. Nevertheless, where there is a direct choice based on technical and economic factors, tubular members will generally prove to be easier to protect from corrosion. This arises from their lack of edges and overall smoothness of lines. Additionally, coatings such as tapes can be used with much greater ease on such members.

Edges are always difficult to protect (see Figure 10.8). Tests by the authors have shown that the reduction of film thickness on the edges of typical steel sections can be as much as 60%. Improved coating thickness will be achieved if they are contoured to a smoother shape. Although

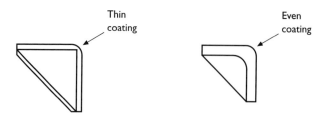

Figure 10.8   Contouring of sharp edges.

additional stripe coatings can be applied at edges, they are, by their nature, features that can easily be damaged.

Generally, large flat areas are much easier to protect than a range of sections of different shapes with random orientations of plane. As fabrication and design have developed, structures have tended to provide cleaner lines. It is interesting to compare the two bridges across the Forth in Scotland. The rail bridge requires almost constant painting with its complex lattice girders, whereas the more modern road bridge is much easier to maintain. Provided they are cleaned flush with the main surface, welds generally provide cleaner lines than bolts or rivets. Complex structural geometry provides problems for satisfactory coating and any improvements that can be made by the designer will improve the durability of protection and ease of maintenance.

10.9   Tanks

Many of the features already discussed can be conveniently illustrated by examining possible approaches to the design of water tanks. Although they have been simplified, the two sketches of tanks in Figure 10.9 demonstrate some of the fundamental points to be covered.

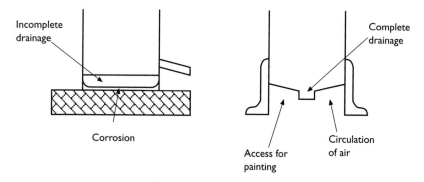

Figure 10.9   Design of tanks.

(i)   Access for protection of the underside.
(ii) Drainage at the lowest point.
(iii) Welded structure for smoothness.
(iv) Circular construction for ease of maintenance.

It would be unreasonable to ignore the additional fabrication costs that might be involved in the 'improved' design, but from the standpoint of corrosion protection it is undoubtedly superior.

10.10   Fabrication and construction

During fabrication additional items may be welded on to the main elements, such as brackets and hooks to assist with the handling and erection of the steelwork. These can be advantageous from the protection standpoint because they can assist in the avoidance of damage to the coating.

Generally, their presence on the steel sections do not cause any problems so may be left in place and painted. This may be satisfactory if the welds are properly cleaned (see below) before painting. Sometimes they are flame-cut to remove them or if small, mechanically removed. If these broken surfaces are not ground to a reasonable finish, coatings applied to them are likely to fail prematurely.

Welds. Weld spatter is the term given to the small globules produced during welding which stick to the steel in the vicinity of the weld. These provide points for pinpoint rusting of coatings applied over them, and should be removed.

Flux residues remaining from welding may be alkaline and so will affect coatings. These residues should be removed.

Discontinuous or 'skip' welds are frequently used for situations where continuous welding is not considered necessary. Although satisfactory from an engineering point of view, they are difficult to coat satisfactorily particularly when used for lap joints; they should be avoided.

Butt joints are easier to protect than lap joints, which present a crevice situation on one side which will inevitably promote coating breakdown.

Generally, welds are easier to coat than nuts and bolts. However, unless they are prepared properly before painting, undercutting, flux and spatter can all cause problems.

Nuts and bolts. With so many sharp edges and the ease with which paint coatings applied to them can be damaged, nuts and bolts can be a problem. Often, the thickness of paint applied is less than on the main structural elements. It is not uncommon to see fasteners rusting and causing rust streaks on an otherwise well-protected structure. Furthermore, the presence of a joint when steel is bolted adds to the difficulties of providing sound protection, because it may act as a crevice.

10.11   Corrosion of steel in contact with other materials

Where steel-framed buildings are clad with masonry and similar materials, these materials can crack if they are in direct contact with steel which corrodes. Unlike heavy steel sections which are able to resist the pressure of the rust, brittle materials cannot do so and consequently are physically damaged. It may, therefore, be advisable to paint the steel if there is any likelihood of moisture reaching the surface through inadequate joints, leaking roofs, etc. It may be preferable to design the building so that there is no direct contact between the steel and other materials used in the construction.

Some porous materials can promote corrosion of steel if they are in direct contact with it, e.g. wood, particularly if it becomes damp. Again, it is advisable to paint both contacting surfaces.

Further reading

British Standards Institution (1969). *Recommendations for the Design of Metal Articles that are to be Coated*, BS 4479.

British Standards Institution (1986). *Method for the Determination of Bimetallic Corrosion in Outdoor Corrosion Tests*, BS 6682.

British Standards Institution (1984). *Commentary on Corrosion at Bimetallic Contacts and its Alleviations*. PD 6484: 1979 (Confirmed 1984).

Chandler, K. A. and Stanners, J. F., *Design in High Strength Structural Steels*. Iron and Steel Institute Publication, No. 122, London, 1969, pp. 127–33.

Chandler, K. A., Design on steel structures for the prevention of corrosion. *Br. Corros. J. Supplementary Issue*, **10** (1968) 42–7.

Shrier, L. L. (Ed.), *Corrosion*, Vol. 2. Newnes and Butterworths, London, 1976, Section 10.3.

Chapter 11

Maintenance painting

11.1 Introduction

Maintenance of protective systems on structures is an essential part of the overall strategy of protection. Maintenance costs enormous sums of money throughout the world, yet in many cases scant attention is paid to the overall problems of repainting. Maintenance painting represents by far the largest part of structural steel painted, yet the understanding of the basic processes and the technology available are far less advanced than for painting new steel. Often an almost fatalistic attitude is adopted, yet, with the correct approach, considerable economies are possible. There are examples of structures where 30 coats of paint have been applied over the years; this can be checked by microscopic examination of detached paint flakes (see Figure 11.1). Such areas generally have not corroded but, on

Figure 11.1 Cross-section showing multiple paint layers.

the same structure, other parts have rusted quite badly because virtually no paint is present on the steel. It is possible that at the next repainting, a thirty-first or thirty-second coat will be applied to the sound areas and they will continue to protect the steel, while the corroded parts will, after wire-brushing and application of a three-coat system, be rusting again within a year or so.

The situation may be slightly more serious because the amount of corrosion may have reached a stage where the structural stability has been affected and welding of additional steel may be required. Although thick coatings generally will protect the steel, internal tension arising from the number of coats applied over the years may lead to cracking and flaking; there is, therefore, a limit to the number of coats that can be applied and, eventually, the whole coating may have to be removed.

The above scenario does occur but it would be unfair to maintenance engineers to suggest that it is typical of all structures and plants. Nevertheless, unless certain clearly defined steps are taken, it is often difficult to avoid this situation, particularly when the repaint is being carried out largely for cosmetic reasons and the overall appearance is considered to be more important than the protection of the steel from corrosion.

There are two main reasons for carrying out maintenance repainting:

(i) Preservation of the structure's integrity so that it can perform its function satisfactorily over its design life without the need for structural repairs.

(ii) To preserve its appearance: this may be a legal requirement in certain situations but generally it is more concerned with maintenance of a satisfactory overall environment and is, of course, taken to be a measure of efficiency and good housekeeping.

Clearly (i) is the more important but (ii) cannot be discounted, particularly where buildings and structures are visible to the public and shoddy appearance has a serious effect on the environment.

It might be assumed that in the process of maintaining the appearance the structural integrity would automatically be covered. However, this is not necessarily so, because appearance is a matter of visual approval often at some distance from the structure and important structural elements may not be apparent to the viewer. Therefore, even where constructions are regularly painted for aesthetic reasons, it is still necessary to take account of 'hidden' steelwork. Some paint types, for example bitumen, often have sufficient elasticity to accommodate the expansion of corrosion products on the steel surface without obvious disruption of the paint film.

Conversely, the structural preservation may be achieved without necessarily providing an overall attractive appearance to the structure itself. A considerable amount of corrosion can occur on some parts of steelwork without affecting the integrity, whereas on other parts attack by corrosion may be critical. It follows, therefore, that a proper plan must be formulated based on the requirements. If appearance is unimportant and the critical areas are properly protected from corrosion, then very limited repainting may be required, particularly on structures with a short life expectation. However, this approach must arise from a decision made on the requirements, not as a result of indecision regarding the overall attitude to maintenance. Furthermore, an element of monitoring should be incorporated to ensure that corrosion is maintained within acceptable limits.

In some situations, painting can be avoided altogether by adding suitable corrosion allowances.[1] In fact, this may be essential if the design involves spaces where access for repainting is impossible.

Having decided the reasons for regular repainting and the requirements involved, there should be a clearly formulated plan of action to ensure satisfactory and economic maintenance procedures. Undoubtedly this is best carried out at the initial design stage but revision of the plan may be necessary during the lifetime of the structure because of changes in the use and purpose of the structure, alterations in the environment and developments of new materials and techniques for coating steelwork.

11.2 The general approach to maintenance painting

It must be said at the outset that the efficiency of maintenance treatments will be directly related to and affected by the initial protective system and the standard of maintenance treatments over a period of years.[1] It is much easier to plan maintenance on a new structure that has been adequately protected to meet the service conditions. On older structures, the level and cost of maintenance will be determined by the history of painting over its lifetime. If there has been a long record of neglect or if the original protective systems and surface preparation of the steelwork were inadequate for the service conditions, no amount of touching up will solve the technical and economic problems posed.

In such situations, only the complete cleaning of all corroded steelwork and the virtual repainting of the structure will provide a long-term economic solution. Unfortunately, in many situations it will not be possible to blast-clean the steelwork so other less effective cleaning methods will have to be used. This will result in shorter periods between repainting and an increase in the overall costs of maintaining the structure. This is particularly disadvantageous for structures with long design lives. Typically, highway bridges may be designed to last for over 100 years and a proper approach to maintenance is essential if costs are to be controlled. On other structures designed for, say 25 years, the problem is less acute and while the appearance may be poor, the structural integrity may be unaffected. It is, therefore, important at the outset to determine the reasons for repainting and to formulate a clear strategy for the maintenance of structures. The *ad hoc* approach often adopted is unsatisfactory, both from the technical and economic standpoints.

It is worth making a point regarding maintenance budgets because in times of financial stringency they are the ones most likely to be cut and are always subject to reductions to an unrealistic level for the work required. To some extent, attitudes are changing because the importance of asset preservation is better appreciated nowadays. In situations where immediate savings are required, it may be quite reasonable to reduce maintenance budgets, including those for repainting. However, the remaining part of the budget should be spent on the critical parts of the structure to ensure maximum benefit from the expenditure and this is not usually done. In fact, it can only be done if maintenance is properly planned so that individual areas can be assessed.

11.3 Planning maintenance

Planning is always necessary, particularly in difficult situations such as on offshore structures where repainting must be carried out over a limited period. However, planning is being considered here in the sense of a long-term commitment to maintenance. This will involve decisions regarding the acceptable criteria of breakdown of paint coatings before repainting is carried out.

It is not unusual for Codes of Practice, etc., to specify that the ideal time for maintenance painting is when the painted steelwork shows a percentage, say 5%, visible rusting. This is a legacy from the old days when the surface preparation of the steelwork was weathering and wirebrushing. In those cases 5% visible rusting was probably just the tip of the iceberg and there would be significant rusting under the remaining paintwork that appeared to be sound.

The objective of modern-day maintenance painting should be to repaint when only the minimum (and easily carried out) surface preparation is necessary. Dry blast-cleaning, wet abrasive blast-cleaning and even UHP water jetting will generally be more difficult, even impossible, to carry out in the maintenance situation.

The condition of painted surfaces can be divided into the following categories:

Surface condition 1. The top coats with loss of decorative appearance by fading, chalking (see Chapter 13), possibly slight cracking or checking of the top coat only, no visible rusting or deterioration of the substrate.

This is the ideal surface for maintenance, requiring only that any surfaces still glossy should be abraded and any loose, powdery deposits be removed by brushing or low-pressure water jetting.

Surface condition 2.   The major part of the paintwork similar to condition 1, but visible rusting occurring in vulnerable areas, for example, where there is water entrapment.

Too often when the visible rust areas represent a small percentage of the total, the standard of surface preparation is governed by the larger area of reasonably sound paint. The result is normally that the rusted areas are inadequately treated, fail again prematurely and consequently set the pattern for the maintenance painting cycle and possibly even threaten the integrity of the whole structure. Any significant rusting of the steel, however small in area, requires some form of blast-cleaning or UHP water jetting, if it is not going to become the weak link in the system. In these cases it is necessary to ensure that the existing paint is sound enough to withstand the 'feathering' needed to adhere the new paint to the old.

Surface condition 3.   This is where there is pinpoint rusting as opposed to rust blistering. This is caused by inadequate thickness of paint, almost certainly below that originally specified or recommended by the paint supplier. Generally, this does not indicate that the paint coating is undermined by corrosion and washing/brushing, plus abrasion, may be sufficient surface preparation.

Surface condition 4.   This is where there is apparently random corrosion blistering, cracking or flaking down to the steel. This generally means that the entire surface has to be cleaned down to bare metal, preferably by UHP water jetting or wet abrasive blasting.

Although improved hand-cleaning methods, including use of special discs, are available as alternatives to blast-cleaning, they are comparatively expensive in the time required for use and are unlikely to provide cleaning to the same standard. The above discussion has been concerned with conventional paint coating systems of up to about $200\,\mu$m total thickness. Thicker coatings, possibly over 1 mm in thickness, do not generally fail in the same way. Often the adhesion between the coating and steel weakens, leading to flaking over comparatively large areas rather than the local spread of rust. The criterion for maintenance with such coatings may be different and will be determined by factors other than the amount of rust formed on the steel surface.

Irrespective of the particular criterion chosen, it is clearly necessary to inspect structures so that maintenance can be properly planned and this will be considered below.

11.4   Inspections and surveys for maintenance

Some form of inspection is carried out on all structures but from the standpoint of corrosion and coatings it is often of a somewhat cursory nature, frequently being carried out from a point some distance from the steelwork to be examined. In some cases, this is the only practicable way of inspecting a structure because of problems of access. In these situations, the use of binoculars and viewing from different positions may provide useful information. However, wherever possible, close contact inspection is necessary to determine the extent and requirements for maintenance. Undoubtedly, proper surveys are the most satisfactory way of assessing the situation. Obviously, there is a cost factor involved in carrying out such surveys, particularly for access to allow close examination of the coating. Consequently, this level of examination may be limited to certain parts of the structure, to reduce costs to a reasonable level. The experience and competence of the surveyor will clearly be important in providing an overall view of the maintenance requirements on parts that have not been physically examined. The purpose of a survey can be summarised as follows:

   (i)   To assess the extent of coating breakdown and rusting over the structure as a whole.
   (ii)   Based on assessment (i), to determine the critical areas for maintenance and the overall methods to be used for the surface preparations and re-coating.
   (iii) To provide a basis for the preparation of a sound specification for the re-coating work.
   (iv)   In conjunction with (iii), to obtain tender prices for the maintenance work.

The survey should cover both the deterioration of coatings and actual corrosion of the steel. In some situations, corrosion can be measured by simple direct methods, using calipers and gauges. Sometimes other methods, such as ultrasonics, may be required. It is, of course, usually necessary to remove rust before making such measurements, although ultrasonic probes are available with which such removal is not necessary.

Often the amount of corrosion cannot be determined because the original thickness of the members is not known. In such cases it is still possible to measure the thickness of steel remaining, which is the essential requirement for determining whether it is structurally stable. The survey also provides opportunities to determine whether specific features of the design are promoting corrosion of the steel or breakdown of the coatings, e.g. water traps or crevices; recommendations for remedial action can be made to alleviate such problems.

11.4.1   Survey procedures

The procedures will be determined to some extent by the nature and siting of the structure. Where access is comparatively easy, a full survey can be carried out without too much difficulty. However, where the structure is in continuous use, e.g. a highway bridge, then both the costs and difficulties of a full survey may be such that only a limited survey can be carried out. Its effectiveness will depend very much on the expertise of the surveyor. Such surveys require careful planning to ensure that a reasonable overall assessment is made, because the most inaccessible parts may be those that will require most attention during maintenance.

The engineer responsible for the structure should discuss the work with the surveyor to ensure that a cost-effective plan is produced. The main options open to the engineer are:

(i) A full survey of the whole structure. This will not always be practicable and will be the most time-consuming and expensive option. Where the deterioration of the protective coating is not uniform, it may be the required option, if long-term protection is to be achieved. The type of structure will be a determining factor, as will the overall history of coating maintenance. Only close examination will determine the exact nature of a surface defect. Often what appears to be a sound coating when viewed from a distance proves to be in the early stages of failure, e.g. there may be a lack of local adhesion that can spread to large areas over a period of time.

(ii) A limited survey is more common, particularly where access is difficult. Such surveys should be carefully planned to ensure that representative areas are examined to provide a reasonable assessment of the whole. Certain areas are always more prone to coating breakdown and corrosion than others and where practicable these should be included.

11.4.1.1   Planning the survey

The engineer must decide the extent of the survey and the requirements for inspection. The survey should be carried out as near to the time of maintenance painting as is practicable. It should not be carried out more than a year beforehand since additional breakdown may have occurred in that time so the survey may then not provide an accurate assessment of the requirements. In the case of limited surveys, the actual areas to be examined must be defined and the requirements specified. The type of access for each area must be decided, with agreement on the testing to be carried out *in situ* and the recording required. Additionally, any laboratory testing should be specified.

The areas must be designated in some suitable way with numbered location and with sketches where required. Although each structure requires separate treatment, the overall approach is well illustrated by considering a bridge.

Planning a bridge survey. The UK Highways Agency has its own specifications and standards for highway bridges and these include survey requirements which must be followed. However, in the case of other bridges and similar structures, which do not necessarily follow the Department's procedures, variations may occur.

The engineer should supply drawings of the structure so that the areas to be surveyed can be suitably marked. Alternatively, the surveyor may prepare simple sketches for this purpose.

Bridges can be divided into well-defined areas:

(i)   Main span(s).
(ii)   Span supports, e.g. piers and abutments.
(iii) Superstructures, e.g. parapets, towers (on suspension bridges), cables, etc.
(iv)   Special areas, e.g. bearings and joints.

SAMPLE AREAS
Main span. On main spans, for a limited survey, sample areas of reasonable size should be selected at right angles to the main beams so that all beams are included. The widths of band may vary but should not be less than 0.5 m. For each beam this would provide a number of different areas for examination. Other transverse elements such as stiffeners would also be included. For a box girder, there would be only one sample area on each side and on the soffit, making a total of three. The insides of box girders should be treated in the same way and should include elements such as stiffeners and diaphragms. Welded and bolted areas should be included in sample areas.

The number of sample areas will be determined by the size and nature of the bridge.

Span supports. Sample areas can be chosen at intervals at right angles to the main direction of the support as horizontal bands round the support.

Superstructure. Parapets can be sampled in a manner similar to that for the main span. Other features such as cables require individual decisions for adequate sampling.

Special areas. The selection of special areas will depend on the bridge design but would include bearings, expansion joints and special features that had not been included in the other sample areas.

11.4.1.2   Inspection and testing

The type of inspection and requirement for testing should be agreed for each sample area. The types of observation and tests are considered below.

(i)   Observations

(a) Type and amount of coating breakdown by visual assessment. This must be carried out to provide the maximum amount of information both for the maintenance requirements and as an indication of the performance of the coating. There are a number of published methods that can be used but the essential requirement is for all those concerned with the maintenance procedures to be clear what the observations mean. Vague comments such as 'some blistering and slight rust' can be ambiguous. The recording methods may be based on ISO 4628. Paints and Varnishes. Evaluation of Degradation of Paint Coatings Designation of Intensity, Quantity and Size of Common Types of Defects.

Part 2: Blistering
Part 3: Visible Corrosion
Part 4: Cracking
Part 5: Flaking
Part 6: Chalking

These standards contain diagrams with numbered ratings which can be used to provide a fairly rapid and accurate assessment of a painted surface. However, useful though these methods are, it is often advantageous to include a description and photograph, particularly where the appearance does not closely relate to the standard photograph. If deep pitting is evident the depth should be estimated and included in the observations. A more accurate indication is provided if a pit gauge is used.

(b) An indication of obvious surface contamination should be provided.

(c) Areas showing abnormal or severe breakdown should be reported as these may require special attention during maintenance.

(d) Where breakdown is associated with specific design features or occurrences such as dripping of water onto an area, this should be reported.

(e) Where appropriate, the surveyor should use a suitable magnifier, preferably illuminated ($\times$10 to $\times$20 magnification), to obtain further information, e.g. on checking and surface contamination.

(ii)   Non-destructive testing. The thickness of the coating can be measured with suitable dry film thickness gauges. These are described in Chapter 9, with an indication of methods of calibrating and usage.

(iii)   Destructive tests. Adhesion can only be tested by destructive methods. Various test methods are available. The method in which a 'dolly' is stuck onto the painted surface and then pulled off using a calibrated pull-off adhesion tester provides a quantitative reading. There are a number of proprietary makes of adhesion testing instruments that can be used under site conditions. They use various methods to exert the pull-off force. The limitations for their use for surveys are: the difficulty of scoring around the 'dolly' under site conditions, the need to wait for the adhesive to cure and the number of tests carried out in one batch is determined by the number of 'dollies' available. The X-cut or 'St Andrew's Cross' test (see Figure 11.2) can be carried out very quickly over a large area and, although not quantitative, in expert hands can give a very good indication of the adhesion and brittleness of the coating.

BS 3900 Part E6, the cross-cut test, is a straightforward test for panels in the laboratory but is difficult to carry out correctly on site (see Figure 11.3).

ASTM D3359-83 Method A is simple and capable of providing the required information. It is carried out using a sharp blade, e.g. a 'Stanley' knife and cutting through the coating to the metallic substrate, preferably

Figure 11.2   St Andrew's cross-cut test.

Figure 11.3   Abraded patch and St Andrew's cross-cut test area.

with one stroke. The cut, approximately 50 mm long, is then repeated across the original cut at an angle of approximately 30° to form a St Andrew's Cross. Adhesive tape is then applied over the centre of the cross and pulled sharply from the surface. The maximum area of detachment of paint from the position of the crossed cuts is reported. The greater the length, the poorer the adhesion.

Abrasion of the coating down to the steel exposes all paint layers (see Figure 11.4).

Further tests using the end of the knife to detach more of the coating will, in experienced hands, reveal additional information that will indicate whether the paint can be mechanically cleaned or locally blast-cleaned or must be removed over a large area. The surveyor should report the results of these tests and where appropriate indicate the state of the surface from which the paint has been detached, e.g. whether clean, rusted steel or primed steel.

(iv)   Paint flakes. Paint flakes, when examined microscopically, can provide useful information such as thickness of individual coats in a system. Special thickness measurement instruments that use an inclined or V-cut can be used but it is not always easy to determine individual coatings.

Paint flakes should be stored in individual bags with proper identification.

(v)   Site analytical tests. Tests for surface contamination are detailed in Section 9.5.1.3.

Figure 11.4   Abraded test area.

(vi)   Solvent swab test. The surface of the paint is rubbed with a cotton wool swab wetted with xylene. If the top surface dissolves readily it indicates that the paint is of the non-convertible type, such as chlorinated or acrylated rubber. This information will influence the choice of coating for repainting.

(vii)   Laboratory tests. Certain tests may require to be carried out in well-equipped analytical laboratories. The test samples will be collected by the surveyor, identified, and then sent to a suitable laboratory for analysis. Other tests are more straightforward and may be carried out by the surveyor with comparatively simple equipment. An experienced surveyor should be able to examine detached paint flakes provided he has access to a binocular microscope with a magnification of ×30 to ×45.

High magnifications do not have sufficient depth of focus for the roughness of a paint flake to be examined. Paint flakes are generally not mounted and polished like metallurgical specimens because the polishing process tends to destroy the surface of the paint. Examination of cross-sections of paint flakes gives useful information on individual coating thicknesses and adhesion between coats. Also, the underside of the flakes can be examined for contamination.

Ideally, the microscope should have a camera attachment so that monocular photographs can be taken. Since the enlargement of the magnification of the final prints will be different from the microscope enlargement this should be determined so that direct measurements can be made from the photographic prints.

The presence or absence of lead pigments may govern the choice of surface preparation method. Other environmental factors, e.g. vicinity to a river or public place, may also require the analysis for other toxic substances such as cadmium and chromium.

In the laboratory, a range of techniques can be used for the comprehensive analysis of paint (see Section 16.6).

The routine determination of water swab samples will include chlorides, sulphates, pH and total dissolved solids. The engineer should specify whether analysis is required for other suspected contaminants.

11.4.1.3 Access for surveys

Arrangements for access must be made in advance of the survey. Failure to do this may lead to expensive delays. Scaffolding is rarely justified except possibly at some areas of special interest. Ladders are the cheapest form of access and may be suitable for the examination of limited areas. However, they are not really suitable for the type of examination required in surveys and moving them can also be time consuming, particularly where they have to be lashed to members for safety reasons. Access should be by mobile lifting platforms or hydraulic hoists for most parts of a structure but other methods may be required for areas that present difficulties of access. The surveyor has to carry out a number of tests, so satisfactory access is important.

11.4.1.4 Recording of survey

It is convenient to record on printed sheets which cover the main items to be included in the report. These would include:

- Identity of sample area
- Film thickness measurements
- Breakdown of coating (type and quantity)
- Swab tests
- Adhesion
- Paint flakes and other samples taken
- Photograph identification (when taken)
- Comments on contamination of surface, exposed substrates and any areas requiring special attention (see Figure 11.5)
- Weather and date of survey.

If sketches assist, then they should be included.

Photographs will often be of assistance to engineers, particularly where

Figure 11.5   Corroded special areas.

the breakdown is of an unusual nature, or where the state of the area is difficult to describe. Where reports are to be photocopied, additional colour prints should be prepared to be included in all reports.

Administrative details will also be included to cover the location and type of structure, name of surveyor, analytical laboratories, etc.

11.4.1.5   Recommendations

On the basis of the survey, recommendations can be made on surface preparation and repainting of the structure. On a large structure the recommendations will vary depending on the state of the coating at different parts. Additionally, any other recommendations that will be of value to the engineer should be included, e.g. design features that have promoted corrosion.

11.4.1.6   Comments on surveys

Surveys can be quite expensive and are carried out primarily to determine coating maintenance procedures. However, if carried out regularly, surveys provide a 'log book' of coating performance. This will be useful for determining suitable coatings for other structures and will indicate likely problem areas so that subsequent surveys may be more limited in scope with an overall saving in cost. On parts of the structure where the breakdown is such that complete removal of the coating is to be recommended, only limited detail of the paint breakdown is required. The general methods of a survey should be used even if the overall examination is carried out on a limited scale.

11.4.2 Feasibility trials

The Highways Agency (UK) advises a feasibility trial after its bridges have been surveyed, before a specification for maintenance painting is prepared. This is a sound approach because it is often difficult to determine both the extent of the surface preparation requirements and the difficulties in achieving them. Furthermore, this type of trial provides an opportunity to check compatibility of coating systems. It also ensures that contractors who tender for the work will be aware of any difficulties involved in the maintenance painting and will, therefore, be able to provide tenders that meet the requirements of the client with a minimum of disagreement. Although such trials add to the overall cost, they may prove to be economic on surveys of structures other than bridges.

11.5   Maintenance procedures

The two main aspects to be dealt with are (i) surface preparation and (ii) repainting. The survey prior to maintenance should provide sufficient information to determine the type and level of surface preparation and painting required. Even the most limited survey is advisable to ensure that a reasonable specification can be produced. Without this, maintenance repainting may prove to be a difficult, and often unsuccessful, operation.

11.5.1 Surface preparation prior to repainting

As already noted, maintenance of paint coatings is most economically achieved by repainting before there is any serious deterioration of the coating or rusting of the steel substrate. In such a situation, the procedures are fairly straightforward on conventional coatings used for land-based structures. On more specialised structures, e.g. offshore structures and dock gates, special procedures may be required and these will often be specific to the particular situation, although the same approach will usually be adopted.

(i) Where coatings are firmly adherent with no signs of incipient breakdown such as blistering, they should not be removed as they will provide a sound basis for the maintenance coats. Generally, further coats can be applied without difficulty after the surface has been thoroughly washed to remove contaminants. However, some coatings, e.g. two-pack epoxies, harden to an extent where some abrasion of the coating is required before repainting to ensure good adhesion. Even oleo-resinous paints may require some abrading to remove gloss. On small areas, power-operated discs may be satisfactory to provide the abraded surface, but on larger areas light blast-cleaning will usually be more economic.

(ii) Where there is some minor deterioration of the paint coating but no rusting of the steel substrate, then all loose paint should be removed by the methods noted above, i.e. by power-operated tools, UHP water jetting or blast-cleaning. The choice will be determined by environmental and economic factors. The surface must be washed before further coatings are applied.

(iii) If the deterioration has reached a stage where the coating is failing and the steel is rusting, the procedures will be determined by the extent of the breakdown. The only really effective way of cleaning steelwork where there is a considerable area of rusting is by blast-cleaning, but this is not always practicable. It may not be possible, for environmental reasons, to use open blast-cleaning, and vacuum methods, i.e. with recovery of the abrasive, may be uneconomic, particularly from the point of view of time required for cleaning. In such situations, power tools may be the only acceptable method,

It follows that if there is a considerable area of rusting to be dealt with and blast-cleaning or UHP water jetting is not practicable, then the maintenance coatings applied to such surfaces will tend to provide comparatively short lives.

The coating breakdown may well be quite severe at welds and a decision must be made regarding the cause of this breakdown. It may be necessary to grind weld areas to avoid a repetition of such failures after repainting.

The requirements for surface preparation prior to repainting can be summarised as follows. Suitable action must be taken to provide a firm base for future coats of paint. The effectiveness of this operation will to a large extent determine the life of the coatings applied during maintenance. If rust and poorly adherent paint are the surfaces to which new paint is applied, then poor performance will be inevitable. Further requirements concerning paint compatibility are discussed below.

(iv) The above comments have been concerned with paint films but metal coatings also eventually require maintenance and this is usually achieved with paint coatings. Hot-dip galvanised coatings may be used bare, i.e. not painted, and over a period of time the zinc will corrode away, usually not evenly over the whole surface, leaving a mixture of zinc, alloy layer and rust. Undoubtedly it is preferable to paint galvanised surfaces before any significant rusting of the steel substrate occurs. At this stage it is comparatively easy to remove loose zinc corrosion products by manual cleaning methods and then to apply a paint system.

If the galvanised steel has been painted and no rusting is occurring, the treatment will be similar to that for paint coatings applied directly to steel. There is usually no requirement to remove areas of sound galvanised surface even if the paint has flaked or weathered away. All loose paint should, of course, be removed. In situations where a great deal of rusting of the steel has occurred, blast-cleaning is probably the only satisfactory way of treating the surface prior to repainting. Sherardised coatings should be treated in a similar way to those that have been hot-dip galvanised. For small areas of breakdown, power-operated abrasive wheels can be used, but this is uneconomic for large areas. Sprayed metal coatings present a more difficult problem, particularly if they have been painted, a procedure rarely adopted nowadays for zinc.

Some authorities consider that once the underlying sprayed zinc coating has corroded under the paint film, then the only course of action is to blast-clean the surface with sharp abrasives to remove the whole of the coating system, i.e. paint and zinc. Others consider the light blast-cleaning to remove the paint and surface zinc corrosion products is a satisfactory procedure, provided dry-blast methods are used. Wet-blasting of zinc sprayed coatings is not recommended except where they are to be completely removed. This is because of the difficulties of ensuring that the remaining zinc coating is completely dry before repainting. Again, where previously non-painted zinc sprayed coatings are to be painted, it is preferable to remove zinc corrosion products by dry brushing methods. Although painting of zinc sprayed coating is not considered good practice by many authorities, it may be necessary to apply paint over badly corroded zinc coatings if the removal of the remaining zinc coating by blasting is impractical. In such situations, power-operated tools can be used to remove zinc and corrosion products, but a layer of the sprayed coating will remain on the steel surface provided rusting has not occurred.

Sprayed aluminium coatings are generally less of a problem than those of sprayed zinc, but in situations where they have virtually corroded away they require to be removed by blast-cleaning. Other methods such as the use of power tools are unlikely to be effective in removing the whole coating, although they can be used for removing surface corrosion products.

11.5.2 Painting

The effectiveness of maintenance painting will be determined by the substrate to which the paint is applied, the compatibility of the material used in relation to the coating on the structure and the overall thickness of the final coating. The importance of preparing the surface properly, both the steel and retained coatings, has been discussed above. Although the necessity to provide a sound base for maintenance coatings is obvious, it is sometimes difficult to achieve under practical conditions, and in such situations it is unlikely that the coatings will provide long-term protection. If they are applied to steel that is rusty, or if chloride and sulphate salts remain after blast-cleaning, then further corrosion under the paint film is probable, with blistering, cracking and eventual flaking. Again, if the retained paint coating has poor adhesion to the steel, then it may well flake off carrying the maintenance coating applied to it.

Compatibility between the coatings on the structure and those applied during maintenance is essential. For example, although solvent-evaporating types of paint can be applied to epoxies, the reverse situation may lead to problems, because the strong solvent in the epoxy will soften the chlorinated-rubber binder. As noted earlier, where paints have hardened during service it may be necessary to abrade them to provide a key for the maintenance coatings.

The choice of coatings will depend upon a number of factors. Traditionally, oil-based coatings with mild, slow evaporating solvents have been used for maintenance, because of their tolerance of minor weaknesses in the existing film and to small levels of contamination. The requirement for faster drying, higher film build, plus environmental legislation requiring reductions in the levels of volatile solvents has encouraged paint manufacturers to produce high solids, low solvent formulations. These can be applied to the older type of weathered coatings without any sign of immediate incompatibility. Paint manufacturers claim such will up-grade the coating system from single-pack to two-pack. However, even those two-pack materials described as 'surface tolerant' can produce weaknesses in existing coating systems that were not previously evident. Even coatings claimed to be able to be applied over damp surfaces do not perform well when there is excessive moisture, condensation or when the existing coating is cohesively weak and retains moisture after the surface-tolerant material has cured. On general land-based structures it is usually advantageous to apply the same types of paint, provided they have given sound protection. However, there is often a requirement to up-grade the overall protective system using more durable coatings. In such cases, checks for compatibility are essential and this should preferably be carried out on the structure. Generally, a number of different types of area have to be dealt with as follows:

  (i)   Intact paint coating of acceptable thickness: usually, apart from washing down with clean water, no further action is required.
  (ii) Reasonably intact coating but with slight blistering or flaking: after removal of all the loose paint coating, the surface is washed to remove contamination and undercoat(s) and finishing coats are applied to provide the required thickness.
  (iii) Areas where there has been rusting of the steel or where the coating

has been removed prior to repainting: a full protective system including, where appropriate, an inhibitive primer, should be applied to the appropriate thickness.

Careful application to ensure a smooth final coating is necessary and the finishing coating is generally applied over the whole structure, after the various areas have been patch painted. Metal coatings are treated in the same general way, but it is a more difficult operation, particularly where there is a variation in the state of the metal coating on different parts of the structure, and it may be useful to obtain specialist advice.

11.6   Environmental conditions during repainting

Unlike the initial protection of steelwork, which can be carried out in a shop with reasonably controlled conditions, the only environmental control usually possible when carrying out maintenance painting is the choice of an appropriate time of the year for the work. Sometimes for operational reasons, maintenance painting has to be carried out under less than ideal weather conditions. The effect of rainfall and condensation on surface preparation and paint application are obvious, but cold temperatures, extreme humidities (high and low) and air movement also affect different coatings in different ways.[2]

Coatings that dry by solvent evaporation, such as vinyls, acrylics, bitumens, etc. are affected by the cold because all solvents have a lower vapour pressure as temperatures are reduced, thereby slowing the evaporation from the film. Solvent entrapment in primers and undercoats can cause subsequent blistering or loss of adhesion in the top coats. With waterborne coatings, the more humid (moisture-laden) the air is, the less able it is to hold water evaporating from the film, the cure is slowed down and the longer the coating stays vulnerable to rainfall, etc. High humidity can also affect coatings that cure by oxidation, for example alkyds. Different formulations are affected to different extents and the paint manufacturer should be consulted about the maximum humidity recommended for their material.

Low temperatures can also inhibit the cure of oxidation curing coatings, but particularly two-pack materials that cure by chemical reaction, such as unmodified epoxies. Normally these will not cure at all below 5°C and should preferably only be applied above 10°C. Some modified epoxy coatings used for maintenance coating are claimed to cure down to 0°C, albeit slowly. The use of isocyanate curing agents can lower this threshold. Care is also necessary when applying paint at normal ambient temperatures, but where the paint is itself excessively cold due to poor storage conditions or a heavy steel structure is excessively cold due to the heat-sink effect. High winds can also lower temperatures below the ambient due to the wind chill effect and they can also cause dry spray (see Chapter 13) and increase the risk of overspray.

It is also important to realise overspray can travel surprisingly great distances, hence containment must be effective to ensure claims are not made by third parties, because of contamination of their property by this airborne overspray. To provide a controlled environment for work and reduce dust contamination, containment of the painting area is now common (see Figure 11.6).

Ventilation is an important factor inside containments to protect the health of the operators, remove solvent vapours, provide adequate visibility for performing the work and reducing contamination of freshly blasted or painted surfaces by abrasive particles. A general guideline is to provide one complete change of air every three minutes during the blasting operation. For ventilation during the painting period, both the explosive limits (LEL) of the solvent vapour and the threshold limit value (TLV) of the airborne toxic material must be considered. Using containment under adverse weather conditions may well need a combination of heating, ventilation and dehumidification. Details are beyond the scope of this book and, if required, experts should be consulted.[3]

Figure 11.6 Containment of blasting and painting operations in a sensitive area.

11.7   Health and safety matters

Health and safety requirements for maintenance painting are the same as those described in Chapters 3, 4, 5 and 9. In addition, it is particularly important that any contractors, inspectors, surveyors, etc. are made immediately aware of the safety requirements for a specific site. Also, since surveyors often work on their own, there should always be an additional person in sight of – and in contact with – the surveyor, to assist if an accident or dangerous situation occurs. This applies in all situations on site, but particularly those where the surveyor is working in any area which may be regarded as an enclosed space.

References

1.   Iron and Steel Institute, Sixth Report of the Corrosion Committee, Special Report No. 66, London, 1959.
2.   Hare, C. H., *J. Protective Coatings and Linings* (Jan. 2001) 27–40.
3.   Jacobi, K. and Brown, R., *Protective Coatings Europe* (Jan. 2001) 17–21.

Control methods other than coatings

Although most structural steelwork is protected from corrosion by coatings, other control methods are also adopted, in particular cathodic protection which is considered below. Methods such as inhibition, water treatment and air conditioning are used in special circumstances and reference is also made to these in this chapter. The other system of controlling corrosion without the use of coatings is to employ low-alloy weathering steels; these steels are also discussed in this chapter.

12.1 Cathodic protection

Cathodic protection has only a limited application as a method for preventing the corrosion of steelwork but it is used for many important structures, particularly offshore, and so is an important method of corrosion control. The concept of cathodic protection is straightforward, but in practice specialist advice should be sought to obtain economic protection. In this chapter only the broad principles will be considered.

An essential point concerning cathodic protection is that it can be used only where steel is immersed in water or buried in soil of suitable conductivity. It is not a method that can be suitably employed with steel exposed in the atmosphere. Although attempts have been made to develop coatings with suitable conductivity so that cathodic protection could be used in the atmosphere, these have not proved to be of practical value for steelwork. The method has been used to protect steel reinforcement in concrete but the conditions are different from those for structural steelwork. Certain metal coatings, particularly zinc, do provide a form of cathodic protection to steel, but this is not the primary reason for using them.

12.1.1   Basic principles

The basic principles of corrosion are considered in Chapter 2 and the main points can be summarised as follows:

(i)   Corrosion at ambient temperatures is electrochemical.
(ii)   The process can be sub-divided into two reactions: one at the anode and the other at the cathode.
(iii) For such reactions to occur there must be a suitable electrolyte present.

For steel (or strictly iron) the two reactions can be depicted as follows:

at the anode $2Fe \rightarrow 2Fe^{2+} + 4e^-$
at the cathode $O_2 + 2H_2O + 4e^- \rightarrow 4OH^-$

As can be seen, there is an electrical balance because all the electrons released in the anodic reaction are consumed in the cathodic reaction. If electrons are supplied to the iron from an external source such as by impressing a current, then the anodic reactions will be suppressed and the potential of the iron will be lowered. If it is lowered sufficiently, then no current will flow between the anodes and cathodes on the iron surface, so corrosion will cease. This is the basis for cathodic protection and provided sufficient current is supplied, corrosion will be prevented. If insufficient current is applied, there will be partial protection and the corrosion rate will be reduced below what would be anticipated in the absence of cathodic protection.

12.1.2   The application of cathodic protection

To achieve cathodic protection it is necessary to supply an external current so that no local currents flow on the steel surface, i.e. to supply electrons to the steel. In order to achieve this it is necessary to set up a cell with an auxiliary anode, the steel being the cathode. Additionally, an electrolyte is necessary to ensure that the electrochemical cell can operate. There are two ways of achieving these requirements in environments such as water or soil, where there is sufficient conductivity to allow proper operation of the cell:

(i)   By application of a current from an external source using an inert anode, i.e. impressing the current. This is called the impressed current method.
(ii)   By using an active anode of more negative potential than the steel and electrically connecting it to form a cell. Electrons pass to the

steel which becomes the cathode so preventing or reducing corro-sion. In this method the anode, e.g. zinc, is 'sacrificed' to protect the steel and it is called the sacrificial anode method.

Both methods are widely used and, in practice, have relative advantages and disadvantages.

12.1.3   Sacrificial anode method

Zinc, aluminium and magnesium all have more negative potentials than iron or steel and are used as anode materials to provide full or partial cathodic protection to steelwork. Over a period of time they corrode and so must be replaced at suitable intervals.

The anodes are manufactured into suitable shapes and sizes and are electrically connected to the steel in various ways, which will not be con-sidered here. However, the method of connection must be such as to ensure good electrical contact.

In practice, the exact composition of the anode material is important because it must fulfil certain requirements. Clearly, it must maintain a sufficient negative potential to ensure that cathodic protection is achieved, but it must also continue to corrode during use. If the anode becomes passive, i.e. develops a protective surface film, it will not operate properly. Additionally, it must develop a high anode efficiency. To achieve these requirements, certain small alloy additions may be made, e.g. indium to prevent passivity in aluminium anodes. Alternatively, some elements may be detrimental, e.g. iron in zinc anodes, and they must be maintained below a certain level. The quality control of anode composi-tions is an essential requirement if sound cathodic protection is to be achieved.

12.1.4   Impressed current method

In this method, again in the presence of a suitable electrolyte, current from a DC source is delivered through an auxiliary anode so that the steel becomes the cathode of a large electrochemical cell, i.e. the potential is reduced to a level where corrosion does not occur. A number of different materials have been used as auxiliary anodes, including high-silicon irons, graphite and lead alloys. A long-life anode material can be produced by coating metals such as titanium with platinum.

12.1.5   Choice of method for cathodic protection

In some situations both methods may be employed. Generally, however, depending on the particular circumstances, either the sacrificial anode or impressed current method is used. There are relative advantages and disadvantages with both methods. These are summarised below.

The impressed current method. The main advantages are that there is better control with this method to provide the performance required, that fewer anodes are required and that the high driving voltage provides efficient protection of larger structures. Against these factors, there are disadvantages compared with the sacrificial anode methods:

(i)   Supervision of operation needs to be at a higher level.
(ii)   A source of power is required.
(iii)   Although capable of good control, over-protection can result with poor control, which may cause problems with coatings and high-strength steels.
(iv)   It is possible to connect the electrical circuit incorrectly so that corrosion rather than protection is induced.
(v)   In some marine situations, physical damage of the anodes may be more likely.

The sacrificial anode method. The main advantages are that initial costs are lower and the installation is comparatively straightforward and usually additional anodes can be added if required. The disadvantages include the following:

(i)   There is a limit on the driving voltage and this is generally lower than with the impressed current method.
(ii)   The available current is determined by the anode area and this may lead to the need for the use of a considerable number of anodes.
(iii)   In some soils, the low conductivity of the environment may be a problem.

12.1.6   Practical applications of cathodic protection

Cathodic protection can be used only where there is an electrolyte of suitable conductivity. The main areas of use are as follows:

(i)   Water-immersion (exterior surfaces): ships, offshore structures, marine installations, submarine pipelines, dry docks, etc.
(ii)   Underground or buried steelwork (exterior surfaces): pipelines, sewers, water distributors, underground tanks, etc.
(iii)   Interior surfaces of tanks, condensers and heat exchangers, etc.

Cathodic protection, if properly applied, should be capable of stopping all forms of corrosion including those arising from pitting, crevices and bacterial activity. Generally, the most economic employment of cathodic protection will be in conjunction with suitable coatings. The method ensures that corrosion does not occur at damaged bare steel areas or at pinholes or pores in the coating, so the costs of operating the system are less than would be the case with uncoated steel.

There is, in fact, one important exception to this general approach. Most offshore structures are not coated in the fully immersed regions and protection is achieved by cathodic protection alone. There is some divergence of view regarding this approach because one large oil company does coat the immersed area of its platforms. Clearly, this is an economic calculation based on the cost of coating as opposed to the costs of operating the cathodic protection system with more anodes. It should be added that the immersed steelwork that is initially uncoated does form a calcareous deposit arising from the action of the cathodic protection and the reaction of salts in seawater.

It is not intended in this book to discuss the design of such systems in any detail but rather to indicate the areas where cathodic protection can be beneficially employed and the factors involved in the choice and design. In broad terms, the following points must be taken into account:

(i)   Total superficial area to be protected.
(ii) Estimate of current requirements.
(iii) Determination of resistivity (conductivity) of the environment.
(iv) Assessment of electrical current requirements for the system.
(v)   Selection of the most suitable method, i.e. impressed current or sacrificial anode method, material and number of anodes required and calculation of the type of anode.
(vi) Cost.

It is not possible to measure the current necessary to achieve cathodic protection because the original anodic and cathodic areas are present on the same steel surface. To overcome this, the potential of the structure is measured by means of a suitable reference electrode. A number of different electrodes are used and when quoting potentials it is necessary to refer to the actual electrode used to measure them. The most commonly used reference electrode is copper/copper sulphate ($Cu/CuSO_4$), especially for soils. Silver/silver chloride ($Ag/AgCl$) is often used for immersed situations; unlike the $Cu/CuSO_4$ electrode it does not require a salt bridge. Calomel (saturated potassium chloride (KCl)) electrodes are generally reserved for laboratory work and for calibration purposes. Zinc can also be used. It is not particularly accurate but is cheap and may provide long-term reliability as a semi-permanent fixed reference electrode.

The potentials at which cathodic protection will be obtained in different environments are based on both theoretical considerations and practical experience. For example, in seawater at 25°C, the potential at which full protection of steel will be obtained with respect to the Ag/AgCl reference electrode is considered to be −0.8 V. In polluted water a more negative potential may be required and in the presence of sulphate-reducing bacteria, −0.9 to −0.95 V with reference to Ag/AgCl is considered necessary. The current requirements can be calculated theoretically if certain assumptions are made. In practice, the current density required will be influenced by a number of factors, e.g. seawater velocity, oxygen concentration, electrical resistivity of the environment, presence of bacteria, etc.

For off-shore structures, the various classification societies, e.g. Lloyds, or national authorities, lay down required standards and, of course, these have to be adhered to. Generally, the design of a suitable cathodic protection system is a specialised matter. The number and positioning of the anodes, the operating controls and determination of suitable monitoring procedures are matters that will determine both the cost and efficient operation of the system.

There are a number of European and corresponding British Standards covering various aspects of cathodic protection.

12.1.7 Coatings and cathodic protection

The combination of coatings and cathodic protection provides maximum corrosion protection for immersed steel structures. However, compatibility between the coating and the cathodic protection is essential.

With the impressed current method, if any part of the steel surface is polarised to potentials more negative than − 1100 mV (Cu/CuSO$_4$) there is a risk of over-protection which may cause disbondment of the coating. This is partly due to the evolution of hydrogen gas and partly due to the build-up of high alkaline deposits.

The extent to which a type of coating can resist disbondment can be evaluated by standard laboratory tests. Most epoxy systems have good resistance to such disbondment.

12.2   Conditioning of the environment

Corrosion has been defined as a chemical or electrochemical reaction between a metal or alloy and its environment. Generally, steel is coated to prevent or reduce reaction with the environment, but in some circumstances it may be preferable to treat the environment to reduce or prevent corrosion. The approach depends on whether the environment is primarily air or an aqueous solution, so the two will be considered separately.

12.2.1   Treatment of the air

Most steelwork is exposed to air but generally there is no way in which the environment can be treated to reduce external corrosion, except in the sense that buildings and structures can be sited in less corrosive parts of a complex. However, where the air is enclosed, e.g. inside a building, or in a 'box structure', it is possible to take action to reduce the corrosiveness of the air. As noted in Chapter 2, the main factors that determine the corrosivity of air are the presence of moisture and pollutants such as sulphur dioxide or contaminants such as chlorides. The simplest method of treating the environment is to remove moisture. Often complete removal of moisture is not required because, by lowering the relative humidity to a suitable level, corrosion becomes inappreciable. Generally, a level of about 50% relative humidity is satisfactory, although in some cases, e.g. in the presence of chlorides, it may be necessary to reduce it below that figure.

Relative humidity is influenced by temperature and often it can be reduced by heating rooms, or air-conditioning equipment can be installed to remove moisture from the air. Sometimes, in fairly air-tight enclosures such as box girders on bridges, the humidity can be controlled by using desiccants such as silica gel or activated alumina. Silica gel is considered to be effective for 2–3 years if used at the rate of $250\,g/m^3$ of void, provided the space is well sealed and manholes are kept closed.[1] There is a change of colour when silica gel has lost its effectiveness and it is necessary to incorporate a system by which this can be checked. The desiccant can be heated to remove moisture and re-used. Where steel sections are completely sealed, a small amount of corrosion may occur because some moisture will be trapped inside the box or tube, but once this has reacted with the steel no further serious corrosion is likely. Corrosion can be avoided by purging air spaces and filling them with inert gases. This method is employed with certain holds on ships.

Air is often treated with desiccants to remove moisture but volatile corrosion inhibitors are also used to retard or prevent the corrosion of steel surfaces in packages. This is a comparatively expensive method and is not used for large steel sections but may be adopted for smaller steel components.

Volatile corrosion inhibitors (VCI) are used for this purpose. The term 'vapour phase inhibitor' (VPI) is also sometimes used but VCI covers all types of this class of inhibitor. The mechanism will not be considered in detail but these inhibitors are chosen to provide different degrees of volatility, which then react in the vapour phase with any moisture in the package to provide an inhibitive solution at the steel surface. The most common inhibitors are dicyclohexyl ammonium nitrite (DCHN) and cyclohexylamine carbonate (CHC). CHC is more volatile than DCHN and more soluble in water. Both types may attack copper alloys and other materials so must be used with some caution on assemblies.

The VCIs are available commercially, usually under proprietary names, in the form of impregnated wrapping papers and as granules in porous sachets. Proprietary products may contain a mixture of the two inhibitors, sometimes with additions of other inhibitors.

12.2.2 Treatment of aqueous solutions

Although steelwork is used in seawater and river water, there is usually no practical way in which the waters can be treated to prevent corrosion. In soils some treatments are possible but generally they rely upon the use of non-corrosive backfills in contact with pipes.

There are methods of treating water to reduce its corrosiveness. These include the removal of oxygen, a method commonly used for boiler waters, and making the solution alkaline. These are not applicable to general structural steelwork, although there may be situations where water is contained within constructional members. For example, with some buildings the fire protection system relies upon water being pumped through tubular members and control of alkalinity may be a suitable way of preventing internal corrosion. A common method of controlling corrosion in aqueous situations is to use inhibitors. These are chemicals added to the water in suitable concentrations which retard either the anodic or cathodic action at the steel surface and, based on this, are commonly called anodic or cathodic inhibitors. Typical examples of the former type are chromates, nitrites and phosphates, all of which can be considered as acting to reinforce oxide films on the steel. Cathodic inhibitors are generally considered to include silicates and polyphosphates but in some cases the inhibitors may have an influence on both anodic and cathodic reactions. Inhibitors are important in the oil industry in both the extraction and refining processes, but these are specialised topics which will not be considered here.

12.3 Alloy steels

The corrosion resistance of steels can be markedly improved by adding other metals to produce alloys. The most resistant of the common steel alloys is stainless steel. It is a good deal more expensive than ordinary steel and, although widely used in process plant, is employed to only a limited extent in structures, mainly for fasteners in particularly aggressive situations and sometimes for bearings. It is more widely used on buildings for cladding, balustrades, doors, etc.

Although there are a number of different groups within the overall classification of stainless steel, the one most commonly used in buildings and structures is *austenitic stainless steel*, so described because of its metallurgical structure. In fact, steels with 12% or more of chromium fall into the category of stainless steels but the common austenitic types contain over 30% of alloying elements, 18–20% chromium, 8–10% nickel and about 3% molybdenum.

The other group of alloy steels that have been used for structures and buildings are much lower in alloy content, only about 2–3%. These are called 'weathering steels', the best known of which is the US Steel Corporation version 'COR TEN', also produced under licence in other countries. Unlike stainless steels they have been used for structural members as well as cladding for buildings.

12.3.1   Stainless steels

These steels owe their corrosion resistance to the formation of a passive surface oxide film, basically Cr_2O_3.

12.3.1.1   Corrosion characteristics of stainless steels

The austenitic stainless steels are virtually uncorroded when freely exposed in most atmospheric environments. The 304 series, without molybdenum additions, may exhibit rust staining arising from slight pitting but the actual loss of steel by corrosion is negligible. The 304 steels are attacked to a greater extent in marine atmospheres because of the presence of chlorides, and this may lead to a rust-stained appearance but again produces little loss of metal. The 315 and 316 steels perform well even in marine atmospheres and often under immersed conditions. However, in some immersed situations corrosion can occur, particularly in stagnant conditions where marine growths can form. Such organisms shield the steel from oxygen so that breakdown of the passive film is not repaired. Any area where the film cannot be repaired is a potential site for pitting. Such situations as overlaps and crevices may provide conditions where pitting may occur. This is not likely to be serious in most atmospheric conditions but may be more severe under immersed situations. Pitting occurs to a much greater extent on stainless steels than on carbon steels. This arises from the presence of the very protective film, which becomes cathodic to any small breaks where local corrosion occurs. In the presence of an electrolyte, the corroding area, i.e. the anodic part of the cell, is in contact with a large cathodic area, which intensifies the local corrosion. Since the passive film is very adherent at the edge of the local anodic area, corrosion tends not to spread sideways but rather to penetrate into the alloy, i.e. to cause pitting. Such pitting can be serious if comparatively thin sheet material is used as a pipe for transporting liquids, because eventually the steel is perforated by the pitting, allowing escape of the liquid. In most situations where stainless steel is used for structures, this is not such a serious problem, but care should be taken with the design of stainless steel fabrications, particularly where they are exposed to chlorides which are the species most likely to cause pitting. Marine situations are obviously affected by chlorides, but attention should also be paid to the effects of de-icing salts when stainless steel is used on bridges.

These steels are often used as components for structures and buildings but they are also used for architectural panels. Generally, 316-type material is employed for this purpose and care must be exercised during construction to ensure that mortars and cements do not come into contact with the panels. Problems of pitting can occur, particularly with chloride-containing concretes, especially if they are allowed to set and are not immediately removed.

In cities and large towns where stainless steel may be used for cladding, the accumulations of dirt, particularly if not exposed to rainfall, can lead to local breakdown of the passive film and it is advantageous to wash the steel down regularly.[2]

12.3.2 Low-alloy weathering steels

In the early 1970s a large number of bridges as well as other structures and buildings were constructed from these steels, the best known of which was called 'CORTEN', mainly in the USA but also in many other countries, including the United Kingdom. Small additions of alloying elements such as copper, nickel, chromium and somewhat higher amounts of silica and phosphorous than in ordinary steels resulted in an alloy content of only 2–3%. This had the effect of reducing the corrosion rate in air compared with that of unalloyed steel. Furthermore, although initially weathering steels rusted in a similar manner to ordinary steels, after a period of some months the rust became darker and more adherent than conventional rust. Considerable test work on small panels throughout the world confirmed the advantages of these steels provided they were freely exposed in air at inland sites. Their performance compared with ordinary steel showed less improvement when exposed close to the sea, and if they were immersed in water or buried in soil their corrosion rate was similar to that of ordinary steel.

In practice there have been disappointments with the use of these steels; their appearance is variable depending upon orientation and the loose powdery rust is a nuisance and can stain adjacent areas.

Although there probably is a place for weathering steels in certain situations, the design of structures and buildings must take into account the corrosion properties of the steels. This has not always been done in a satisfactory way. A paper by Tinklenberg and Culp[3] sums up what is probably a fairly representative view of many bridge authorities in the USA. The authors say:

... in 1977, a comprehensive evaluation of weathering steel was started. This investigation identified a number of problem areas. These included salt contamination, crevice corrosion, pitting, millscale, accumulation of debris, the capillarity of the rust by-products and the potential of corrosion fatigue. When it was determined that these structures had to be painted and that other equal strength steels were available at a lower cost, the initial reasons for selecting weathering steel were no longer valid...

References

1. *Protective Coating of Iron and Steel Structures Against Corrosion*, BS 5493. British Standards Institution, London, 1977, p. 84.
2. Chandler, K. A., *Stainless Steels*. Iron and Steel Institution Publication 117, London, 1969, p. 127.
3. Tinklenberg, G. L. and Culp, J. D., *J. Protective Coatings and Linings*, **1**(1) (1984) 26.

Chapter 13

Coating defects and failures

13.1   Introduction

A coating failure can be considered to have arisen in any situation where the coating system has failed to attain its potential life. Often, the failure cannot be attributed to a single cause but occurs because of a general lack of attention to a number of facets involved in the coating process; typically, selection of the wrong system for the environment, inadequate surface preparation and poor quality control of application. Failures of this type have serious economic implications because a reduction in the anticipated life involves a considerable increase in costs for maintenance. Any organisation faced with a continuing failure to achieve the coating lives that can be reasonably anticipated in a particular set of circumstances would be well advised to examine its own processes for selection, specification and quality control.

When a failure occurs within, for example, a year of application, a serious and costly situation will arise and immediate action is required. The cause must be established if at all possible so that it will not be repeated when the remedial work is carried out. Another reason for the investigation of such a failure arises from the costs involved and the requirement to establish some responsibility for meeting them. The costs of this type of remedial work can sometimes be many times those for the application of the original coating system. Generally, new access scaffolding is required and the conditions for re-coating may be very much worse than those for the original work. It should also be appreciated that remedial work of this nature is unlikely to achieve the original standards possible under the controlled ambient conditions in a shop.

It is important to be aware of the causes of such failures so that they can be avoided in future. They can be categorised in relation to the stage of the coating process at which lack of attention to some requirements eventually led to the problem.

13.2   Surface preparation

Lack of adequate surface preparation is probably the cause, either directly or indirectly, of a majority of coating failures. Failures arise from surface preparation procedures in a number of ways but most of them are related either to adhesion or under rusting of the coating. Some of the common causes of such failures are:

(i)   Use of steel which is excessively laminated or has other surface defects arising from manufacture.

(ii)   Use of steel which has corroded and pitted to a considerable extent prior to blast-cleaning. It is difficult to clean such steel to a satisfactory standard and generally it is not suitable for structures where a high level of protection is required.

(iii)   Incorrect blast-cleaning profile for the particular coating: this is more commonly a problem with high profiles which are not satisfactorily covered by paint at the peaks. However, with very thick coatings, a very low profile may not provide the conditions for sound adhesion, and this may lead to failures.

(iv)   Failures often arise from the surface preparation used to clean welded areas (Figure 13.1). Even where the main steelwork has been blast-cleaned, the welded areas are often only wirebrushed, so pro-

Figure 13.1   Corrosion at weld, largely due to roughness of the weld profile.

viding a poor surface for paint application. Additionally, welds may provide a rough surface equivalent to a coarse profile, again leading to coating breakdown. Premature failure of coatings often occurs on or near bolts because of inadequate protective treatments (Figure 13.2).

(v) After blast-cleaning, the steel must be protected to a degree sufficient to prevent rusting until the full coating system is applied. Generally, a blast primer of some type is used and this is satisfactory for a short period. However, if the steel is exposed outdoors, particularly in an aggressive environment for a long period, corrosion may well occur and salts will form. If this steel is painted without further thorough cleaning, premature breakdown of the coating can be anticipated. This is true, even for the 'surface-tolerant' types of paints and coatings. Even where a full priming coat is applied, problems can occur if there are long periods of storage outdoors. Similar problems can arise where there are long delays in cladding and roofing buildings so that inadequately protected steel, which will eventually be exposed internally, is exposed to exterior conditions for prolonged periods.

(vi) Where steel is to be exposed to aggressive situations, and particularly where it is to be immersed, the standards employed for

Figure 13.2   Breakdown on bolts due to inadequate coating thickness on edges.

blast-cleaning for coatings exposed in milder conditions may not be sufficient to ensure adequate cleanliness, e.g. the presence of soluble chloride salts after blast-cleaning may lead to localised failures.

13.3   Coating materials

If coating materials are defective in some way, then this will clearly lead to problems and proper control procedures are necessary to avoid this situation. However, failures can arise from incorrect use of materials that are basically sound:

(i)   Inadequate mixing of paints, particularly two-pack materials.
(ii) Excessive use of thinners or use of the wrong type of thinners.
(iii) Use of paint after the expiry of the stated shelf life.
(iv) Using two-pack materials beyond their pot life.
(v)   Use of paint not suitable for the ambient conditions experienced, e.g. paints suitable for use in temperate climates may be unsuitable in very hot climates.
(vi) Selection of a coating material not suitable for the environmental conditions of exposure, e.g. oleo-resinous paints for steelwork that is cathodically protected.

13.4   Coating application

The application techniques for all coating materials have a significant effect on performance. Often poor application procedures result in a reduced overall life but these can also be the cause of early failures. Some typical causes are listed below:

(i)   Operators with insufficient experience of a particular type of coating material may produce what appears to be a sound coating. However, it may be defective and cause problems later. Typically this occurs with fast-drying paints if the spray gun is too far from the surface. This means that particles of paint will have time to dry out and not be wet enough to flow into a continuous film. This is called 'dry spray'. Coating over such a porous, powdery layer with a wet film of paint can make this difficult to detect from the finished film. Nevertheless, it remains an inferior product.
(ii) Most defects with coatings with a solids content of 95% or greater occur during application. High-solids, multi-component coatings based on epoxies or urethanes are available in a range of viscosities. Those that can be applied by normal brush or spray methods can be

improperly mixed so that air is introduced and this can cause pores, pinholes, etc. in the cured film. Excessive mixing of such materials can also create so much heat that the pot life is shortened considerably. Some high-solids materials can only be applied by hot-airless spray equipment, or through dual component feedlines (see Chapter 5). If the heating is incorrect the coating will neither mix nor flow correctly and the applied film remains tacky and soft. Similarly, if the ratio of base resin to curing agents is not correct, soft patches, due to incorrect stoichiometric mix ratio, will result.

(iii) The condition of the equipment is equally important. If it is poorly maintained or of the wrong type, although the coating material will be applied apparently satisfactorily, it will not necessarily cure or dry to provide a suitable protective film. Poor intercoat adhesion or solvent entrapment may well occur, leading to early failures.

(iv) Poor application procedures can result in excessive discontinuities in the film, e.g. pinholes and 'vacuoles' (see Figure 13.3). These may not be apparent from a visual examination but can lead to rusting and coating breakdown, particularly under immersed conditions.

(v) Unsuitable ambient conditions, whether of temperature or relative humidity, may cause problems leading to lack of cure of two-pack

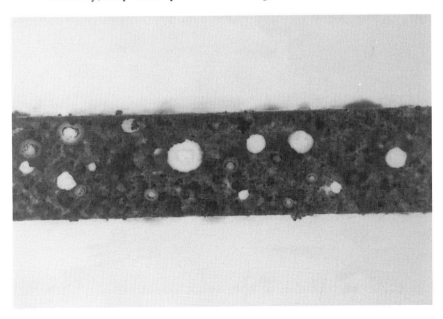

Figure 13.3   Cross-section of paint film showing presence of 'vacuoles'.

materials or adhesion problems arising from moisture on the surface.

(vi) Some paint coatings must be overcoated within specified time limits. Failure to observe these requirements can lead to poor intercoat adhesion.

(vii) Generally, the application of an inadequate thickness of paint will result in a reduction of the coating life. However, with some paints excessive thickness of coating can lead to problems such as entrapment of solvent, producing discontinuities in the film.

(viii) Application of water-based coatings in conditions of high humidity so that the coating remains vulnerable to damage by rainfall for an excessively long period (see Figure 13.4).

Many paint failures arise from poor application which affects adhesion and the film properties. These defects are not always immediately apparent, so careful control of application processes is essential.

Materials other than paint can fail to protect steel adequately if they are not properly applied. For example, the manufacturers' instructions must be followed when applying tapes to ensure that the steelwork is correctly wrapped. Corrosion can occur under badly applied tapes and may not be evident for a considerable time. Sprayed metal coatings

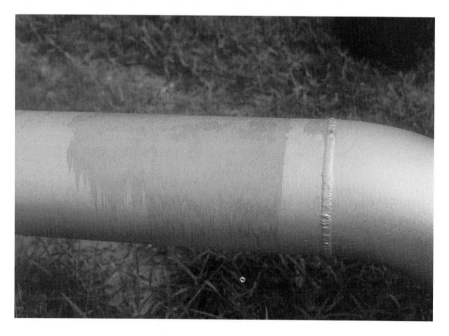

Figure 13.4 Water-based coating damaged by rain 36 hours after application in high humidity conditions.

are also subject to poor application arising from the nature of the equipment and the skill of the operator. Factory-applied processes, e.g. electroplating, hot-dip galvanising and plastic coatings, all require careful control, which is usually in the hands of the companies producing the coatings. Failures can, however, occur on all these materials if they are not properly applied.

13.5   Transport and storage

The use of correct handling techniques will reduce damage during transport and storage to a minimum, but it is inevitable that some damage will occur. It should be noted that excessive damage to a coating is not necessarily a sign of excessively rough handling but may signal that the coating as a whole is soft and has poor adhesion to the substrate, due, for example, to entrapped solvent.

Although mechanical damage to a coating during fabrication, handling, transport and storage is strictly a failure, it is usually obvious and is repaired. However, despite repainting, these damaged areas can be a source of weakness in the system, particularly if rusting has occurred and inadequate cleaning methods are employed before repainting. Mechanical damage can be particularly troublesome on factory-applied coatings because it is generally not possible to repair the coating to the original standard of that applied in the works. Special kits of repair materials are often available for plastic coatings and these are usually superior to touch-up with paints.

13.6   Types of coating defects

As already observed, there may be a reduction in life without any obvious visual defects in the coating, and a careful laboratory investigation may be required to determine the cause of failure. In many cases, however, visible effects are manifested. In the UK, a coloured, visual guide to coatings and application defects has been published.[1] Some of the more common types will be considered below.

13.6.1   Adhesion loss (flaking, peeling, etc.)

Loss of adhesion is generally visible as the lifting of the paint from the underlying surface in the form of flakes or scales. If the cohesive strength of the film is strong, the coating may form large shallow blisters. Excessive mechanical damage during handling can often be a symptom of poor adhesion. The following points are the main causes of loss of adhesion:

(i)　Loose, friable or powdery active or inert materials on the surface before painting, i.e. rust, dirt, zinc salts (Figure 13.5), dry spray, millscale, chalking.

(ii)　Contamination, preventing the paint from 'wetting' the surface, i.e. oil, grease, silicones, amine bloom, plasticiser migration.

(iii)　Surface too smooth to provide mechanical bonding, i.e. galvanising, aged polyurethane or epoxy coatings (Figure 13.6), too shallow blast profile.

(iv)　Application of catalysed materials in too advanced a state of their curing, i.e. in excess of pot life.

13.6.2　Bacterial or fungal attack

Bacteria or fungi can thrive on a dirt layer on the surface of a paint film without detriment to its protective properties. In some cases, however, bacteria can use certain components of a coating as their food source. The result can be a rapid failure. Munter[2] quotes the case of the use of coal-tar epoxies under sewage conditions. Coal-tar epoxies with amine-type curing agents have excellent resistance but coal-tar epoxies with polyamide curing agents are rapidly attacked and fail quickly.

Figure 13.5　Loss of adhesion due to painting over zinc corrosion products.

Figure 13.6   Adhesion failure.

13.6.3 *Bleeding*

Bleeding is the brown staining of a coloured topcoat due to the migration of bituminous material from an undercoat such as a coal-tar epoxy, asphalt, etc.

13.6.4 *Blistering*

There are two forms of blistering; one arises within the coating itself (Figure 13.7) and the other is caused by corrosion of the substrate (Figure 13.8). Blistering within coatings is generally caused either by solvents which are trapped within or under the paint film, or by water which is drawn through the paint film by the osmotic forces exerted by hygroscopic salts at the paint–substrate interface. The gas or the liquid then exerts a pressure and, if it becomes greater than the cohesive strength of the paint film, the blisters break.

Solvent entrapment can arise from 'skin curing' of the top layer of the coating. Overcoating or immersing the coating before the solvent has had an opportunity to evaporate also causes entrapment within the coating. The measurement or evaluation of the degree and size of blistering can be carried out by the use of standard photographs or diagrams, such as those published by ASTM, or BS 3900 Part H1 (ISO 4628/1).

Figure 13.7 Blister from within the coating film.

Figure 13.8 Blister from corrosion of the substrate.

Coatings applied to steel generally fail eventually by disruption of the paint film due to the large volume of corrosion products accumulating at the coating–metallic interface. If the thickness of the paint coating is initially inadequate then the corrosion will be of an overall nature, but when corrosion arises from water and aggressive ions being drawn through the film by the osmotic action of soluble iron corrosion products, then the attack will start from corrosion pits. This can be overall or confined to isolated areas. Corrosion does not automatically spread under paint films if the adhesion is good.

The blistering of paint films on steel in seawater involves the operation of corrosion cells on the metal surface. Iron dissolves at anodic areas and hydrogen evolves at the cathodic areas, leaving an accumulation of sodium hydroxide in the cathodic blisters. Anodic blisters that develop before the appearance of rust are small and filled with acid liquid; they readily fracture and become the seat of anodic pits. Paints based on linseed and other drying oils are attacked by alkalis and, therefore, such paints are especially prone to blistering under immersed conditions.

13.6.5   Blooming or blushing

Blooming (often called blushing in the USA, although in the UK this term is confined to this phenomenon on lacquers that cure by evaporation, e.g. nitrocellulose) is a deposit like the bloom on a grape which sometimes forms on glossy finishes, causing loss of gloss and dulling of colour. It is generally caused by high humidities or condensation during the curing period. This occurs with amine-cured epoxies, in particular when they are applied at relative humidities in excess of 85%. It is not generally considered detrimental to the coating but can reduce the adhesion of subsequent overcoats. It can normally be removed by wiping the surface with a rag saturated in the appropriate solvent.

13.6.6   Chalking

Chalking is the formation of a friable, powdery coating on the surface of a paint film caused by disintegration of the binder due to the effect of weathering, particularly exposure to the actinic (photochemically active) rays of the sun and condensation from dew.

Different binders react at different rates; for example, epoxies react quite quickly whereas acrylics and polyurethanes can remain unchanged for long periods. Chalking of epoxies, however, is generally considered to be a surface phenomenon only and therefore not harmful, except to the decorative appearance. Oil-based paints, however, are thin enough to be eroded away so that eventually the undercoat is 'grinning' through. In all cases it is generally considered the most acceptable form of failure since the surface preparation for subsequent maintenance consists only of removing loose powdery material and it is generally not necessary to remove the old coatings completely.

13.6.7   Cissing, crawling and fisheyeing

Cissing, crawling and fisheyeing are all forms of surface defect where the paint has not 'wetted' the surface correctly and the wet paint recedes from small areas, leaving either holes of various shapes and sizes or attenuated films. The cause is generally oil, grease or silicones on the surface to be painted, or water or oil in the paint during application. A typical source of such contamination is the air supply to the spray gun. These defects can also be caused by the use of the wrong solvent.

13.6.8   Cobwebbing

Cobwebbing is the formation of thin, spider-web-like strings during the spraying of fast-drying, solvent-evaporation type coatings, such as chlorinated-rubber or vinyls. It is caused by the too rapid evaporation of the solvent, such as when the application is carried out under high ambient temperatures.

13.6.9   Cracking

Cracking may be visible or require the use of some form of magnifier to detect it. There are various types of cracking and a range of terminology is employed to describe them.

Hair cracking. Fine cracks which do not penetrate the top coat; they occur erratically and at random.
Checking. Fine cracks which do not penetrate the top coat and are distributed over the surface giving the semblance of a small pattern.
Cracking. Specifically a breakdown in which the cracks penetrate at least one coat and which may be expected to result ultimately in complete failure.
Crazing. Resembling checking but the cracks are deeper and broader.
Crocodiling or alligatoring. A drastic type of crazing producing a pattern resembling the hide of a crocodile.

In addition, there is a special form of cracking called 'mud cracking' which on a small scale resembles the appearance of dried-out mud and is an effect specific to overthick applications of zinc silicates (Figure 13.9).
All paint films, but especially coatings containing solvents, are subjected

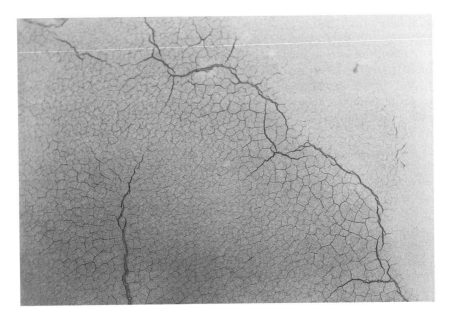

Figure 13.9   'Mud cracking' of over-thick zinc silicate primer.

to an internal stress which may have the effect of reducing their tolerance to external stresses. If a coating cracks spontaneously, this is because the internal stress has increased to a value greater than the tensile strength. Internal stresses are dependent upon plasticisation, pigmentation, ageing and the conditions under which the coating was applied and cured.

In some instances, e.g. some anti-fouling paints, water can act as a plasticiser and the films remain flexible under immersed conditions but crack when dried out. In other cases, e.g. oil-based paints, water and ageing can gradually leach out the plasticising elements and the film becomes brittle with age. In yet other cases, e.g. chlorinated-rubber, the plasticiser elements migrate under the influence of heat. Paint films can be sufficiently flexible for most purposes but not when applied over a softer or more flexible undercoating; for example, alkyd paints, which are applied over bitumen coatings generally crack in the particular form known as crocodiling.

13.6.10   Complete or partial failure to cure of two-part materials

Catalysed materials such as two-pack epoxies and polyurethanes will sometimes dry without the addition of the catalyst. However, such a film is incorrectly cured and will not give the service intended. The paint film is generally softer than the fully cured coating and may have a tendency to sag.

Even if the two materials are added together, they must be adequately mixed and in the correct proportions, or again the polymer formed will not be the correct one and will be less durable. Also, there might be a tendency for the mixed coating to have a shorter pot life than expected and it may even set up in the pot or in the spray lines. ('Set up' is the term used for conversion of a liquid paint to a gel-like consistency, or for thickening, i.e. increase in viscosity.) With the exception of the isocyanate-cured epoxies, the two-pack epoxies will not cure adequately at ambient temperatures below 5°C.

13.6.11   Dry spray

This occurs when, during the spraying of paint, the particles hitting the surface are insufficiently fluid to flow together to form a uniform coating; the result is a powdery layer. The defect is generally caused when fast-drying materials such as zinc silicates or two-pack epoxies are sprayed with the gun too far away from the surface, or with very strong cross-currents of wind.

13.6.12   Fading

Changes in colour of paints are mainly due to ultraviolet degradation of the coloured pigments or dyes. Selective deterioration of one pigment can cause a complete colour change, for example, from green to blue, but generally the effect is of the brighter colours fading and becoming dull with time.

13.6.13   Lifting or pulling up

This is the softening and expansion of a previously applied paint when a new paint with strong solvents is applied over it. The effect is often similar to that obtained with 'paint removers'. It can be prevented by careful choice of overcoating paints to ensure that their solvents are compatible, or in some instances by allowing the previous coat to dry and harden to an acceptable degree.

13.6.14   Orange peel

Incorrect spray application, such as with the gun too close to the surface, or with air pressure too low for proper atomisation, can produce a surface effect that causes a paint film to resemble the skin of an orange. Before the coating has 'set' it is sometimes possible to brush out the excess paint, otherwise it generally means waiting until the coating has cured, sanding down, and applying a further coat.

13.6.15 Pinholes and holidays

Pinholes are minute holes formed in a paint film during application and drying. They are often caused by air or gas bubbles which burst forming small craters in the wet paint film which fail to flow out before the paint has set. Some confusion arises from the term 'holiday', which is sometimes defined as skipped or missed areas left uncoated with paint, but is often used to indicate pinholes.

13.6.16 Pinpoint rusting

Pinpoint rusting, consisting of small isolated spots of rust often forming a directional pattern associated with the blast-cleaning or paint spray operation, is caused by the high peaks of the blast-cleaned surface not being adequately covered by the paint film. If it is caught at an early enough stage the remedy can often be no more than sanding down and application of more paint.

13.6.17 Runs and sags

A run is a downward movement of a paint film over an otherwise flat surface, which is often caused by the collection of excess quantities of paint at irregularities in the surface, e.g. cracks, small holes, etc. The excess material continues to flow after the surrounding surface has set. It is unwise to consider such defects as affecting only the appearance and therefore being unimportant for industrial applications since, due to surface tension effects, the perimeters of such areas can be accompanied by pinholes or holidays.

Runs and sags in a paint film application can be symptomatic of more than just over-thick application or excess thinners. With two-pack catalysed materials it can also indicate failure to add the catalyst correctly or the use of a paint beyond its shelf life.

13.6.18 Saponification

In painting practice this refers to the decomposition of an oil-containing binder by reaction with alkali and water. The effect occurs particularly on new concrete, or with immersed coatings used in conjunction with cathodic protection.

13.6.19   Skin curing

Skin curing is the term used to describe the situation in which the top surface is cured faster than the body of a coating. This normally results in a soft cheesy type film. The effect can be produced by curing of paint in direct strong sunlight or by forced heating of the top surface, particularly with thick films which cure by chemical reaction, e.g. two-pack epoxies. However, it can also occur to a different extent with films that cure by oxidation or by evaporation. The effect may be particularly pronounced on heavy steel sections where the substrate may be at a significantly lower temperature than the top of the paint surface. The effect is further accelerated when the paint surface is black.

To prevent skin curing it is advisable to avoid direct heating of two-pack epoxy and polyurethane surfaces and also painting and curing in very strong sunlight.

13.6.20   Spot-blast boundary breakdown

During maintenance painting, when spot blast-cleaning is carried out, it is necessary to 'feather back' the adjacent, and apparently sound, adjacent paint surface. However, this partial blasting operation often damages the old paint more than is obvious to the eye and, in particular, weakens its adhesion. Consequently, on exposure the paint film is penetrated by moisture, etc., at this point and eventually fails by a particularly localised form of corrosion blistering.

Figure 13.10 shows a typical example where a spot-blasted weld plus new paint and the surrounding, untouched, old paint are in good condition, but corrosion blistering has occurred in a clearly defined area, where the new and old paint have overlapped.

13.6.21   Thickness faults

Edges of steel sections are not only generally subject to more mechanical damage than other areas, but will also generally have thinner paint films. The attenuation at an edge depends on its radius and the type of coating, but reductions of 60% are not uncommon. Nuts and bolts and threaded sections also provide many edges and corners that are difficult to coat adequately.

Weld areas are often the first part of a structure to show failure. This may be due to contamination from welding rods or from soaps used for weld tests, but is generally due to inadequate preparation in the form of lack of removal of sharp edges, undercuts, etc., which then protrude into the paint film thickness. Blast-cleaning alone is seldom adequate for such areas.

Figure 13.10   Corrosion blistering at overlap between old and new paint at a spot-blasted area.

13.6.22   Uneven gloss

A patchy appearance to a glossy finish can be due to inadequate sealing of a porous substrate, moisture in the film, changes in temperature during application or paint applied over an adequately cured coating.

13.6.23   Undercutting

Undercutting is corrosion that penetrates laterally under a paint film. It can start from a break in the film or from an edge. It may also arise from corrosion products that have not been removed adequately before coating. Corrosion occurs rapidly under such conditions because the rust retains moisture and, coupled with any soluble iron corrosion salts, forms an ideal corrosion cell.

The measure of a good protective coating is its ability to withstand or minimise undercutting. This is almost always related to its adhesion properties. The greater the adhesion, the less tendency for the coating to be undercut.

13.6.24 Wrinkling

Wrinkling of a paint film as it dries is characteristic of oil-based paints applied in too thick a film. In these cases the surface of the film cures by oxidation from the atmosphere but the lower layers of the paint take considerably longer to cure.

References

1. *Fitz's Atlas of Coating Defects*. Pub. MPI Group.
2. Munger, C. G., *Corrosion Prevention by Protective Coatings*. NACE, Houston, p. 354.

Chapter 14

The selection of coating systems

14.1 Introduction

The selection of the most suitable system of corrosion protection for a specific situation is not necessarily a simple matter. Many factors may be involved in the selection process and the final choice may well be a compromise. In particular the following information must be considered:

(i)   The conditions that have to be resisted by the chosen system, e.g. immersed sea conditions.
(ii) The ability of the various systems to resist such conditions.
(iii) For coatings able to provide the necessary protection, the nature of other problems that have to be considered to provide an economic solution.

When a suitable system has been selected, engineers frequently ask 'How long will it last?' This is clearly an important question. After all, if an alloy is selected for a particular purpose, you would expect to be provided with information on, say, its tensile strength. Information on durability of coatings is not so readily available. Although some types of coating would be expected to last longer than others in a particular set of circumstances, the coating material is only one factor in determining the life of a protective system. It is therefore worth summarising some of the factors that influence the durability of coatings:

(i)   Standard of surface preparation of the steel.
(ii) Application procedures.
(iii) Thickness of coating.
(iv) Design of structure.
(v)   Local variations in the environment.

If all other requirements, such as surface preparation, are met, then clearly the coating material will play a major role in determining durability.

However, without testing a particular material in the specific situation where it is to be exposed, only a broad view of the durability can be expressed. For example, a generic type of coating supplied by one company will not necessarily perform in the same way as that from another. Consequently, actual 'lives' cannot be predicted. Generally, a broad indication is the best information likely to be offered.

There is general agreement on many aspects of steel protection, e.g. the importance of a high standard of surface preparation. However, there are often wide divergences of opinion when it comes to selecting protective systems for specific situations. Even where considerable attention has been given to the matter of choosing the most suitable system, e.g. on offshore structures, there is rarely agreement on the overall virtues of one particular coating system. To some extent this is understandable because specifiers have had varying experiences with coating systems, and the high level of salesmanship of paint companies and other coating suppliers must also be taken into account. Generally, the system will be chosen from one of the following groups:

(i)   Paint coatings.
(ii) Metallic coatings.
(iii) High-duty or special coatings such as neoprene.

Paint coatings are usually, but not always, the cheapest initially. However, taking into account the aggressiveness of the conditions to which the steelwork will be exposed, and maintenance, the most economic may be within any of the above groups or even a combination of different types. Under immersed or buried conditions, cathodic protection may be used, with or without coatings. For other situations, alloys may be selected and methods of corrosion control such as inhibition may be preferred to coatings. However, this chapter is concerned only with the selection of coating systems.

Many factors will be taken into account when determining the most suitable coating system for a particular situation. Sometimes a single requirement may determine the type of system. Usually, however, the various factors tend to result in a compromise choice depending on their relative importance and the views of the specifier, which may well be determined by personal experiences, good or bad, of a coating system.

A limited number of well-recognised factors will generally determine the broad choice of coating. Then other factors will be considered in the final selection.

14.2   Factors influencing the selection of coating systems

(i) The environmental conditions are clearly of fundamental importance. For aggressive situations only highly resistant coatings will be considered, whereas for mild conditions less resistant and usually cheaper coatings may be preferred, not only because the coating materials are cheaper but because, generally, other parts of the coating process, e.g. application, are also more straightforward. (Environmental conditions are considered in detail in Section 14.3.)

(ii) Access for maintenance is clearly an important factor, particularly if there is only a limited period during which maintenance painting can be carried out, e.g. on an offshore structure. If access is both difficult and expensive, this will obviously influence the nature of the coating system to be selected. On the other hand, if access is straightforward, although it may still be desirable to choose a high-quality system, it will not be an economic necessity from the standpoint of maintenance costs.

(iii) Apart from the problem of access, the requirements for maintenance will vary and, depending on the pattern, it may be preferable to apply less resistant coatings which have advantages, e.g. in application, with repainting at regular intervals to ensure a sound protective film. In other situations there will be a preference for long periods between maintenance.

(iv) The importance of the structure will clearly influence the choice of coating system. Structures are not all equally important with respect to problems arising from maintenance, planned or unplanned. The financial losses ensuing from the requirement to take a structure wholly or partly out of service will vary considerably. Structures considered important in the sense that problems will be costly to rectify or will produce unwelcome publicity will generally be coated to a higher standard than others.

(v) The facilities to apply the selected coating must be available within the time-requirement of the project. Where the coating system is considered at an early stage, then the required arrangements for coating application can usually be organised, though sometimes at a considerable cost. However, the choice of, e.g., a galvanised coating would be influenced by the knowledge that the nearest plant was 500 miles from the fabrication shop. Again, there is a limit to the size of section that can be galvanised and this would have to be taken into account in the design of some structures. Although galvanising has been taken as an example, the situation can arise with all shop-applied coatings and many other coatings which require specialist application.

(vi) The choice of the coating may well be influenced by the problems of transportation to a site some distance from where the coating is applied. The resistance of the coating to mechanical damage and the ease and standard of repair of such areas will be important factors in the selection process.

(vii) When cathodic protection is employed, then the system chosen must be of the non-saponifiable type.

(viii) The ability to carry out maintenance without undue problems is important and it may be preferable to choose coatings that will not require special operations, such as blast-cleaning, to ensure good intercoat adhesion when repainting.

(ix) Ease of application both initially and at maintenance may influence the choice of coating system, particularly if the application is to be carried out at a site where there is a shortage of skilled operators.

(x) Cost is always important and usually plays an important role in the selection of coatings (see Section 14.5).

(xi) The choice of the coating may be determined by its ability to resist certain chemicals or solvents, e.g. for tank linings.

Many other factors can be taken into account. Some, although not important technically, may nevertheless affect the approach to selection, so all of them must be considered at an early stage. Typically, if the requirement is for a glossy blue finish, a range of materials can be discounted. Even if such a material can be applied as a finishing coat to a system of more durability but less pleasing appearance, overall maintenance requirements will be affected by the need to retain the cosmetic appearance.

Experience is clearly an important element to be taken into account when selecting systems, provided it actually relates to the situation in hand. Differences in climatic conditions, particularly temperature and degree of wetness, may have a considerable effect on the performance of a coating. With metal coatings the life may be directly related to the conditions of exposure and these must be known in some detail before relating the probable performance at one site to that obtained at another.

14.3   Selection of coatings for specific environments

Environments can be categorised in different ways but the broad classifications in Table 14.1 are commonly used. Sometimes descriptive terms such as 'Rural' are used and these have been included. These categories can be deceptive, so-called rural areas can be in a wet, marshy environment or windward of pollution from industry, not necessarily near-by. The time that a surface stays wet is an important factor. In Europe for example, surfaces facing north will stay wetter longer than those facing south. This particularly affects metals, such as iron and steel substrates and metallic coatings such as zinc and aluminium, but also plays a part in the deterioration of paint coatings. A particularly difficult environment is where there is

Table 14.1 Types of environment

Type	Description
Atmospheric exterior	
Slight pollution inland (rural)	Areas without heavy industry with low pollution levels
Polluted inland (semi-industrial)	A reasonable level of pollution arising from industrial processes sited some distance away. Suburban areas of large cities
Highly polluted inland (industrial)	High pollution near certain industrial processes, airborne chemical particles and sulphur dioxide
Chemical pollution inland (chemical)	Near specific chemical process plants with severe pollution by specific corrosive species
Coastal non-polluted (marine)	Near the coast away from industry but with chloride contamination
Coastal polluted (industrial–marine)	As above but with additional pollution from industrial or chemical processes
Atmospheric interior	
Very warm	Virtually no condensation, typically inside a modern office building
Dry, some condensation	Not always heated, so some condensation. Typically general storage building
Prolonged condensation	This covers a range of environments inside buildings where industrial processes take place. May be very corrosive to steel if chemical fumes persist. Typically a pickling shop
Immersed	Although seawater is generally more corrosive than river water, from the coating selection viewpoint they can be considered as one environment
Splash zone between low and high tides	Particularly corrosive to steel. More problems than with fully immersed because cathodic protection not effective
Soil burial	A range of soils of different corrosivity but generally high-duty coatings and/or cathodic protection used

a corrosive atmosphere, such as salt spray onto surfaces that are not regularly washed by rainfall. The soffit of a motorway bridge where the road is salted in winter is a typical example. Even interior surfaces are not exempt. ISO 12944 'Corrosion Protection of Steel Structures by Protective Paint Systems, Part 2 Classification of Environments', gives the corrosion

Table 14.2   Paint performance and steel corrosion in different environments

Type of environment	Sheffield	Calshot
	Exterior/polluted	Coastal/non-polluted
Corrosion rate of bare steel (μm/year)	109	28
Life to first maintenance of oleo-resinous paint system (years)	6	6

Data from Sixth Report of the Corrosion Committee, Iron and Steel Institute, Special Report No. 66, London, 1959.

rate of steel in thickness loss in microns per year as 50 for both the interior of rooms with high humidity and for an urban exterior.

When considering the aggressivity it should be appreciated that the environments do not necessarily have the same effect on bare steel and coatings (see Table 14.2). This is old data from the time when industry in Sheffield made it one of the most corrosive areas in the UK. Now in common with most industrial areas in Europe, acid pollution in the atmosphere is considerably reduced, but the implication of Table 14.2 remains relevant.

Environments that would not be considered particularly corrosive to steel may cause early breakdown of some coatings. Even the colour of the paint may influence its performance in some environments. For example, a problem with skin curing of black coal tar epoxy on piling being installed in the Middle East was alleviated by application of a temporary coating of white emulsion, thus reducing surface temperatures by 5°C. Ultraviolet rays from the sun have no significant effect on steel corrosion but can cause breakdown of organic coatings.

The corrosiveness of an environment to steel is important because, if the coating breaks down to allow rusting of steel, the rate will influence the course of further coating breakdown.

14.4   Types of coatings

Hundreds of different coating materials are available and these can be incorporated into an almost indefinite number of coating systems. However, the various coatings fall within fairly easily defined groups. Although there are many differences within the groups, there are even greater differences between groups. Consequently the first stage of selection is to choose the types of coating that are likely to be suitable in a particular set of circumstances.

To assist in the selection process, the various coatings are discussed in detail in Chapter 4 for paint coatings, Chapter 6 for high-duty coatings and Chapter 7 for metallic coatings.

14.5   Costs of protective systems

Some form of economic assessment is required to provide a basis for the selection of protective systems. The initial cost is important but proper comparisons can be made only by taking into account the total costs over the life of a structure, i.e. all the maintenace costs added to the initial cost. In practice, comparative protective coating costs are meaningful only over a period of about 30 years, so for structures with long design lives, e.g. highway bridges, there is little point in assessing costs over a period of 100 years.

To achieve genuine comparisons, the costs are usually based on Net Present Value (NPV). The calculation is based on the standard formula for compound interest

$$\text{present value (PV)} = \frac{\text{future value (FV)}}{(1+i)^n}$$

where i is the interest rate expressed as a decimal, e.g. $15\% = 0.15$, and n is the number of years. This formula is the basis of 'discounting', i.e. future value can be discounted to present value. If the interest rate is, say, 10% then if a sum of £1000 is required in 3 years' time, the present value would be:

$$\frac{1000}{1.1^3} \quad \text{or about £750}$$

In other words, if £750 is invested now at 10% it will provide the £1000 required for maintenance in 3 years' time.

The calculation can be carried out for the lifetime of the structure by adding all the maintenance period costs together. If, additionally, the initial cost of protection is included, then the formula for calculating the total cash required for protection, including investing for maintenance, i.e. net present value (NPV), is:

$$\text{NPV} = C + \frac{M_1}{(1+i)^{n1}} + \frac{M_2}{(1+i)^{n2}} + \dots$$

where C = initial cost, M = maintenance cost in year n_1, n = number of years to each maintenance period and i = interest rate.

In practice it is difficult to make accurate assessments of NPV because of the variables such as inflation and interest rates, which cannot be forecast with any certainty. On the technical side, it is not always an easy matter to predict the time period between maintenance paintings. Many other factors can also make long-term forecasting difficult. These include the following:

(i) Over the lifetime of a structure, changes may be made that influence the costs of maintenance. Further services may be required, parts may be added or alterations may be made, and these may involve increased costs arising from access difficulties.

(ii) The actual use for which the structure was designed may change, so altering the maintenance requirements.

(iii) The environment may alter for a variety of reasons, not least the construction of other plants in the vicinity, which may increase the overall aggressivity of the conditions that the coating has to withstand. Furthermore, this may increase overall costs because of the requirements to prevent damage to other structures adjacent to the one being repainted, e.g. from blast-cleaning operations.

(iv) The design life of a structure is based on certain long-term assumptions which may change with time. This may involve the closure of plants, making a structure obsolete; alternatively, the life may be increased well beyond the initial intention.

(v) Changes in the nature of the area in which a structure is sited may lead to requirements for a higher standard of appearance, so leading to increased maintenance for purely aesthetic reasons.

(vi) During the lifetime of a structure, new techniques for surface preparation and coating application may be developed. These, and the introduction of new coating materials, may alter the approach to maintenance and result in a complete re-assessment of the protective requirements.

The overall economic assessment may be outside the control of the specialist concerned with steel protection because accountants will decide the requirements for the return on investment. Their techniques will be determined by, and their decisions will be made on, facts and data not necessarily available to the coatings specialist or engineer concerned with the project. Methods such as the 'internal rate of return' may be used to determine the advantages to be gained on different levels of investment. It may be decided that little financial benefit will accrue from a higher initial cost of protection despite the longer-term benefits of lower maintenance requirements.

The engineer or coating specialists should offer clear alternatives with a reasonable assessment of costs and, at least for the initial protection, these can be determined with some accuracy.

14.5.1   Calculating the costs of alternative protective systems

The basic economic assessment concerns the overall comparison of different systems for protecting a structure. The number of systems suitable for a particular situation will vary. In a rural atmosphere almost every possible system could be considered, whereas for the immersed part of a jetty to be cathodically protected only a limited number of types of system would be technically feasible. In general the cost of coating materials is a small part of the overall cost and in the majority of cases it pays to use the best materials available.

In calculating costs, technical and 'common sense' aspects should not be ignored. The costs that are relevant are those quoted, not general costs provided by a manufacturer or supplier at a time when he is not directly involved in the project.

The main costs to be considered are for:

(i)   Surface preparation
(ii)   Coating application
(iii)   Coating material
(iv)   Overheads

If a paint system is taken as an example, then (i)–(iii) above can be established reasonably easily. The overheads cover tools and equipment, supervision, administration, inspection, scaffolding, delay costs and disruption costs. The delay and disruption costs may be greater for one system than another and in some cases may be the fundamental factor in choosing a system.

The actual costs will be obtained directly from contractors where work is offered to tender or where particular firms are invited to offer quotations for the protective system. In such cases, the overhead costs such as tools, equipment and supervision will be included in the overall price. As the cost will be based on a specification, this must be prepared so that disputes are kept to a minimum, otherwise costs may well escalate. Where the work is being carried out by direct labour, all the overhead costs must be included in the calculations. Each of the main factors will be considered briefly.

14.5.1.1   Surface preparation

The cost of surface preparation will largely depend upon the condition of the surface. Table 14.3 shows an example of the amount of abrasive used on different surface conditions to achieve particular visual standards of cleanliness. More importantly from a cost point of view this also relates to time spent. It must also be remembered that this cleanliness is only to visual standards. For initially rusted and pitted surfaces, there is a need for chemical cleanliness and this will probably add extra cost.

The cost of manual cleaning has not been included because in practice it is so variable, depending upon the thoroughness of the application and the result achieved. In the literature many surveys claim that the cost is about half that for blast-cleaning. Others, particularly in the USA, claim that to obtain the highest standard of visual cleanliness with power tools is about the same cost as achieving SA2$\frac{1}{2}$ by abrasive cleaning.

Table 14.3   Surface preparation related to initial condition of steel

Kilogrammes of abrasive used per 10 m^2 by dry blasting

Initial condition	Visual standard of cleanliness		
	Sa3	Sa2$\frac{1}{2}$	Sa2
Loose millscale or light rust or no pitting	4.06	3.85	2.03
Tight millscale or general rust or light pitting	5.62	5.33	2.79
Painted surface or heavy rust or moderate pitting	7.27	6.89	3.64
Thick paint or rust nodules or heavy pitting	10.15	9.64	5.07

14.5.1.2   Coating application

In broad terms, the labour costs for application are in the ratio of about 1:2:3:4 for airless spray (the cheapest), air spray, roller and brush. On the other hand, paint 'losses' for the methods are about 30% for air spray and 5% for roller and brush, so additional paint costs must be balanced against labour costs. Further additional costs may arise with spraying, e.g. for masking and protection of adjacent steelwork.

Paint and sprayed metal coatings are usually costed by area, whereas hot-dip galvanising is costed by weight of steel. The type of paint influences the application costs. Generally, alkyd paints would cost 15–20% less to apply than the two-pack types of paint.

14.5.1.3   Coating materials

The cost of metal coatings can be calculated fairly easily, although hot-dipped galvanised coatings are usually specified as weight per unit area and sprayed coatings by thickness. For paint coatings, prices should not be compared by the price per unit volume of wet paint because of the variations that occur in solids content. *The factor of importance is the dry film thickness/unit area covered.* Paints generally have a solids content of 40–50% (vol/vol) but this can be much lower with some resins; on the other hand, with solventless paints the solids are virtually 100%.

Covering powers are quoted by paint manufacturers and these usually include an allowance for losses during application. Such losses arise in a number of ways, e.g. by overspraying and paint remaining in the can or application equipment at the end of the day. The losses depend on the efficiency of the applicator but some are inevitable. Furthermore, it is not possible to achieve the exact thickness of paint coating specified, so it will, in practice, be slightly thicker, hence requiring additional paint. Rough surfaces also require additional paint to give adequate coverage.

14.5.1.4 Overheads

Overheads may cover a considerable percentage of the total cost of painting. They will, of course, be included in a tender for a job. They are important because if they are more or less fixed, irrespective of the paint system to be used, then in calculating the cost per unit of life of a coating system, the longer-life and usually more expensive paints will show to advantage.

14.5.1.5 Other aspects of coating economics

Many of the costs involved in coating can be assessed fairly accurately but others are more difficult to calculate. The type of paint may have an influence quite apart from its durability. Typically, paints which dry slowly may add to the overall cost in the applicators' shops or on-site. Many epoxy paints will not cure below about 5°C, so delays may arise in some circumstances. There may be a critical time period for overcoating certain types of paint; this may cause problems and add to the overall cost. Some types of paint systems require careful handling to prevent damage during the full-drying period of the film. Stacking of steelwork can be a problem with some paint films because they tend to stick together if not fully dry, e.g. solvent evaporation types. When determining comparative costs prior to selection of the coating system, specific figures must be used rather than general costs per unit area. These costs vary considerably with the type of work and must be taken into account. The geometry of the steelwork, e.g. simple surfaces in tanks compared with lattice girders, and the problems of access, will have a marked influence.

In 1991 in the UK, the SCI Study Group on Maintenance Cost of Bridges obtained some sample costings for maintenance painting a steel bridge from one major painting contractor. The costs are now out of date, hence factors are used to provide a guide for this single instance, indicating relative costs and some of the factors that need to be taken into account.

1 *Cost of access*
    (a) Provide suspended scaffolding, double-boarded with polythene membrane in between and sheeted-in using tarpaulins to soffit and sides of bridge.   FACTOR *1.0 × PER M²*

(b) Provide all necessary heating and lighting within the scaffold area.

FACTOR *0.4 × PER M²*

2 *Cost of blast-cleaning*

This will be dependent upon the substrate encountered.

(a) Remove unsound metal coating down to clean steel; as for Sa2½.

FACTOR *3.0 × PER M²*

(b) Remove all existing unsound two-pack epoxy to clean steel as Sa2½.

FACTOR *2.4 × PER M²*

(c) Remove unsound paint down to clean steel as for Sa2½.

FACTOR *1.6 × PER M²*

3 *Cost of grit and debris removal*

(a) Remove non-toxic waste at cessation of each shift to skip.

FACTOR *0.1 × PER M²*

(b) Remove toxic material to sealed skip, stored in locked compound overnight and carted to licensed tip.   FACTOR *0.35 × PER M²*

4 *Paint application costs*

(a) Prepare and apply by airless spray conventional five-coat alkyd system to 325 microns min. dry film thickness.

FACTOR *1.6 × PER M²*

(b) Prepare and apply by airless spray five-coat micaceous iron oxide/silicone alkyd system to 325 microns min. dry film thickness.

FACTOR *2.4 × PER M²*

(c) Prepare and apply by airless spray four coats of 2-part epoxy to 50 microns min. dry film thickness.   FACTOR *4.0 × PER M²*

5 *Closure costs*

Where it is necessary for work to be carried out over the highway, road management costs would be in the order of

FACTOR *0.6 × PER M²*

Notes

The above costings are for a typical steelbox/composite type structure, over a highway where the soffit height does not exceed 10 m from the carriageway. Access charges could be reduced if the headroom were, say, only 6 m, but the traffic management costs would probably increase by an amount similar to the savings.

On a bow string structure the above costs would be increased by 75–100% (except access costs would only increase by approximately 10%).

Work done during possession periods over a railway would increase the above costs by approximately 100% (based upon a four-hour possession).

The additional costs incurred for work over waterways would equate to road management savings, resulting in no overall increase, with the exception of disposal costs and these would increase by approximately 100%.

The final cost has many variants, not least market forces.

Periods of low activity can alter costings by as much as 100% and this position can quickly reverse if there is an over-supply of activity in the market place. The above figures are offered as an average in normal market conditions.

14.5.2 Initial costs of protective systems

The *Steelwork Corrosion Protective Guide* (see Further reading) published jointly by steel, paint and metal coating interests, provides costs for a number of commonly used systems. The actual costs are now out of date but the broad comparisons are included in Table 14.4. Clearly there can be widespread variations in practice depending upon the type and situation of the structure. Nevertheless, the table provides some indication of initial costs and also requirements for deciding whether coatings are to be applied in the shop or on site.

Table 14.4 Comparative costs of protective systems

Surface preparation[a]	Protective system[b]	Thickness (μm)	Site- or shop-applied	Comparative cost
MC	Zinc phosphate/alkyd	35	Site	x
	HB zinc phosphate/alkyd	75		
	HB alkyd	60		
BC Sa2½	Blast primer	20	Shop	1.26x
	HB zinc phosphate alkyd	60	Shop	
	HB alkyd	60	Site	
BC Sa2½	Two-pack zinc-rich epoxy	25	Shop	1.58x
	HB zinc phosphate/alkyd	75	Shop	
	MIO/alkyd	50	Site	
	MIO/alkyd	50	Site	
Pickle	Hot-dip galvanisation	85	Shop	1.16x
Pickle	Hot-dip galvanisation	85	Shop	
	T wash		Site	1.7x
	HB chlorinated-rubber	75	Site	
BC Sa2½	Two-pack zinc-rich epoxy	25	Shop	
	Two-pack MIO/epoxy	85	Shop	
	Two-pack MIO/epoxy	85	Shop	1.7x
	HB chlorinated-rubber	75	Site	
BC Sa2	Hot-dip galvanised	140	Shop	1.6x
BC Sa3	Sprayed aluminium or sprayed zinc, two coats of sealer	150	Shop	2.2x

a MC = manually cleaned. BC = blast-cleaned.
b HB = high build. MIO = micaceous iron oxide pigment.

Further reading

Useful publications on the costs and economic assessment of protective coating systems, include the following:

Appleman, B. R., Economies of corrosion protection by coating. *J. Protective Coatings and Linings* (March 1985) 26–33.

Brevoort, G. H. and Roebuck, A. M. (1979) The simplified cost calculation and comparison of paint and protective coating systems. Expanded life and economic justification. *Corrosion 79*. Paper No. 37. NACE, Houston, 1979.

Federal Highway Administration, Coatings and corrosion costs of highway structural steel. Report No. FHWA-RD-79-121, Washington, 1980.

Mitchell, M., Selecting protective coatings based on cost/performance analysis, *Corrosion Management* (Jan/Feb 1999) 14–20.

National Association of Corrosion Engineers, Direct calculation of economic appraisals of corrosion control measures, RP-0Z-72, NACE, Houston.

Roberts, J. W. and Davis, L. H., Evaluation of bridge corrosion cost model. *NACE Conference*, 1982, Paper 140.

Roebuck, A. H. and Brevoort, G. H., Coating work costs – the selection and justification of coatings and galvanising, *Materials Performance* (Oct 1984) 46–52.

Sisler, C. W., *Materials Protection*, **9** (1970) 23.

Steelwork Corrosion Protection Guide: External Environments. Available from Paintmakers Association of Great Britain Ltd., London.

Chapter 15

Protective systems for different situations

15.1   General steelwork exposed to the atmosphere

The choice of coating system will be determined to some extent by the type of structure, its relative importance and the environment of exposure. Metal coatings will often provide long maintenance-free lives in mild environments: hot-dip galvanising, sprayed zinc and sprayed aluminium have all been used for this purpose without additional paint coatings. It should, however, be appreciated that the corrosion of zinc is closely related to the amount of pollution in the atmosphere. Fortunately in the UK at least, there is an overall reduction in atmospheric pollution, and metallic coatings can last significantly longer without the protection of paint, although weathered metallic coatings may be painted purely for aesthetic reasons.

Sprayed zinc coatings do not fall into the same category as hot-dip coatings so far as painting is concerned. The porous nature of the sprayed coatings leads to problems of satisfactorily cleaning them after a period of exposure. Consequently, there is a strong possibility of failure if the paint coating is applied to weathered zinc-sprayed steelwork. If sprayed zinc coatings are to be used they should be sealed. Metal coatings, particularly hot-dip zinc, are useful for rails and balustrades on structures and are recommended as a coating for fasteners.

Most steelwork is protected with paint. In many ways paint is a more flexible material and can be applied more easily than metal coatings, particularly on-site. The facilities for applying paint are also far more widely available than for other types of coating. Compared with bare metal coating, paint also has the advantage of being available in a wide range of colours and can be selected to provide a good cosmetic appearance. However, metal coatings also have advantages over paint coatings. Their lives are more predictable, they generally have better abrasion resistance and no drying time is involved, which makes for easier handling. They are also more resistant to damage and zinc in particular provides protection of the steel at scratches. Hot-dip coatings also have an important advantage which is often overlooked. They do not require the somewhat complex process of coating which is required for paint. Therefore, inspection is easier and generally defects in the coating can be readily observed.

Generally, for atmospheric structures exposed inland or in non-industrial conditions on the coast, oleo-resinous types of paint, e.g. alkyd, modified phenolic and epoxy esters (particularly for priming coats), are used. Silicone alkyd top coats are used in some situations to provide better gloss retention. Two-pack epoxies and urethanes are now being used extensively for aggressive environments.

The overall thickness of the coating influences the 'life' of the system, and it is one of the many improvements associated with modern paint systems that most can be applied with a high build, thus reducing the number of coats needed for the optimum thickness.

Consideration will now be given to a few typical structures exposed in the atmosphere to indicate specific problems that may arise.

15.1.1 Bridges

The design of the bridge has an important influence on the long-term maintenance requirements and sometimes on the selection of the initial protective coatings. The straightforward constructions with beams and girders (continuous or box girders) commonly used for highway bridges are easier to protect than the more complex truss and lattice type. Suspension bridges add further problems because of the requirement to protect the cables.

Bridges will be exposed in virtually every type of environment from the mild rural atmosphere to the industrial–marine situation where, apart from chlorides, industrial pollution may be intense, depending upon the industries in the proximity of the bridge. Large areas of steel are more easily coated than the smaller ones used for trusses, with their greater number of edges and joints.

The design of the bridge structure influences coating performance in other ways. Because of the construction with a large deck area, rain does not flow from the whole structure as with, for example, a gantry, but tends to collect and run away at select points such as expansion joints and at the ends of the bridge. Often the water collects in an open area of steelwork, promoting coating breakdown. Again, the undergirders of a bridge may be almost permanently sheltered from the sun and may remain damp for long periods. Paintwork near bearings also tends to break down more rapidly than the freely exposed facias. An additional hazard on many road bridges is the widespread use of road salts, which can lead not only to the premature breakdown of coatings but also to accelerated corrosion of the steel and increased maintenance costs.

Very serious corrosion of bridges has occurred in some parts of the world but, in view of their importance, a high degree of attention is usually paid to the selection of coating systems. Furthermore, increased attention to maintenance requirements is being given to most bridges. Modern bridges are usually blast-cleaned to take a variety of coating systems. In 1997 the UK Steel Construction Institute published Technical Report 241 'Durability of Steel Bridges. A survey of the performance of protective coatings'. The purpose of the Report was to present views and findings of a survey about the durability of corrosion protection systems applied to primary structural members of bridge steelwork.

Information on such durability is a key factor in assessing the 'Whole Life Cost' as required by the UK's Highways Agency. Among the conclusions of the Report is the fact that, according to those responsible for arranging maintenance work, the majority of existing structures are now only undergoing major maintenance at intervals in excess of 20 years, although the views expressed by painting contractors and paint manufacturers were more conservative.

The existing coating systems for bridges included:

1   Aluminium metal spray (no longer a Highways Agency Approved System).
2   Oleo-resinous (no longer a Highways Agency Approved System).
3   Acrylated rubber.
4   High-build epoxy polyurethane.
5   Aluminium metal spray + acrylated rubber.
6   Aluminium metal spray + high-build epoxy polyurethane.

Alternative systems proposed by paint manufacturers included:

1   Zinc-rich epoxy/MIO epoxy with polyurethane or acrylic finish.
2   Glass flake epoxy with polyurethane or acrylic finish.
3   Aluminium metal spray + polysiloxane.
4   Solvent-free polyurethane.

Non-highway bridges, including a large number of rail bridges, are protected with a wide range of coatings. The general requirements are similar to those for road bridges. Where colour is not important, micaceous iron oxide (MIO) pigments are often used for the weathering coats.

15.1.2 Buildings

Steel-framed buildings are often fully encased with masonry or brickwork, in which case the selection of the coating systems will be determined by the protection afforded by the encasement. It is often assumed that no water will penetrate encasing materials, but roofs leak and joints are not always watertight. If the designer is satisfied that there will be no ingress of water onto the steelwork, then both the surface preparation and the coating system can be at a low level, e.g. wirebrushing and a suitable primer with possibly a finishing coat. Even in this case it may be advantageous to apply a resistant coating, e.g. a thick bituminous coating at the foot of columns, because if there is any leakage water will tend to collect there. If there is any doubt regarding ingress of water then a full protective system should be applied. Where access for inspection is not possible, this should also be taken into account when selecting the coating system.

Although there are no basic problems in selecting coating systems for the main steelwork elements in buildings where the steel is freely exposed, difficulties may arise where some form of fire protection has to be incorporated into the overall coating system. Fire protection is a specialised topic and it will not be discussed here, but problems may arise inside buildings where the steel is protected for fire protection rather than corrosion prevention. Some methods of fire protection completely enclose the steelwork and it may be assumed that this will also act as a protection against corrosion. This is not necessarily the case. In dry, warm conditions, corrosion protection does not have to be of a high quality, but in damp situations, moisture may well penetrate to the steel, so a reasonably protective coating system should be selected. Inside buildings the protective system will be determined by the conditions and may have to be of the highest quality, e.g. in electroplating shops. In some chemical atmospheres paints will not provide sufficient protection and specialised coatings may be necessary, e.g. glass-fibre-reinforced plastics. However, in warm, dry conditions conventional alkyd systems will provide adequate performance although, where colour is not a determining factor, hot-dip galvanised coatings should be considered.

The conditions may be worse than had been anticipated because of combustion products from equipment or, in warehouses, for example, because doors are left open for prolonged periods. In buildings where, because of the nature of the equipment used or because of the type of manufacturing operation in progress, e.g. food processing, flakes of paint or rust are unacceptable, thorough surface preparation may be essential.

When selecting coatings for the interior steelwork of buildings, the possibilities of delays in roofing and cladding should be taken into account. Coating systems suitable for dry interiors will not necessarily withstand the exterior conditions prevailing before the building is completed. If this extends for some months, repainting may be required earlier than anticipated. Similarly, long-term protection of steelwork due to delays in construction will lead to the same premature breakdown.

15.1.3 Storage tanks (exterior)

Tank designs range from simple cylindrical shapes to complex spherical structures with architectural connotations. The exterior system will be determined by the environment and other factors considered in Chapter 14, but, because of their design, tanks often provide special problems. Generally, they are welded from a number of plates and, because of the size of many of them, they cannot easily be fabricated to the finished product in the shop. Consequently, there is often a good deal of welding on-site. To ensure sound performance of the exterior coating, all these welds should be blast-cleaned before painting. As this represents a large percentage of the total area it is often more economic to carry out the cleaning and painting on-site rather than in the shop. To some extent this will influence the choice of protective systems, particularly if the tank is being constructed close to other plants or buildings.

For water tanks, the American Waterworks Association lists a number of paint systems which are considered suitable for exterior protection. These include a three- or four-coat alkyd system, a two-coat alkyd system with a silicone alkyd finish, a three-coat vinyl system and a zinc-rich primer. These systems will cover the requirements for most environments where a specific colour is specified for the weather coat. However, if the tank contains running water colder than the ambient temperature there is a likelihood of condensation on the outside surface and alkyd-based systems are likely to blister. For tanks containing materials other than water, the exterior coatings must take into account possible spillage and, wherever practicable, should be chosen with this in mind.

15.2 Offshore structures

The selection of systems for offshore structures is not a straightforward matter and there is no general agreement on the most suitable systems for the different parts of the structure. Although the choice of suitable systems to withstand the various conditions encountered may not be difficult, the problems involved in application, storage and handling may be immense. Generally, a short delay in commissioning a platform designed for oil or gas production can cost millions of pounds, so it would be misleading to suggest that in all circumstances the protective coatings are the main priority, despite the aggressive nature of the environments that the steelwork has to withstand.

Maintenance is a problem with offshore structures where accommodation for operatives is limited and the time available for painting covers only a few months of the year. Furthermore, the costs of access are much greater than for onshore structures. A range of coatings has been developed mainly to take account of the difficulties involved in protecting these structures.

Most large platforms are the responsibility of organisations with technical expertise available, but in the future a number of smaller structures are likely to be constructed for purposes other than oil production. It is, therefore, worth considering the approach to protection of structures built off the coast. An offshore structure has to resist different environments: atmospheric, immersed and splash zone, and sometimes, depending on the type of structure, mud on the sea bed. Each of these will be considered separately.

15.2.1   Atmospheric zone

The approach to coating selection is similar to that for other atmospheric marine environments. Generally, chemical-resistant coatings such as epoxies, urethanes, chlorinated rubber and vinyls will be selected. The selection of a particular system is determined by the organisation's experience and view of the coatings available. As experience is gained at a particular geographical location, e.g. the North Sea, changes are made in the system, so that specifications for new structures are often significantly different from those used for the older ones. All steel is blast-cleaned to at least Sa2½ and sometimes even to Sa3. A blast primer will generally be used to protect the steelwork after blast-cleaning, although the system primer may be applied immediately, depending on circumstances, such as the requirement for welding.

The following systems indicate the approach taken by different organisations. They are typical but other systems have also been used.

(a) Vinyl system (3–4 coats)   200–250 μm
(b) Zinc phosphate pigmented two-pack epoxy primer, two-pack epoxy (2 coats)   300 μm
(c) Inorganic zinc silicate primer, two-pack epoxy (2 coats)   325 μm
(d) Chlorinated-rubber system (3–4 coats)   250–300 μm

The dry film thicknesses indicated are for the complete systems.

Although two-pack epoxy materials are very resistant to the environment, they tend to chalk fairly quickly, so in some situations a final coat of urethane has been applied. Again, because epoxies may require some abrasion to ensure sound adhesion when repainting, chlorinated-rubber finishing coats have been considered as a way of easing maintenance repainting. In the authors' view, this is not a very satisfactory approach because of the problems that may arise if epoxy coatings have to be reapplied at some time. This may entail the complete removal of the chlorinated-rubber coating at the areas to be repainted, to avoid the possibility of attack by the solvents in the epoxy.

15.2.2   Immersed zone

The main control method in this zone is cathodic protection, generally without the use of coatings, although some organisations do specify coal-tar epoxide coatings. The decision whether or not to use coatings is an economic one. Where coatings are used, fewer anodes will be required to protect the steel and they would be expected to last longer. On the other hand, the cost of painting the large areas of submerged steelwork is not inconsiderable and adds to the difficulties of planning and programming the construction work.

15.2.3   Splash zone

This is the most aggressive of the three environments considered and because of the conditions, where seawater spray is continually wetting the area, is the most difficult to maintain in a satisfactory manner. Generally, coatings similar to those used for the atmospheric zone are employed in the splash zone, usually at a greater film thickness. This is not completely satisfactory but is usually considered to be a practical way of dealing with the situation. Sometimes the steel thickness is increased to act as a 'corrosion allowance' and to allow for abrasion and wear; it is then usually coated with the same system as the rest of the structure. Other more resistant coatings are also employed, for example on the hot riser pipes, which are particularly prone to attack in this area. The following coatings have been used on different structures:

(i)   Corrosion-resistant sheathing; usually a nickel–copper alloy such as Monel 400. This is fitted round the tubular member and welded into place to seal gaps between the sheathing and steelwork. This is an expensive form of protection and is not widely used except on hot risers. The sheathing can also be torn by impact and seawater can then penetrate between it and the steelwork.

(ii)   Thick rubber or neoprene coatings up to 15 mm in thickness.

(iii)   Various forms of polymeric resins and glass-flake-reinforced polyester material.

In principle, special tapes could be used on tubular members but, so far as the authors are aware, these have not been used to any extent in practice.

15.3   Ships

Ships can be considered in the context of structures although some of the problems concerned with the protection and maintenance of the steelwork are different from those encountered in static marine structures. In particular, there are two major differences from the approach to protection that might be afforded to piers and jetties: (i) ships can be dry docked to allow access to the underwater parts for maintenance, and (ii) the speed and fuel costs are markedly affected by marine fouling that collects on the hull, so additional coatings must be applied to provide anti-fouling properties. Nevertheless, in broad terms, ships are protected in much the same way as other structures and the same general principles apply. In fact, the widespread use of blast-cleaning for structural steelwork owes much to the shipbuilding industry, which was among the first to appreciate the need to provide automatic cleaning facilities as part of the construction process.

15.3.1   Surface preparation

Ship plate is commonly blast-cleaned in automatic plants before fabrication. It is then usually protected with a suitable blast or prefabrication primer, which is quick-drying and can be 'welded-through'. Such primers are often applied in an automatic spraying unit direct from the blast-cleaning operation which is carried out in a centrifugal type machine. Often, a pre-heating plant is used to warm the steel before entry into the blast-cleaning plant to improve adhesion of the primer, particularly in cold weather.

Generally, the standard is to Sa2½, or the equivalent. For some coatings Sa3 may be specified. As shipyards are on the coast, particular attention must be paid to the effects of chloride, which may contaminate the surface in the form of NaCl (sodium chloride) or produce chloride-containing corrosion salts during the storage of the ship plate.

15.3.2   Areas to be protected

Protective treatments vary with the particular areas of the ship under consideration. The main areas are as follows:

(i)   Underwater plating: ships' bottoms, which are completely immersed in service.

(ii)   Boot-top plating which is somewhat similar to the splash zone and parts may be more or less completely immersed, depending on the loading of the ship.

(iii) Topside and superstructure: exposed to the atmosphere and affected by sea salts.

(iv)   Cargo holds and tanks: carrying a variety of different materials.

(v)   Ballast tanks: containing seawater.

(vi)   Accommodation units which on exterior surfaces may be exposed to conditions similar to those of the topside.

(vii) Machinery, rails, etc., on decks.

(viii) Machinery, internal, which will have to withstand oil, grease and seawater from leaking pipes.

In the next few sections a broad outline of the types of coating systems used will be given. However, selection of protective coatings for ships is a specialist undertaking and the specification will usually be prepared by those who are knowledgeable in a particular field, e.g. naval vessels, oil tankers or small work boats. Each of these has different requirements.

15.3.3   Underwater plating

The conditions that have to be resisted by the underwater plating include seawater, which is a high-conductivity electrolyte, abrasion and the possibility of damage, particularly in harbours. Furthermore, as most ships' hulls are cathodically protected, the coating must be resistant to saponification, so excluding conventional oleo-resinous paints from consideration. The following types of binders are generally used for underwater plate: coal-tar epoxy, chlorinated-rubber, vinyl and bitumen or pitch. To differentiate these coatings from the anti-fouling coatings, they are commonly termed 'anti-corrosive or A/C compositions'. The thickness applied will depend upon the type of ship and the conditions it will have to withstand and can vary from 200 to 600 μm total film thickness.

Although, with modern painting, high-build coatings can be applied in one application, it is usual to apply a multi-coat system to avoid the possibilities of pinholes in the coating. Some authorities consider that zinc-pigmented primers should not be employed for underwater service while others consider them to be suitable; there is conflicting evidence on this aspect and advice should be sought from paint manufacturers when selecting primers.

15.3.4   Anti-fouling paints

As already noted, the attachment of fouling to ships' hulls impairs the operational efficiency of the ship so special coatings are applied over the anti-corrosive compositions. Most anti-fouling coatings no longer employ toxic compounds of tin, mercury or copper due to their adverse effect on the environment, these being replaced by proprietary organic materials.

The anti-fouling paints work by providing a concentration of material toxic to foulings in the vicinity of the painted surface. Provided this is maintained at a suitable concentration, all fouling organisms will be killed. Such paints must, therefore, be formulated so they can supply the toxins at a suitable rate while at the same time supplying sufficient coating life to allow for reasonable periods between dry dockings. Two main types of coating are used. In the one commonly used for many years, the paint matrix remains unattacked and the toxin particles are leached out at the required rate. A more recent development is a coating with a soluble matrix which is often described as a 'self-polishing' type. In this type layers of the paint containing the toxin are removed to provide the necessary concentration of poison. This has an advantage because suitable coloured layers can be incorporated to provide an indication of the level of toxin left in the paint. Furthermore, there may be some advantage in the continuously smoother surface produced with this type of coating.

New, semi-soft silicone based anti-fouling compositions are being trialled. These contain no toxins nor leachates, but rely on the low surface energy of the silicone polymer, to ensure that foulings have no key and simply slip away from the surface, when minimal build-up has occurred. The performance of anti-fouling coatings may be affected by the action of the anti-corrosive compositions and tie coats over which they are applied. It is, therefore, advisable to contact the coating manufacturer to ensure compatibility of all the coats in the system. It is usual to apply a barrier coat over the remains of any old anti-fouling coats, e.g. aluminium-pigmented bitumen, when applying further anti-corrosive coatings during maintenance painting.

15.3.5 Boot topping

As the boot-topping area is similar to the splash zone, high-quality protective coatings must be selected. On some ships the decorative aspects may be important, so coloured finishes are employed. Cathodic protection is not effective in this area because it is often not immersed in the sea. Nevertheless, the coatings for this area are often similar to those selected for the underwater plates although epoxy polyamides may be used in preference to coal-tar epoxies.

15.3.6 Topsides and superstructures

In many ways the approach to the selection of coatings for these areas is similar to that for general structures exposed to atmospheric marine conditions. The decorative aspects may be important, particularly for passenger ships; because of the moist atmosphere the coatings have to withstand, small spots of rust generally produce a considerable amount of rust staining. Particular attention should be paid to edges and welds to ensure coating integrity.

Silicone alkyds, acrylics and epoxy/polyurethane systems are currently favoured with three-coat systems yielding dry paint film thickness of $200\,\mu$m for the former and $350\,\mu$m for the latter.

15.3.7   Steel decks

Steel decks are difficult to protect satisfactorily because of the combined effects of corrosion, abrasion wear and the requirement, for safety reasons, of non-slip properties. The non-slip requirements are generally obtained by using specially formulated paints containing grit or by adding suitable grit to the finishing coat. Sprayed metal coatings have also been used on some decks.

Although alkyds, chlorinated-rubber and bituminous coatings have been used for weather decks, where there is heavy traffic epoxies and urethanes are commonly preferred. Zinc silicates have also been used but are not suitable where the cargoes are of an acid or alkaline nature. The coating thickness will depend on the conditions of use but may be $450\,\mu$m or even more.

15.3.8   Machinery, pipes, etc.

Often these are supplied fully coated or at least primed and it is difficult to generalise on the selection of coatings because the conditions vary so much. Coatings similar to those used for the superstructure are suitable.

Machinery in the bilge area and below deck level may have to withstand oil, grease, steam and seawater. Careful attention is required to such areas and the application of a high-quality coating such as an epoxy during ship construction will be beneficial. However, because of conditions, difficulties of access for repainting and the inevitable damage that will occur, the problems may be of design more than of protection.

15.3.9   Cargo and ballast tanks

These tanks may have to withstand very aggressive conditions, not only from the nature of the cargoes carried but also from seawater ballast and washing-out. Although crude oil leaves a wax-type film on the surface of the tank and may provide some protection, it is not continuous. Consequently, any seawater ballast can result in local attack and pitting. Some crude oils contain sulphur compounds which may react to form acids which will, of course, attack the steel. Refined oil products provide little protection. The coating or lining for the tank must be selected to withstand the corrosive effects of the cargo and a wide range of coatings is used, depending on the requirements.

However, it is generally possible to find an epoxy system with a carefully selected curing agent which, providing it can be applied correctly, will withstand most cargoes.

Where ballast tanks are permanently filled with seawater, cathodic protection is a suitable control measure. Where the tanks are out of ballast for reasonably long periods, a joint cathodic protection–coal-tar epoxide coating is often used.

Owing to the problems of protecting cargo tanks, methods other than conventional coatings are widely used. These include:

(i)   Dehumidification to remove moisture from empty tanks or above oil cargoes.
(ii) Inhibitors.
(iii) Oil films floated on the surface of the ballast water.
(iv) Injection of inert gases to remove oxygen.
(v)   Wax and grease coatings.

15.3.10   Freshwater tanks

Tanks for non-potable water can be coated with bituminous solutions or coal-tar epoxies, usually to a thickness of 150–200 μm. However, a non-tainting coating must be used for drinking water. This may be a suitably formulated bituminous paint or certain epoxies. Clearly, the regulations governing coatings for water of drinking quality must be complied with.

15.4   Chemical plants

The approach to the selection of coating systems will be determined by the nature of the plant and the processes being carried out, which will influence the local environment in a number of ways. Emission of polluted gases, although generally subject to government regulations in most countries, still occurs and leakages from joints are by no means unusual. The coatings to be chosen must, therefore, take into account the actual conditions that are likely to be encountered in practice. No matter how carefully coatings are selected initially, if there is not a properly organised maintenance programme then serious problems will almost certainly arise in some plants.

On many structures, such as highway bridges, the amount of corrosion occurring where coatings break down is not likely to cause serious structural problems if maintenance is delayed for a year or two. The maintenance costs will increase but in most environments the loss of steel by corrosion will be acceptable. This is equally true of many chemical plants but, with others, spillage and leakage of chemicals if not suitably resisted by appropriate coatings will result in serious corrosion, often necessitating the replacement of sections of the plant.

For steelwork exposed in corrosive areas, blast-cleaning is essential before application of paint coatings. If the steel has been stored in the plant for any length of time, the rust will contain corrosive salts and these will not be removed by hand-cleaning methods such as wirebrushing, so premature breakdown of coatings applied over the rusty steel will occur. In some situations, thick grease paints may be applied over rusted steel but, generally, these will not be used on plants where personnel can come into contact with the surface. In the most corrosive areas where acidic or alkaline fumes are present or where the corrosive fluids are in contact with the steel, even thick paint coatings may not suffice, and glass-reinforced plastic materials or other high-duty coatings discussed in Chapter 6 may be required. Such coatings must be applied with great care because, if corrosive solutions are able to penetrate damaged areas of the coating, serious corrosion can occur. For the average type of chemical environment, the high-resistance paints such as epoxies, urethanes and vinyls will generally provide adequate protection provided they are applied to a reasonably high film thickness of at least $200\,\mu$m.

In America, zinc silicate coatings are widely used as primers with systems based on the above binders. This primer is not so widely used in many other parts of the world, where a primer based on the main binder used for the undercoats and finishing coats is used. A particular problem occurs with the exterior surfaces of machinery and equipment, e.g. pumps and valves, which are often supplied with the manufacturer's coating system. This may be satisfactory in some situations but often the coating is not adequate for the conditions. It is virtually impossible to obtain equipment painted to the client's specification, so additional coats should be applied where this is practicable. Alternatively, the coatings should be removed by blast-cleaning, where this will not cause damage to the equipment, and repainted to a satisfactory standard.

Tapes can be used to protect pipes but care must be taken to ensure that the steel is well cleaned before wrapping and that the manufacturer's procedures are correctly followed. Regular examination is advisable because, if there is ingress of moisture to the steel, quite severe corrosion may occur without any obvious deterioration of the tape. Pipes are usually colour coded and this must be taken into account. Often the conditions are sufficiently corrosive to preclude the use of any coating, in which case suitably resistant alloys must be used. However, for many situations in chemical plants where there is unlikely to be condensation on the surface, alkyd or acrylic paints will be suitable; the choice will depend on the environment and the type of plant.

It is always advisable to carry out an inspection of newly painted steelwork within a few months of application. Small areas of damage should be touched up immediately after erection but further damage tends to happen during the first few months as additional services are added. If, at this inspection, there are clear signs of early coating failure, suitable action should be taken before this spreads over a larger area of the plant.

15.5   Oil refineries and installations

Oils are used as temporary protectives for steel so it might be assumed that protection of steelwork in oil installations would not be a problem. In fact, crude oil can be very corrosive and often other corrosive chemicals are produced in these plants. Furthermore, there are considerable temperature variations on the exterior surfaces of different parts of the plant arising from the process operations, with consequent expansion and contraction of the surfaces. To meet these and other requirements, the external steelwork should be blast-cleaned, generally to $Sa2\frac{1}{2}$, and epoxies or similar resistant coatings applied to surfaces exposed to the atmosphere. Coal-tar epoxy or modified, tar-free epoxies, can be used for any underwater or buried steelwork, probably with cathodic protection. Hot-dip galvanised coatings may be considered for less corrosive areas where colour is not important.

15.6   Sewage systems

The major corrosion problem with the immersed parts of sewage systems occurs when hydrogen sulphide is generated. The gas is produced bacteriologically in the sewage, accumulating in slimes and sludges where anaerobic conditions exist. The hydrogen sulphide tends to be absorbed onto surfaces and is oxidised to form sulphuric acid. The problem is accelerated in hot climates or where industrial acid wastes are also discharged into the sewer. Correct design of the system, such as ensuring that the sewage stays oxygenated as long as possible and that the sewer is well ventilated, is the best approach to corrosion control. However, most metals in the system will require some form of protection by coatings. Such coatings should have a high water resistance and good resistance to impact damage. The coating should also be capable of patch repair.

For the interior of pipelines, PVC linings are increasingly being used. If *in situ* coating is required, the choice is often an amine cured coal-tar epoxy. The use of solventless coal-tar epoxies is considered to have advantages. This would reduce the possibilities of pinholes but requires sophisticated equipment and skilled operation to apply. The method allows the application of very thick, quick-drying films of paint but this also emphasises the need for a high standard of surface preparation, notably surfaces free from contamination and rough enough to provide an adequate key.

For the steelwork in aggressive atmospheric environments, two-pack epoxies are preferred and a cosmetic finish with excellent weathering properties can be obtained by making the final coat one based on a two-pack polyurethane. Repair or repainting of such systems requires, at least, sweep blasting or abrading of the surface. For steelwork in a relatively non-aggressive atmosphere, but in damp conditions, vinyl systems are suitable. These are available in light decorative colours and if in good condition can be repaired or repainted without extensive surface preparation.

15.7   Sheet piling

Sheet piling is used for retaining purposes and is not always coated. If, for example, it is used as temporary piling during excavation prior to foundation construction, it probably will not be painted. Generally, where it is driven into soils it is not coated except under conditions that are particularly corrosive. In such situations cathodic protection may be used.

One of the most aggressive environments for piling is on the coast, where it may be used for harbour and dock installations or on beaches as groynes. The conditions involve four zones similar to those on an offshore structure; atmospheric, intertidal (splash zone), immersed in the sea and buried in the ground. Generally, problems arise at the intertidal and immersed areas. The protection of the area exposed to the atmosphere is not usually a problem and it can be repainted without great difficulty. The part buried in the soil does not usually require a coating but the immersed and intertidal zones are generally coated. Apart from the difficulties of maintenance painting, these areas are prone to damage whether in harbours or on beaches. One approach is to use thicker steel to provide a 'corrosion allowance', possibly with a coating. Alternatively, or additionally, cathodic protection may be used for the immersed zone.

There is no general agreement on the most economic coating system for sheet piles. A number of comprehensive test programmes have been conducted on a variety of coatings but it is difficult to draw any firm conclusions from the results. They vary, depending on the particular test programme. Some coatings have performed well in most tests, e.g. neoprene, but this would be considered too expensive for most situations. Sprayed aluminium coatings have also performed well in a number of tests. On the more conventional coatings for this purpose, coal-tar epoxies have generally performed well. In one series of tests the performance of these coatings was inferior when applied to bare blast-cleaned steel compared with application to a zinc silicate primer. On the other hand, an organic zinc-rich paint was much less effective under immersed conditions.

Isocyanate-cured pitch epoxies have performed well in some tests but the selection of suitable coatings will be influenced by the likelihood of damage. In many situations cathodic protection would be advisable. Often the reverse of the pile in seawater conditions can be protected without difficulty, but cathodic protection can be applied to both sides of sheet piles.

15.8   Jetties and harbours

In some ways, the problems are similar to those experienced by sheet piling in the sense that the steelwork has to withstand the conditions produced by the four zones: atmospheric, splash, immersed and buried or mud-line.

The selection of suitable coatings for the atmospheric zone is not a problem and will be determined by the factors considered for other situations, e.g.

  (i)   Difficulties of access for maintenance.
 (ii)   Periods required between maintenance repaintings.
(iii)   Difficulties arising from loss of use of facilities during maintenance.
 (iv)   Appearance, including colour.
  (v)   Geographical situation.
 (vi)   Likelihood of damage arising from the use of the facilities, e.g. ships damaging paintwork.
(vii)   Whether the facilities are part of an industrial complex with pollution of the air in addition to airborne sea salts.
(viii)   Possibility of abrasion from industrial processes.

Metal coatings, particularly hot-dip galvanised steel (painted) in the atmospheric zone and sealed sprayed aluminium coatings in the immersed zone, may be considered in some situations. However, both metals act sacrificially in seawater and problems may arise if they are connected to large areas of adjacent bare steel or poorly coated steel, because the metal coatings will corrode rapidly to protect areas of bare steel. Generally, resistant coatings such as epoxies and chlorinated-rubber will be considered for long-term protection of steelwork exposed in the atmosphere, although conventional oleo-resinous coatings will be suitable for many situations provided maintenance painting is carried out regularly.

For the immersed and splash zones, coal-tar epoxies or urethane pitches will generally be selected, preferably with cathodic protection. Where long-term protection is essential in some situations, solventless epoxies and other special materials applied as thick coatings may be preferred.

The selection of coatings for the interior surfaces of tanks and the external surfaces of pipelines is discussed in Chapter 6.

15.9   Steel in reinforced concrete

Two of the materials most commonly used for structures and large buildings are steel and reinforced concrete. This book is concerned primarily with structural steel but many of the problems that arise with reinforced concrete result from the corrosion of the steel reinforcements or rebar as they are commonly called.

Cases of corrosion of rebar with subsequent serious deterioration of concrete structures have been reported over many years. However, in the last 10–15 years the problem has taken on a new significance both in buildings and structures. Particularly severe corrosion has occurred on many large bridge decks in North America.

For many years the general view was that, unlike steel structures, those of reinforced concrete were not subject to significant corrosion. However, in many situations this has proved not to be so and attention is being focused on the problem.

15.9.1   Concrete

The corrosion of rebar is basically the same as for any other steel, but to appreciate the particular problems that occur with concrete it is necessary to look at some of the properties of this material. Concrete is produced from the following constituents:

  (i)   Cement.
 (ii)   Fine aggregate, e.g. sand.
(iii)   Coarse aggregate – gravel or crushed rock.
 (iv)   Water.

Additionally, other materials called admixtures may be used to modify the properties of the concrete. Of the various constituents, cement is the most important from the corrosion standpoint.

15.9.1.1   Cement

Cement is produced by mixing suitable components such as limestone and clay and heating them to about 500°C. This produces a series of reactions that result in a mixture of calcium silicate, aluminates or aluminoferrites. When these are mixed with water to produce mortar or cement the important aspect so far as corrosion is concerned is the alkaline nature of the final product.

15.9.1.2 Admixtures

Admixtures of inorganic or organic materials are added immediately before or during mixing to modify the properties in some way, e.g. to increase or decrease the setting time. Of these, calcium chloride, used as an accelerator is the one that most influences corrosion because of its chloride content.

15.9.1.3 Basic chemistry of concrete

In the setting of concrete, as the aggregates are basically inert they have little direct influence on the concrete. The exception is when aggregates contain chloride, e.g., sand from near the coast. The main reaction of interest is that between the cement and water which provides material for the 'cementing' together of the constituents and produces the density, hardness and strength of the concrete.

When water is added, the reactions with cement are fairly complex and are described in detail in text books on concrete. Basically a series of hydrates are formed, e.g. calcium silicate hydrate; calcium aluminate and calcium aluminoferrite hydrates. The reactions, which may continue in wet concrete for some time, lead to a hardening of the concrete.

The important aspect of these reactions from the corrosion standpoint is the alkalinity produced both in the mass of the material and in the moisture remaining in the pores and capillaries. Additionally, chlorides, if present, react with calcium aluminate and calcium aluminoferrite to form insoluble chloroaluminates and chloroferrites, in which chloride is found in a non-active form. In practice, the reaction is not complete, so some active soluble chloride will remain in the concrete. Nevertheless, concretes or cements with a high proportion of these phases reduce the availability of chlorides, which, as will be discussed later, is a primary cause of steel corrosion.

15.9.1.4 Properties of reinforced concrete

Although strong in compression, concrete is weak in tension, so steel reinforcement bars or meshes are used to improve the tensile strength. Steel does not normally corrode in an alkaline environment; however, in some circumstances, which will be discussed in detail later, problems can arise if certain corrosive species are available to react with the steel. The properties of the concrete that have the most influence on the diffusion or ingress of these species are its permeability and porosity, which are to some extent controlled by the mix of the concrete.

15.9.2   The corrosion of rebar in concrete

The mechanism of steel corrosion has been considered in Chapter 2, and this is applicable to rebar corrosion. Basically, the process can be divided into reactions that occur at the anode and those occurring at the cathode, with the reactants combining to form ferrous hydroxide, which is oxidised to rust – Fe_2O_3 . H_2O. The corrosion occurs at the anodic areas.

In alkaline solutions, steel becomes passive, i.e. a protective film is produced which prevents these reactions occurring, or if they do occur the corrosion is very slight. As concrete produces an alkaline reaction, little or no corrosion of the steel occurs provided the passive film remains intact. However, certain corrosive species, such as chlorides, can attack the film or diffuse through it, so allowing corrosion to occur. Passivity can also be broken down by the ingress of acid solutions which neutralise the alkalinity.

Processes which lead to the breakdown of the passive film will be considered next.

15.9.2.1   Breakdown of passivity

Passivity can be broken down by natural deterioration of concrete, called carbonation, or by ingress of corrosive species, especially chlorides.

15.9.2.2   Carbonation

Air contains a number of acidic gases, in particular carbon dioxide and sulphur dioxide. Concrete is permeable and allows the slow ingress of these acidic gases, which react with the alkaline components in the concrete, e.g. calcium hydroxide, to form carbonates and sulphates at the same time as neutralising the alkalinity.

These reactions result in what is termed a carbonated layer in the concrete, so called even if sulphates predominate. If the carbonated layer reaches the steel surface then as the environment is no longer alkaline, passivity breaks down and the steel is not protected from corrosion. This situation can arise as follows:

(i)   Over a period of time, carbonation may progress to the point where it reaches the steel reinforcement. The rate of carbonation in most sound concrete is generally low, under $\frac{1}{2}$mm per year, so if the depth of cover is 20mm the reinforcements will be in an alkaline environment for a long time. Where the depth of concrete cover is 50–70mm, then no problems should arise during the design life of most structures provided other factors do not lead to a loss of passivity.

(ii)   Certain types of cracks formed by shrinkage or loading may allow the ingress of acid species and the carbonated layer so formed may reach the steel even though the general cover is adequate. Many factors will determine the probability of corrosion arising from cracks, e.g. their depth and width, whether they move as a result of alternating loads and the degree to which they are self-sealed by corrosion products, debris, etc., so preventing the ingress of corrosive species.

(iii)   Where reinforcement bars or mesh are not correctly placed, local reductions in cover may occur with premature failure of passivity of the steel at a few places. This may be superficial but in some circumstances can lead to more general failure.

15.9.2.3   Chlorides

Chloride ions can penetrate the passive film or may replace hydroxyls in its structure, so destroying passivity, at least locally. If sufficient chloride is present, then general passivity breakdown occurs. Chlorides react with calcium aluminate and calcium aluminoferrite in concrete to form insoluble compounds. However, some active chloride always remains free to reduce passivity and promote corrosion. Chlorides may arise from the material used to make the concrete or from external sources.

Chlorides in concrete. Chlorides may be present in the aggregates if they are taken from marine environments or they may be present in the water if this is of a saline nature. Although problems have arisen from such chloride-containing materials, particularly in the Middle East, there is now a general appreciation of the inadvisability of using such materials for reinforced concrete. Some admixtures, particularly calcium chloride used as a set-accelerator, may also cause corrosion problems.

Chlorides from external sources. The two sources of external chlorides are:

(i)   Seawater which contains a high proportion of chlorides.
(ii)   De-icing salts such as sodium chloride used to reduce the freezing point of ice, particularly on bridge decks.

Chloride concentration. There appears to be no definite threshold level of chloride below which corrosion is unlikely to occur. Various figures have been used, e.g. 0.15% (based on cement content) in a bridge deck has been stated to be safe with no likelihood of corrosion, whereas over 0.3% chloride is a cause for concern. Chloride limits for cement mixes have been set by various authorities, e.g. 2% $CaCl_2$ as a limit, but there appears to be no general agreement on threshold levels.

15.9.2.4   Influence of concrete on corrosion

Corrosion will occur only if the corrosive species reach the steel surface, so the permeability of the concrete has an influence. With sound aggregate, the best path for the ingress of corrosive ions is the cement phase, the permeability of which will be influenced by the water to cement (W/C) ratio. A low W/C ratio reduces permeability, so retarding the ingress of ions to the steel surface.

Other aspects of the concrete mix influence to a greater or lesser extent the overall protective value of the concrete, e.g. aggregate grading, degree of compaction, adequacy of cure. These and the earlier factors considered all affect permeability, which has a direct relation to corrosion. Additionally, high permeability will increase the rate of carbonation.

15.9.2.5   Concrete cover

It has been established in test work and from practical experience that the cover, i.e. the thickness of concrete covering the steel, is of fundamental importance to the protective qualities of concrete. Adequate cover will ensure that carbonation does not reach the steel surface, so decreasing the overall permeability of the concrete. Standards and Codes of Practice, e.g. BS CP114 Part 2, 'Structural use in reinforced concrete in buildings', indicate the correct cover for different situations.

15.9.3   Types of failure with reinforced concrete

The basic problem with the corrosion of reinforcing bars arises not so much from loss of strength but from the bulky corrosion products that are formed.

The rust exerts a considerable degree of force on the concrete in contact with it, possibly as much as ten times the concrete's tensile strength. As the concrete has a low degree of elasticity, the force leads to cracking, delamination and eventual spalling of the concrete. This results in loss of protection to the reinforcing bars and the danger arising from pieces of concrete falling from the structure or building. The various forms of failure tend to be progressive with cracks leading to spalling.

The early stages of corrosion are not necessarily visible, so a structural survey must take this into account. Delamination may not be obvious without close examination. The rust may have forced the concrete away from the reinforcement but the cracks may not be obvious at that stage although the concrete has in fact failed. Often the depth of cover determines whether a crack or delamination occurs.

15.9.4 Corrosion control methods

Control methods should be incorporated before corrosion occurs and a number of approaches have been used:

(i) Ensuring that the concrete is of a high standard with low permeability and with adequate cover to the steel.
(ii) The use of suitable membranes for waterproofing to reduce the ingress of moisture, chlorides, etc.
(iii) Protecting the reinforcements with suitable coatings. Zinc in the form of galvanised coatings and powder epoxy coatings have been used.
(iv) Retarding the ingress of ions into the concrete. Various coatings have been applied to the exterior surfaces of the concrete structure or building to achieve this.
(v) Using corrosion-resistant reinforcements, e.g. stainless steel or carbon steel clad with stainless steel or nickel.
(vi) Cathodic protection has also been used to retard further attack where corrosion of rebars has already occurred.

Other methods have also been considered but have limited practical value. These include inhibition of the concrete and the use of wax beads and other suitable materials to seal the internal pores of the concrete.

15.9.4.1 Coatings

Coatings have been used in two different ways: (a) application to the concrete surface to retard the ingress of corrosive species, particularly chlorides; (b) Coating of the rebars to protect them from attack.

Coating of concrete. Various methods have been used:

(i) Paint coatings such as epoxies, chlorinated rubbers and vinyls.
(ii) Sealers that penetrate the concrete to some extent, e.g. epoxies and acrylics.
(iii) Renderings of thick coatings, e.g. epoxy mastics, applied by trowel. They also tend to fill in minor surface defects.
(iv) Materials that block the concrete pores, e.g. silicates and polyurethanes.

The concrete has to be properly prepared to ensure that the coatings adhere properly. Their effectiveness has not been established and investigational work is in progress to provide long-term data.

Coating of rebar. Two particular coatings have been used for this purpose: (a) hot-dip galvanised zinc; (b) powder epoxies.

(i) *Hot-dip galvanised rebar.* There are differing views on the effectiveness of zinc-coated rebar. Tonini and Cook[1] conclude that zinc-coated bars offer a minimum performance enhancement over unprotected bars of at least four times. Their paper and others produced by 'zinc interests' demonstrate this enhanced performance. However, other authorities are not so enthusiastic. The test methods used to obtain some of the data have been under scrutiny. It is probably fair to say that in cases of loss of alkalinity due to carbonation or low concrete cover, galvanising the rebar is an effective way of retarding the onset of corrosion. However, the reduction in corrosion where high chlorides prevail is less marked. A restriction on the use of galvanised rebar is the general requirement for bending of the bars before rather than after galvanising. Additionally, zinc corrodes in highly alkaline situations so that while the steel may be protected, problems may arise from the zinc corrosion products.

(ii) *Epoxy-coated rebar.* Fusion-bonded epoxy coatings have been used for rebar since the early 1970s. They are similar to those used for pipelines and have been discussed in Chapter 6. They should not be confused with the paints used on concrete surfaces. Epoxy-coated rebars are specified in North America for a number of situations, particularly for bridge decks. They have not been widely used in the UK, partly because of problems with standards. It seems likely that they will be used to a greater extent in future. Problems such as damage to the coating may occur but practical data on their performance are encouraging.

15.9.4.2 Cathodic protection

Where rebars have corroded to produce deterioration of concrete, cathodic protection has been employed as an alternative to concrete removal and repair. The high electrical resistivity of concrete means that specially designed systems are required with specially designed anodes. Wyatt and Irvine[2] have produced a useful review and a book on this topic has recently been published.[3] The basic principles of cathodic protection are discussed in Chapter 13.

References

1. Tonini, D. E. and Cook, A. R., The performance of galvanised reinforcement in high chloride environments. NACE compilation of papers on Rebar Corrosion 1976–1982. NACE, Houston.
2. Wyatt, B. S. and Irvine, D. J., A review of the cathodic protection of reinforced concrete. *Materials Performance*, **26**(12) (1987) 12–21.
3. Berkeley, K. G. C. and Pathmanaban, S., *Cathodic Protection of Reinforcement Steel in Concrete*. Butterworths, London, 1990.

Chapter 16

Testing of coatings

16.1 Introduction

It is virtually impossible to predict the 'life' of organic coating systems from a knowledge of their formulations, even if these were available to users, which normally they are not. A good deal of performance data are available for metallic zinc coatings and, to a lesser extent, for aluminium, so it is possible to provide a reasonable assessment of their performance in many situations. It is not possible to provide similar data for organic coatings such as paint. Even within generic groups, there can be wide differences, so the performance of a paint supplied by one manufacturer will not necessarily be the same as that from others. Additionally, the performance of any paint is significantly influenced by the surface preparation of the steel prior to coating and the application of the paint. Hence, data on lives of coatings have to be obtained in one of two ways: (a) experience of a coating in a situation similar to that where it is to be used, and (b) by carrying out tests.

Experience is important when selecting coatings, but some caution is required. The conditions of exposure, the surface preparation and the type of structure should be very similar when applying previous performance data to predict future performance of a coating. Furthermore, the coating must be very similar and this should be checked with the manufacturer to ensure that the formulation has not been changed in any significant way.

Apart from the 'life' of the coating other aspects of performance, e.g. drying time and hardness, may also be required and again tests have been devised to check on many of these. There are other reasons for testing and generally they fall into one or other of the following categories:

(i) Development of new coating materials.
(ii) Comparative tests to determine the best coatings for a particular situation.
(iii) Quality control measures for coatings and coating materials.

(iv) Specialised tests for determining the composition of paints, particularly for failure investigations.

There is also investigational work or research on the fundamental properties of coatings and their mode of protection. Such work is carried out by large paint and material manufacturers, universities and organisations concerned with research. This is necessary for obtaining a fundamental understanding of coatings and leads to the development of new and improved materials and application procedures. The detailed work and the results obtained are published in journals and discussed at conferences. Generally, the results and data produced are of more interest to those involved in research and development of coatings than to users. The methods and techniques employed in research will not be discussed in any detail here because they go beyond the requirements of testing and form a specialist topic.

Users and specifiers of protective coatings may carry out the tests themselves or use data supplied by others, such as paint manufacturers. These will usually fall within groups (ii) or (iii) above. It is always essential to check that the results of the test are valid for the situation where they are to be applied. This is particularly important when studying the results of tests carried out by other organisations. Typically, if the tests are carried out on coatings applied to cold-reduced steel panels, the comparisons between different coatings may be valid but the actual 'life' obtained may not provide a good indication of the durability of the coating when applied to other types of surface.

The methods of carrying out tests can also be divided into a number of categories:

(i) Tests on the dried film to determine and check the physical and mechanical properties of films, e.g. hardness and abrasion resistance: usually carried out in a laboratory.

(ii) Measurement of properties of the dried film such as film thickness and freedom from pores: often carried out on-site.

(iii) Tests of properties of the liquid paint, e.g. viscosity and specific gravity.

(iv) Laboratory tests to determine the ability of the coating to resist certain environmental conditions, e.g. salt-spray tests. These are sometimes called 'accelerated tests'.

(v) Tests carried out on coated steel in a natural environment often at special test sites. These are generally termed 'field tests'.

(vi) Determination of the performance of a coating on an actual structure. These may be called 'service trials', but equally they may be used to monitor the performance of coatings.

Each of these methods of testing will be dealt with, but, first, some general points concerned with testing will be considered.

16.2 Test requirements

Of the various test requirements the most difficult to achieve is that of durability. Tests of the paint system to be used in the environment where a structure is to be constructed are likely to provide the most realistic results, but the time factor precludes the use of such tests for most situations. Although it is not always a requirement that coatings should be tested to failure, a prolonged period of testing is inevitable, particularly for modern high-performance paints. Attempts have been made to develop more rapid ways of assessing coating performance, but with limited success.

A lack of precision with tests is often excused on the grounds that 'at least it separates the sheep from the goats'. Unfortunately, even this is not necessarily so. For example, sample panels prepared with artificially thick films of paint on small test panels can perform better than in practice and the converse can apply.

Tests on the dried paint film are usually carried out to the requirements of national or international standards, e.g. ISO 4628 or the equivalent BS 3900 (1969) 'Methods of test for paint': this means that the tests are reproducible and so it should be possible for different laboratories to obtain very similar results on the same materials. From this viewpoint, such tests are useful as a method of quality control. However, they are concerned with what might be described as the 'intrinsic qualities' of the paint film under the particular test conditions. It does not follow that a test, e.g. for adhesion, carried out under the standard conditions would necessarily give the same results if carried out on a different substrate. Nevertheless, the advantages to the user of such test methods, which will be considered in Section 16.3, is that they can form a basis for a paint specification. Despite their limitations, they do provide a general indication that a particular batch of paint meets the requirements of the specification or standard concerned. In some cases, the tests may have a practical significance, e.g. scratch-resistance, but it should be appreciated that most such tests are carried out on simple coats of paint rather than on the paint systems used in practice.

In the hands of specialists almost any test will provide useful information, provided the results are correctly interpreted. Problems can, however, arise when tests are standardised and appear in national or international standards. While it is clearly advantageous to have standard procedures and equipment, it cannot be assumed that all tests carried out to the standard will provide the same results on particular coatings or materials. It is not uncommon for results to vary between laboratories. This was demonstrated over 50 years ago when a special series of tests was formulated to compare the results of salt spray and sulphur dioxide tests carried out by experienced personnel in a number of different laboratories.[1] It is important to appreciate the limitations of tests and not to be tempted into directly applying results of simple tests to more complex situations.

Tests on paint films are usually carried out in the laboratory but it is possible to use the test methods on paint films applied to actual structures. Such tests do not always meet the requirements of a standard, which is generally related to a specified set of conditions. Nevertheless, they may provide very useful practical information.

Tests on the paint itself again provide evidence of quality and demonstrate whether it meets the requirements set out, either in standards or in the paint manufacturers' data sheets. Although such tests are usually carried out in properly organised laboratories, some are straightforward, requiring simple apparatus, and can suitably be carried out on-site, e.g. specific gravity tests.

Tests involving the properties of the paint film are those primarily concerned with application and are probably the most important from the users' standpoint. These cover paint film thickness, both wet and dry, sag tests, holiday detection and porosity. All of these have a direct influence on durability in the sense that films containing holidays or applied at thicknesses outside the specification requirements are likely to fail more quickly.

All the above types of test play an important role in ensuring that the quality of the paint, its application properties and its actual application meet the requirements set either by the manufacturer or users of the paint. However, they do not provide direct information on the probable durability of the paint system to be used. There are two general ways of attempting to make such assessments.

The best way of determining the performance of a coating or paint system is to actually expose it to the conditions it will have to withstand in practice. This is generally quite impracticable except where long-term procedures for maintenance painting are involved. In such situations, it may be possible to apply coating systems to the actual structure and then to monitor their performance over fairly prolonged periods. This is a form of service trial and probably is the best type of method for determining paint durability. However, it has limitations, which will be considered later. The more usual way of testing paints in natural environments is by means of what are generally termed 'field tests'. Paint coatings are applied to steel specimens and exposed to natural environments, such as the atmosphere, immersed in seawater, or even buried in soil. Although the environments are obviously only representative they do, at least to some extent, reproduce many of the factors met in practice. These will be considered in Section 16.5, but clearly there are some technical limitations to such tests.

There are other disadvantages with 'natural testing', not least being the costs involved. To carry out a field test correctly requires at least one properly organised test site, and generally more than one in order to cover a range of conditions. The test sites may be some distance from the main workplace, which involves travelling at regular intervals, and usually the site has to be monitored for environmental factors such as pollution, rainfall, etc. The biggest disadvantage of field tests is the time required to obtain results. Although it is not always necessary to expose coatings until they fail, a prolonged period of testing is usually required. These disadvantages and limitations have led to attempts to develop more rapid ways of assessing the performance of coatings. These have followed three directions:

(i) Attempts to simulate the main factors involved in natural exposure by designing either complex testing cabinets or more commonly by developing fairly simple tests such as the 'salt-spray'.

(ii) On the basis of theoretical principles of paint protection, the development of tests to determine characteristics related to performance. Such tests have mainly been based on various measurements of the electrical properties of the film, e.g. electrical resistance or permeability, and have involved a variety of techniques. These have included electrical impedance measurements, differential scanning calorimetric techniques (DSC), resistance measurements, ellipsometry techniques and many other approaches.

(iii) Development of combined laboratory technique and natural exposure tests, where the paint coatings are exposed under natural conditions but the changes in properties are investigated by some of the techniques used in (ii).

Tests to determine the resistance of paint films to a variety of different chemicals and solutions are usually effective if properly carried out, e.g. suitable immersion tests.

Despite the considerable efforts that have been applied to laboratory tests, it must be said that few of them have proved to be particularly useful in predicting the practical performance of paints. This is a sweeping statement but both authors have been involved in the conduct and development of test methods. Based on this experience, they must conclude that while in the hands of specialists all the test methods provide useful information, none of them can be considered to be in the category of standard performance tests. Many of them are useful control tests for quality and reliability of manufacture and others provide a good indication of some general characteristics such as resistance to ultraviolet light. The tests in group (i) are often called accelerated tests, but generally this is a misnomer because it is virtually impossible to determine a factor relating the test results to the performance under practical conditions. Those in group (ii), although considered as test methods in some published papers, are really investigations into the nature of protection by paints. The approach noted in (iii) would probably provide the most useful form of testing, but as yet no standardised procedures have been developed.

Although testing has limitations, it is necessary both for the development of new products and for assessing current paint coatings. In the following sections the test methods and equipment used will be considered.

16.3 Laboratory testing of paint films

Before discussing the various tests available for testing paint films, it is worth considering the preparation of the film to be tested. In standard test procedures it is necessary to have the paint film itself prepared in some standardised way. This leads to certain problems because when paints are applied to a steel substrate their properties and the structure of the dried film are not the same throughout the film. One surface is in contact with, and adhering to, the steel with the nature of the adhesive forces depending upon the state of the steel substrate. The paint surface in contact with the environment, air and water, has different properties, depending upon the internal stresses in the film, and the film itself is not homogeneous. This has been shown by various workers, e.g. Mills and Mayne[2] have detected what they term 'D' areas, which have properties different from the rest of the film. Furthermore, even within the films, the drying process influences the structure, particularly in the period immediately after the film has dried. These effects may persist for some time as the drying process continues. The visco-elastic properties of paint films cause certain problems that do not arise with other materials such as metals. Although metals are not completely homogeneous they can be manufactured into representative test specimens, which provide a reasonable assessment and measure of properties such as hardness, elasticity, etc. It is more difficult to achieve the same degree of objectivity when preparing paint films for testing.

The method of application of the paint has an effect on the properties and structure of the film. It is therefore necessary to standardise the application procedure so far as is practicable. Additionally, the paint must be prepared and sampled to ensure that the tests are carried out on representative samples. Details are provided in most standards concerned with paint testing. These usually cover application by spray, brush and dipping. However, a method using what is generally termed a 'doctor blade' is sometimes referred to and it is worth commenting on this. In this method, a calibrated blade is drawn over paint applied to the surface of a test panel, the excess paint being removed by the blade (Figure 16.1). The advantage of this method is the production of an even wet film, the thickness of which lies within a comparatively small band of values. This ensures that tests can be carried out on films of about the same thickness.

Figure 16.1   Preparing test panels by 'doctor blade'.

This may be convenient and suitable for some test procedures but the method of application bears no relation to those used in practice.

If this method is used for comparative testing of different paints, the results may be misleading because an advantage of some paint films may lie with the thickness of coating produced by methods such as brushing or spraying. For example, paints for brush application require a compromise between good flow characteristics to provide a smooth film without ridges and furrows and sufficient structure to prevent attenuation of the film on contours and edges. Inevitably this will differ with individual formulation.

Many of the paint tests included in standards such as BS 3900 are not of direct interest to most paint users. They include tests for properties such as fineness of grind, wet-edge time and combustibility and will not be considered here. For those who require such tests, the various standards provide a clear description of the methods to be used. Other tests which will be of more interest to the general user are considered below, but the appropriate standards should be studied to obtain the full requirements of the test procedures.

16.3.1   Determination of drying time

The drying time may be defined as the length of time that elapses between the application of the paint and a certain level of hardness and 'dryness'.

There are gradations in the *level of hardness* and varying terminology is used: *hard dry*, dry enough to be handled and for further coats of paint to be applied – sometimes described as 'overcoating time' and 'through-dry'; *surface dry*, indicating that the surface is dry but the rest of the paint film may still be soft, sometimes called 'dust dry'; the other commonly used term, *touch dry*, is self-explanatory.

The tests to determine these various stages are straightforward. The two most commonly used tests are for 'surface' and 'hard dry'. The former can be determined by distributing sand grains or small glass balls (ballotini) of a specified size onto the surface to be tested and lightly brushing them off. If they do not stick to the surface, the paint is surface dry. Traditionally, painters have tested for 'hard dry' by twisting their impressed thumb on the painted surface; movement of the underlying paint indicating that the underlying layers are still soft. Accordingly, this method has been copied as a laboratory test apparatus which faithfully reproduces the action. The thumb test, however, is still a useful guide, particularly at the work face. Stages between these two may be checked in various ways. 'Tack-free' may be considered as the stage where a light weight, about 200 g, placed on a filter paper on the paint film causes no damage.

The drying time is influenced by temperature, which is specified in standard tests and also by the degree of ventilation. Drying time does not involve precise measurement, but is useful and generally appears on manufacturers' data sheets. A test connected with drying time is sometimes specified. This is a pressure test to determine whether two painted surfaces will stick when pressed together. This can occur if steel sections are stacked before the paint is hard enough to resist such pressure. The pressure test will indicate the delay required before sections can be stacked. It is carried out by painting circular discs which are placed together in a suitable apparatus under specified pressures.

16.3.2 Gloss

Gloss is the ratio of light reflected from the surface to light incident on the surface. It is measured as a percentage of light reflected from the sample and compared to a standard reflective surface, usually black glass. There are a variety of instruments available for this measurement.

16.3.3 Hiding power

Hiding power is the property of a coating which enables it to hide the surface over which it is painted. There are a number of test methods but the most familiar is the use of hiding power charts. These are coloured or black and white patterned cards to which the paint is applied and an assessment is made of the extent to which the pattern is obliterated.

16.3.4   Adhesion tests

Adhesion is a property of paint films that is difficult to assess other than in comparative terms. Adhesion has been discussed in Section 4.4.1, and its practical significance on performance cannot easily be established. Where the adhesion is poor in the sense that the paint can be comparatively readily removed from the substrate, this is generally considered to indicate the existence of a potential problem.

Many tests have been devised to test adhesion; some 20 methods have been listed in a publication by Gardner and Sward[3] and this is not complete. Some methods give a numerical measure of adhesion, whereas others are qualitative. Some test the adhesion by direct pull-off of the paint film, whereas others use a shearing mechanism. Almost all measure to a degree some other property such as brittleness or cohesive strength. There is still a good deal to be learnt about adhesion and its influence on the durability and performance of paint films. Therefore, all the available tests are open to some form of criticism but in practice serve a useful purpose if their limitations are taken into account. Some of the more commonly used methods are noted below:

The 'cross-hatch' or 'grid' method is comparatively simple and is probably the most widely used practical test. A series of 11 parallel cuts each, e.g., 1 mm apart, are made through the paint film to the steel substrate. This is then repeated across the first series of cuts at right-angles to them. This results in a grid of 100 squares each 1 mm^2 in area. The cuts can be made separately with a sharp knife or razor blade. However, a specially designed cutter, which forms the parallel cuts in one action, is also available. The adhesion is generally determined by applying a piece of self-adhesive tape to the grid, pressing it onto the paint film, then jerking it away. Alternatively, light brushing may be specified. The number of squares remaining indicates the adhesion, i.e. the fewer removed, the better the adhesion, but both the British and International Standard also categorise the appearance of the cuts even if no squares are detached. The width between the cuts should be increased to 2 mm for films between 50 and 250 μm in thickness.

Although the 'cross-hatch' method is straightforward when carried out in the laboratory, difficulties can arise when it is used on paint applied to structures, so a simpler method may be employed. This is discussed in Section 11.4.1.2. It consists basically of using a sharp knife blade to cut a St Andrew's Cross through the paint system to the steel substrate, then removing the coating by a suitable method to check the adhesion.

Pull-off tests are also widely used, partly because they provide a numerical value for adhesion, although its relation to actual film performance is vague. Various designs of apparatus for carrying out the test are available, but the principle of operation is similar for all of them. A small cylindrical piece of steel of known diameter, sometimes called a 'dolly', is glued to the paint film at right angles to the film surface. It is then connected to a device that exerts a force sufficient to pull off the small steel cylinder. The force required to remove the cylinder is a measure of adhesion if the coating is pulled from the steel substrate. Sometimes the film breaks under the applied force. Depending on where the break occurs, this may be a measure of intercoat adhesion or cohesion. Although the test is simple to operate, it is essential for the force to be applied at right angles to the film. Epoxy or acrylic adhesives are used to attach the dolly to the surface to be tested and to avoid wasting time with valueless adhesive failures, the adhesive must cure for the recommended time. Quick-setting adhesives such as the cyanoacrylates are generally not suitable, as their solvents can affect the coating.

One problem with this test is whether to cut the coating around the circumference of the dolly before the test. Some instruments are provided with a special cutter for the purpose. For coatings with very high cohesive strength, unless this is carried out the test will try to detach a very large area of the coating and the result will be meaningless. However, the very act of cutting the coating can cause micro cracks that will lower the pull-off strength.

A simple method of checking adhesion is to push a knife blade between the paint film and the steel substrate, then to prise off the paint. This is probably the oldest method and is clearly subjective. Nevertheless, in the hands of an experienced operator, it is a quick and useful test. A mechanical form of this method has been used, called the adherometer. The force required to remove the paint film with a small sharp knife can be measured.

16.3.5 Abrasion resistance

Abrasion takes many forms. It can be slow attrition by, for example, sand slurries. It can be a continual polishing action, e.g. the rubbing of dirt-engrained overalls. It can be the abrasive blasting of sand blowing in the wind. It can be a heavy weight dragged across the surface, and so on. It is therefore important to carry out the type of test relevant to the service required.

Abrasion resistance tests are probably more commonly carried out on special paints used for ships' decks and road marking rather than for protective coatings. Various test methods are used but they generally rely upon testing the resistance of the coating to the abrasive action of a material such as sand. The abrasive material is either dropped or blown onto the painted surface. A more convenient and quicker method is by using the Taber abrader (see Figure 16.2). The sample is mounted on a turntable, which rotates under a pair of weighted abrading wheels and the weight of

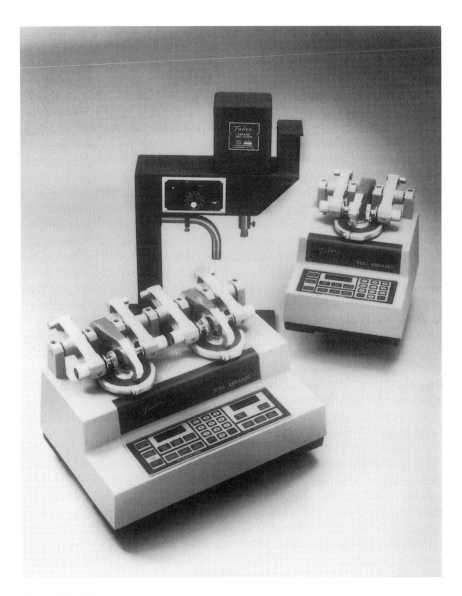

Figure 16.2   Taber abrader.

Source: Sheen Instruments Ltd.

coating lost per thousand revolutions is measured. Although it lacks repro-
ducibility, ASTM Standard Test Method D 4060 is the best known and
most widely specified method.

16.3.6   Physical state of the film

A number of tests are used to determine the physical state of the film, in
particular the hardness and resistance to deformation. Some of the tests
probably cover more than a single property, so they have been grouped
together. Many of these tests are more appropriate to stoving finishes
applied to sheet steel used for domestic appliances and some of them
cannot be carried out directly on paints applied to steel plates. Nevertheless, the properties may have relevance to protective coatings, particularly
where some form of test is included in a paint standard or specification.

The scratch test, as its name implies, is a method of determining the
ability of a paint film to withstand the scratching motion of a sharp pointed
object. The ability to withstand scratching is less important for protective
coatings than for stoved decorative coatings. However, the test itself does
also provide an indication of the hardness of the paint film. Many devices
have been developed for this test, including the use of pencils of different
hardness, i.e. 9H to 6B. The usual instruments for scratch resistance are
based on the use of a hardened steel point attached to an adjustable arm
which is weighted in some way. The point is then moved down the paint
film and the weights are adjusted until the paint film is penetrated to the
substrate. Such devices may be manual or automatic.

The hardness of the paint film can also be measured by various pendulum devices. These are based on the damping effect of the film on the
movement of a pendulum. As the pendulum loses kinetic energy, the
amplitude of the swing decreases. This can be used as a measure of hardness because the hardness of the paint film is related to the energy
absorbed and the time the pendulum continues to swing. Many different
types of pendulum test have been devised but two types are now in
general use, the Persoz pendulum and the König/Albert pendulum, the
latter being considered to be easier and quicker to use. The methods of
use are published in various standards, which include the temperatures
and relative humidity requirements. There is also a limit to the variation of
film thickness allowed for comparing different paint films.

Indentation hardness is commonly used as a measure of the physical or
mechanical properties of alloys such as steel. However, because of the
visco-elastic nature of paint films, the instruments cannot be used in quite
the same way on them and often the time factor must be taken into
account when carrying out the test. Basically, the instruments used for the
test are micro-hardness testers employing an indenter, which may be a ball
or pyramid-shaped, pressed into the paint film under a specified load. The

method is not generally used as a routine measure for protective coatings but the determination of the depth of indentation at various periods does provide information regarding the mechanical properties of the film. In the USA both the Bar Col and Shore Durometer hardness testers are widely used for the testing of thick coatings, plastics and sealants.

There are a number of tests available for determining the flexibility or resistance to deformation of paint films. The two most commonly used are the mandrel test and the Erichsen test. The mandrel test is carried out by bending steel sheet specimens coated with paint around cylinders or mandrels of different diameters. The film is placed in tension, being on the outer face of the steel specimen, and the size of mandrel required to produce cracks in the film indicates its flexibility. The flexibility is greater for films that resist cracking when bent on smaller mandrels. Conical mandrels are also used to allow for a series of tests on one piece of equipment.

The Erichsen test also provides information on film flexibility and resistance to deformation. It is a cupping test in which the coated specimen is placed horizontally on an apparatus so that the paint film is in contact with a hemispherical steel head (20 mm diameter). The head is pressed into the specimen under load until the paint, which is stressed as the steel specimen is itself extended into a hemispherical shape, cracks. The cracks are usually determined by examination with a magnifying device (×50). The criterion is the smallest depth of indentation that results in cracking. The depth is usually measured directly from a scale on the apparatus. This test is widely used for industrial finishes but is probably of less value in determining the properties of protective films because such films are rarely deformed in this way during manufacturing processes. Furthermore, the test is influenced by the degree of adhesion between the paint and the substrate, which is standardised in the test procedures but may not relate to practical conditions for painted structural steel.

A test in this category that is worth noting is the impact or falling-ball test. This consists of dropping a steel ball of standard size and weight from a predetermined height onto a painted specimen. The effect of the impact on the paint film is then determined. This is a useful test for determining the ability of different types of paint to withstand impact, which commonly occurs during the handling and transport of painted steel sections. For use as a rapid indentation test, the steel ball can be dropped onto the reverse side of the specimen, i.e. the non-painted side. The test is sometimes carried out with the panels at low temperature as this is considered to provide additional useful data.

For two-pack or heat-cured materials, such as urethanes and epoxies, it may be necessary to determine a 'degree of cure'. This is quite a difficult quantity to define, even with the aid of sophisticated laboratory equipment, but a 'solvent wipe test' is a convenient and sometimes surprisingly accurate method. It involves rubbing the surface with a cloth impregnated with a particular solvent (as specified by the paint manufacturer) to see to what extent the coating is removed. For thin coatings such as coil coatings, the number of rubs necessary to remove the coating from the substrate is sometimes specified.

16.3.7   Film thickness

The thickness of a paint film is an important property and is generally specified for protective coating treatments. Unlike most of the tests considered above, it is regularly carried out in fabricators' shops or on painted structures. The instruments for determining film thickness are commonly used by operators and others who are not usually concerned with the laboratory testing instruments. As these tests are a vital part of quality control, they are discussed in detail in Chapter 9, dealing with that topic. However, for testing purposes, particular methods may be used that would not generally be applied to quality control work. It will, therefore, be convenient to summarise the various methods of measuring the thickness of paint films.

(i) The average thickness of the paint film can be calculated from the amount of paint used to cover a specific area of steel, provided certain properties of the paint are known.

This method can be used satisfactorily for brush application. It is necessary to know the volume solids or percentage of solvent by volume and to make allowance for the roughness of the surface and loss due to application. The method has a practical use when it is difficult or impossible to measure wet or dry film thickness by the usual instrumentation methods, but it obviously will give no indication of the uniformity of the coating.

The formula used is based on Imperial measurements but can be converted to metric units:

$$t = 1930G\left(1 - \frac{PM}{100V}\right)\Big/ A$$

where t is the film thickness in mils (0.001 in), G = gallons of paint used to cover area A, A = area covered (ft^2), P = weight of a gallon of paint (lb), M = volatile content (per cent by weight) and V = weight per gallon of volatile matter (lb).

The values of P and M can be measured by weighing a known volume of paint and by measuring the loss in weight on drying. V can be measured in the laboratory but it is usually possible to obtain the data from the paint manufacturers. Many data sheets contain the nominal solids content by volume (per cent), so a direct conversion from the weight of wet paint to dry paint can be made.

(ii) Where a small strip of the paint film can be completely detached from the steel, this can be measured by means of a micrometer or, after suitable mounting, with a graduated microscope.

(iii) The most common method of determining the thickness of a paint film on steel is non-destructively by using instruments based on magnetic or electromagnetic principles. A range of such instruments is available and their operation is based on the magnetic attraction between the steel base and the steel probe on the instrument. The paint film acts as an air gap, so affecting the magnetic force, and this can be related to the film thickness. The instruments are generally calibrated on shims of suitable thickness. These instruments are considered in detail in Section 9.5.3.6.

16.4 Testing of paints

The tests considered in Section 16.3 have been concerned with paint films, but there are also tests on the liquid paint that are of interest to the paint user. Of these, only one, the determination of the density, is likely to be carried out in a routine way as a check on the paint itself. Other tests, such as flow time, are really within the province of the paint maker. The determination of flow time is straightforward and is carried out by using a flow cup container with a standard orifice. The temperature of the test is standardised and the cup is filled with paint and the time is measured until the first break occurs in the flow, or until a standard volume, e.g. 50 ml, has been collected.

A sag test is a useful laboratory test to determine the thickness of paint that can be applied on a vertical plane without sagging. The result must be related to ambient temperature. The test is described in BS 3900.

The density of the paint is determined using a small cup generally known as a 'weight per gallon cup', although nowadays densities are expressed in metric terms (kg/litre). The cup, which holds a standard volume of paint, e.g. 100 ml, is filled and a cap with a small hole is placed on top. Excess paint exudes through the hole and is removed. The cup with paint is weighed and this weight, after subtracting the weight of the cup and cap, provides the necessary data to calculate the density:

$$\text{Density} = \frac{w}{V} \text{ kg/litre}$$

where V is the volume of the cup in millilitres and w (the weight of the paint) is in grams.

There are occasions where the volatile content of the paint is required. Provided a controllable oven is available, this can be simply determined. A small amount of paint, e.g. 2–3 g, is placed in a suitable container and weighed. It is then placed in an oven at 105°C for about 3 hours. It is then removed and re-weighed. The container and paint are returned to the oven for a further half hour, removed and re-weighed. If the weight is not the same as that obtained after 3 hours, it is returned and reweighed until the weight is constant. The difference in the original and final weight provides measurement of the volatile content, which can be expressed as a percentage of the weight of the original paint sample:

$$\text{Volatiles (\% by weight)} = \frac{w_1 - w_3}{w_1 - w_2} \times 100$$

where w_1 is the original weight of paint and the container, w_2 is the weight of the container, and w_3 is the final weight of paint and the container. All the weights must be in the same units.

16.5   Laboratory performance tests

The tests discussed above do not provide a direct indication of the durability or protective value of the paints exposed under practical conditions. As already discussed in Section 16.1, attempts have been made to develop laboratory tests that will provide data on the durability of paint coatings. Although there must be doubts regarding many of these test methods, they are widely used, so they will be considered briefly with comments on their limitations.

Two separate aspects of such tests must be taken into account: (i) durability and appearance of the coating itself, and (ii) the effect of any coating breakdown on the corrosion of the steel substrate. The two important factors determining (i) are ultraviolet radiation and moisture, whereas (ii) is mainly influenced by moisture and pollutants and contaminants such as sulphur dioxide (SO_2) and chlorides. There are four types of laboratory test in general use:

(i)   Weathering resistance
(ii)   Salt-spray
(iii) Sulphur dioxide
(iv)   Humidity

16.5.1   Artificial weathering

In their lifetime, coatings are subjected to a wide variety of destructive elements. These include corrosive atmospheres, rain, condensation, sunlight, wet/dry cycling and temperature cycling (see Figure 16.3). There are several methods that attempt to reproduce these and accelerate the weathering process on test panels. Frequently the destructive elements have a synergistic effect on one another and this is one reason why weathering apparatus that concentrate on one element do not correlate well with practice.

Probably the most successful include a cycle of severe exposures, such as

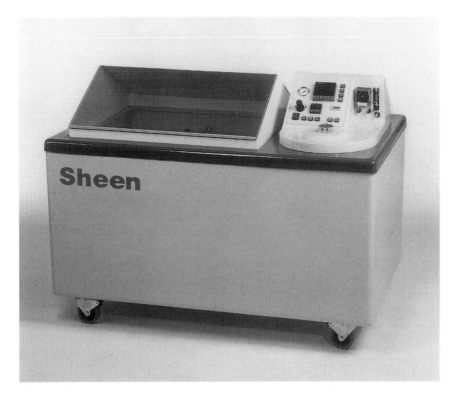

Figure 16.3   Cyclic corrosion testing cabinet.
Source: Sheen Instruments Ltd.

the Prohesion test.[4] Rather than correlate with atmospheric service, such tests are probably more useful for comparing coatings of a similar type. They may make the difference between selecting a coating system that affords effective protection and selecting one that fails. However, it is claimed that ASTM D 5894-96 cyclic test can give good correlation with atmospheric exposure.[5]

16.5.2   Salt-spray tests

These tests are more widely used than accelerated weathering tests and because the salt is based on sodium chloride (NaCl) it might appear that such tests would provide a reasonable assessment of the protective properties of paint films. Although there is some correlation with marine conditions at least with some paint coatings, on the whole the test does not provide a very good basis for classifying the resistance of coatings to natural environments. This has been demonstrated in a number of test programmes, particularly those carried out by the former British Iron and Steel Research Association.[1]

There are a number of variations in the standards for salt-spray and salt droplet tests but they all operate on the same broad basis. A salt solution, usually containing a specified percentage of NaCl or sometimes more complex mixtures, is either sprayed as a mist or atomised into droplets, which reach the painted steel (see Figure 16.4). The specimens then remain damp by virtue of the high humidity in the test cabinet. Various modifications to the salt-spray test, e.g. the addition of acetic acid to accelerate attack, have been used for some purposes. Although the test rapidly picks out defects in the film, e.g. holidays, it cannot be regarded as providing a true acceleration of what occurs in practice. Some types of paint, noticeably zinc-rich primers, can give an outstandingly good performance under very severely corrosive test conditions, whereas in a natural environment which is less corrosive in both duration and degree other priming systems may perform equally well or better. Work in the Netherlands has shown that waterborne coatings (acrylic and alkyd) in the salt-spray test show particularly poor correlation with atmospheric exposure, whereas the cyclic test in the Prohesion Cabinet gave relatively good results.[6]

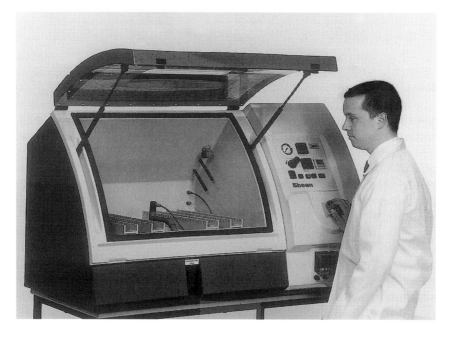

Figure 16.4   Salt-spray test cabinet.
Source: Sheen Instruments Ltd.

16.5.3 Humidity and condensation tests

Humidity tests rely on condensation of moisture on painted panels to assess the resistance of the paint film to water absorption and blistering. Although humidity testing may be useful for testing materials to be used in the tropics and for certain problems concerned with packaging, it seems less useful as a means of assessing paint films. Blistering of paint films is often far more pronounced in humidity cabinets than would be expected from the performance of the same films under natural exposure conditions. The effect of condensation of moisture on paint films is a likely accelerator of breakdown and subsequent corrosion of the steel substrate, but the conditions of most humidity tests do not appear to reproduce natural condensation conditions particularly well.

Other forms of condensation tests have been developed. The 'Cleveland' box consists of an insulated metal tank containing water which is raised in temperature very slightly by the means of aquarium-type heaters. The test panels are supported to make a continuous roof to the tank, with the painted surface facing the water. The slight difference in temperature between the two surfaces of the metal test panel causes light condensation to form on the paint. This test has the advantage that temperatures are closer to those experienced under normal ambient conditions and the condensation formed is similar to atmospheric dew.

16.5.4 Other laboratory tests

Many special tests have been devised to assess particular properties of coatings. Coatings for use under immersed conditions in seawater have been tested in rotor apparatuses. In these either the main tank revolves around the specimens or, more commonly, specimens are rotated in a circular tank containing natural or synthetic seawater. Special rigs have been designed to investigate coatings used under heat transfer conditions such as occur with risers on offshore platforms. Methods have also been developed to assess heat resistance. These special tests have not been standardised and often have been used by only one testing organisation. Many such procedures rely very much on the expertise of those involved in the design and operation of the test and are not suitable for general use.

16.6 Instruments for specialised analysis

A number of instruments and methods are available for tests on coatings and coating materials. These are not used routinely but provide important information about coatings, which usually cannot be obtained by other methods. They are particularly useful for investigating paint failures and for determining the types of coatings present on structures. This may be of value prior to maintenance painting.

Arc/spark spectrograph. This test is used to determine the inorganic elements in a coating. It will detect the presence of metallic elements such as lead, zinc and barium. A small sample of the coating is subjected to an electric arc and the light so produced is transmitted through a prism, which divides the light into a spectrum. This is compared with standards and, as each element has a distinctive spectrum, the presence of individual elements can be determined. On some spectrographs, the comparative analysis is carried out by computer. Generally, the test is qualitative as this is all that is required, but quantitative measurements can be made.

If coatings contain lead this must be taken into account when planning maintenance; this method can be used to determine the presence of metallic pigments.

Atomic absorption. This is also a test to determine metallic and other inorganic elements in a coating and is a quantitative test. The coating sample is dissolved in an acid and then sprayed into a flame. The light is analysed by means of a computer to provide the information regarding the type and amount of a particular element.

Infrared spectrograph. This instrument can be used to analyse the binder or resin used in a coating. Infrared radiation is passed through a sample, which absorbs infrared frequencies according to its composition. The absorption is then analysed to provide the required information.

Gas–liquid chromatography. This method is used for the analysis of organic liquids, and so can be used to determine the types of solvent used in paints.

Differential scanning calorimeter. This is a method of measuring the gain or loss of heat in a chemical reaction and can be used to determine the curing characteristics of heat-cured coatings such as fusion-bonded epoxy.

16.7   Field tests

The application of paint to a steel panel followed by exposure outdoors is one of the simplest and oldest methods of assessing the performance of coatings. However, its very simplicity as a method can lead to problems. Often the problem with field tests, in common with many other types of test, is the lack of a clear aim when formulating the programme.

Many paint companies carry out tests for a number of reasons other than the determination of protective properties. These include the development of new coatings, tests for colour retention, comparative tests on formulations containing different raw materials and others of a similar nature. All these tests of paint properties are important but the techniques required are often different from those used for assessing the protective properties of coatings. Where the properties of the film, e.g. gloss, are being assessed, the substrates to which the paints are applied may not be particularly important. Furthermore, if the tests are basically to determine the decorative effects of the paint film, it is sensible to ensure that a maximum of ultraviolet light reaches the coated specimen. It is, therefore, usual to expose flat specimens at 45° facing south (in the Northern Hemisphere). Generally, steel sheet of comparatively thin gauge is used for such tests. This makes for easy handling and fairly simple wooden exposure racks are often used.

Tests to determine the protective qualities have factors in common with the above procedures but the requirements are not necessarily the same. In temperate climates, moisture is often more important than sun, although this may not be so in other places, e.g. the Arabian Gulf. Consequently, the sheltered surfaces may be more important than those exposed to the sun. Again, north-facing surfaces remain damp for longer periods than south-facing ones (in the Northern Hemisphere) so the orientation of specimens should be considered in relation to this. Rain has a two-fold effect on protective coatings. While it provides moisture, essential for corrosion, it also washes corrosive deposits from the surface. Consequently, coated surfaces sheltered from the rain may perform less well than those that are freely exposed. Furthermore, paints applied to panels of thin sheet steel will generally dry out quicker than on heavy steelwork and the degree of exposure to winds will also affect the duration of periods of condensation.

Vertical and horizontal exposure may provide more useful results than the more common angled exposure. It is easier to compare the performance of coatings applied to standard flat specimens but examination of paint on a structure will show that breakdown often begins at features such as welds, joints, nuts and bolts and faying surfaces, rather than on the main steelwork that is freely exposed. It may, therefore, be preferable to design specimens that contain some or all of these features. Coatings applied to such complex specimens are more difficult to assess than those applied to rectangular sheet or plate, but the information provided may be of more value. A compromise is to use small I-beams. Tests care carried out for a variety of reasons and the procedures will, to some extent, be determined by the purpose of the work. Typical reasons for field tests are:

(i)   Comparison of different paint systems to obtain general information.
(ii)   Comparative tests to determine a suitable coating system for a specific purpose, e.g. a bridge in a given location.
(iii) Tests to assess the effectiveness of different primers in a paint system.

(iv)   Assessment of different methods of surface preparation of the steel-work.

(v)   As part of a research investigation, to determine the effects of factors such as film thickness.

There are, of course, may other purposes for which field tests would be carried out. Once the aim of the tests has been clearly defined, a number of decisions regarding the test procedures must be made. Some of these are discussed below.

16.7.1   Type of specimen to be used for the tests

For tests on coatings for structural steelwork it is preferable to use steel plate rather than sheet for the specimens. This enables the surface prepa-ration of the test specimens to be similar to that to be used in practice: blast-cleaning is more difficult to carry out satisfactorily on thin gauge material.

The specimen size is not, of itself, important but generally larger ones are more representative of practical conditions. Furthermore, the applica-tion of the coating is likely to approach a practical situation as the size of specimen increases. The method of application raises a problem. Contrary to popular belief there is some evidence that different methods of applica-tion produce paint films of different degrees of homogeneity.[7] Most heavy industrial painting is carried out using an airless spray gun but this gener-ally requires too large a sample of paint and has too high an output for practical and economic preparation of test panels. Generally the use of air-spray, rather than brush for the application, and a specimen size of, say, 25 cm^2 provide a reasonable compromise between obtaining a representat-ive area and avoiding problems of handling heavy steel plates.

As noted earlier, inclusion of channels or more complex shapes, includ-ing welded areas and nuts and bolts, may be advantageous in some situ-ations. For special tests a range of differently shaped specimens may be used, e.g. pipes, tubes and small angles. Sometimes very large specimens may be used, e.g. for tests on coatings suitable for sheet piles.

16.7.2   The coating

The coating to be tested will be determined by the nature and purpose of the test. However, there are a number of points worth noting. The thick-ness of each coating in a multi-coat system as well as the total thickness should be measured. Each coating should be applied to the manufacturer's recommended thickness when proprietary paints are being tested. Thick-ness has an influence on performance so this must be taken into account when assessing the results. It is not uncommon for the results of field tests to be more a reflection of coating thickness than the intrinsic qualities of the paints.

It is difficult to achieve what might be termed 'practical application conditions' on test specimens. This arises in part from the size but also because applicators tend to be particularly careful when painting test panels. It is difficult to overcome these problems but, with spray application, it is advisable to place the test specimen on a larger background to ensure more realistic coverage. The use of the same applicator for all the test specimens is also likely to provide a degree of standardisation.

It may be considered appropriate to 'damage' the coating in some way to assess under-rusting at the damaged area. This can be achieved by applying a thin strip of self-adhesive tape to the specimen before applying the coating. This is removed after the coating has dried, so leaving a bare strip. Any residues from the adhesive can be removed by solvent of a type that does not affect the paint. This method is more suited to larger areas. Scratches made after the paint has dried and carried out so as to expose the substrate are widely used. A suitable sharp scribing device is required to ensure that the substrate is exposed, but the method is less reliable than the 'tape' method described above because paint films have differing degrees of elasticity and recovery and therefore the configuration of the scratch mark will vary.

It is usual to coat both sides of rectangular test specimens and all surfaces of more complex test pieces. Generally, the edges do not receive the same thickness of coating as the main areas, so it is advisable to apply stripe coatings to a distance of 15–30 mm from the edge depending on the size of the specimen. These areas are not usually included as part of the test area for assessment purposes.

Specimens are usually painted indoors and ideally this should be carried out under controlled conditions of temperature and humidity in well-ventilated conditions. Generally, temperatures of $23 \pm 2°C$ and a relative humidity of 45–55% are considered suitable. Provided all the coatings are applied over a short period, slight variations in these conditions will not be critical. However, wide variations should be avoided. Specimens should be painted in a vertical position on proper racks positioned away from direct sunlight. Where coatings are applied outdoors, the conditions and time of year should be appropriate to the practical conditions that will eventually be encountered. Suitable covers should be available to avoid rain dropping onto the specimens before the paint coating has dried.

16.7.3   Exposure of specimens

Many field tests are carried out at sites organised for the purpose (Figure 16.5). The essential requirement is a test rack (or racks) to which the specimens are fixed. Depending on the nature, scale and duration of the

Figure 16.5   Exterior paint panel exposure racks.
Source: Q-Panel.

tests, these will usually be made of wood or steel. For the larger test specimens preferred for tests on protective coatings, steel racks will usually be more suitable. The racks are designed to provide the desired orientation for the exposure, e.g. facing south at 45° to the horizontal, and generally are straightforward in design and construction. The main point is to ensure good drainage and to avoid situations where water runs from one specimen onto another. There are a few other points to be noted.

The test specimens must be insulated from each other and the stand. Plastic washers are frequently used to achieve this. Figure 16.6 indicates various methods of fixing specimens to an exposure stand. The stands should be placed on firm ground and on reasonably permanent sites; it is advantageous to place them on a concrete base. Where a series of stands are required they should be placed so that no shadows are cast by one stand on another. Similarly, they should be placed so that buildings, trees, etc., do not cast shadows. If this is impracticable then the stands should be situated so that they all meet the same exposure conditions. The test racks

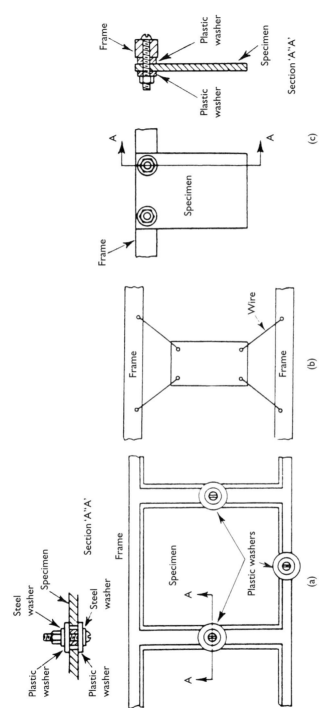

Figure 16.6 Methods of attaching specimens to a test rack: (a) using plastic washers (three washers are also required on the other side of the frame); (b) corner hole suspension using wire; (c) top hole suspension.

should be placed well clear of the ground, preferably at least 1 m above ground level if situated on growing vegetation as this will avoid dampness. For tests on protective systems, the use of H-girders will usually be more suitable. Each girder should be well spaced from its neighbour and laid with a slight slope from the horizontal, at least 1 m from the ground. This will then provide a variety of exposure conditions on the flanges, webs, uppermost and underneath surfaces.

16.7.4   Test sites

Often the test site is an area close to the location of the particular organisation responsible for carrying out the tests. This may be satisfactory for a particular series of comparative tests provided the conclusions drawn from the tests are relevant to the conditions. Generally, though, one test site cannot represent all the conditions likely to be encountered in practice. Various classifications have been adopted to represent practical conditions. The two main factors for tests on protective coatings are the climatic conditions and degree of pollution, although additional factors such as the effects of splash zones where steel is partially immersed in the sea may also have to be taken into account. Sometimes the site is chosen to represent a specific set of conditions.

16.7.4.1   Climatic conditions

Where tests are carried out in a specific country where the climatic conditions do not vary to any great extent, one test site may be sufficient. However, for test programmes with a wider purpose, the effects of climatic variations must be considered. The scope of such conditions has been considered in BS 3900 and should be consulted. The broad categories of climate are as follows:

(i)   Humid tropical: (a) with, and (b) without a dry season.
(ii)   Tropical.
(iii) Very dry: steppe and desert.
(iv)   Warm temperate, subdivided into climates: (a) without a dry season, (b) with dry summer, and (c) with a dry winter.
(v)   Polar.
(vi)   High mountain.

The climatic conditions have a considerable effect on the durability of paint films, in particular ultraviolet radiation and humidity. This influences their protective properties. Pollution also affects the paint film, e.g. the resistance of some binders to specific chemical attack. It also has a significant influence on the corrosion of the steel substrate where coatings have failed to protect, whether through damage, poor films, or local break-down.

16.7.4.2   Atmospheric 'pollution'

Atmospheric pollution arises in various ways but, from the corrosion standpoint, sulphur dioxide resulting from the burning of fuels and industrial processes is probably the most important pollutant. Locally, various corrosive species arise from industrial and chemical processes and they can be aggressive to both coatings and steel, particularly if they are markedly acidic or alkaline.

Near the coast, chloride is the most important contaminant, although being a natural part of the environment it cannot strictly be described as a pollutant. The quantity of dirt particles in the air can also influence break-down of coatings and the corrosion of steel. A dirt-laden atmosphere generally reduces paint degradation by UV but increases corrosion of bare metals.

A number of general categories of atmosphere are recognised although there is no completely satisfactory method of categorising atmospheric environments; often the local variations over a few hundred metres may be considerable. Furthermore, the construction of a process plant in an area can completely alter its nature. Unlike climatic conditions, which do not vary within the broad classifications, atmospheric pollution does not remain the same. Test sites must, therefore, be monitored to ensure that they continue to provide the conditions required for the particular tests conducted at the sites. The generally accepted classification of areas in relation to atmospheric pollution has been considered and the various broad categories are listed in Table 14.1.

16.7.5   Monitoring of test sites

All test sites should have an element of monitoring to ensure that the conditions are known and to allow for variations in test results arising from changes in the environmental conditions. The extent of the monitoring will be determined by the nature of the test programme and the relevance of certain measurements. It is common practice in laboratory tests to include a specimen with a coating of known performance under the test conditions. This can be done on large-scale long-term tests but is not always satisfactory. There is a fairly simple form of monitoring pollution, although it does not provide information regarding sunshine, i.e. ultra-violet radiation, or temperatures. The method consists of exposing small weighed specimens of zinc and/or steel and then, after a suitable period, removing the specimens and, after cleaning off all corrosion products in a suitable chemical solution, re-weighing them. From the initial and final weights, the size of specimen and the density of the metal, the corrosion rate can be calculated. The advantage of the method is simplicity and cheapness. Although it is related more to steel corrosion than to coating performance, it will provide data capable of indicating whether serious changes have occurred in the environmental conditions, i.e. whether the corrosion rate has changed significantly.

There is a range of more complex equipment and apparatus for measuring the environmental variables. Much of this is automatic and so can only be used on sites where a supply of electricity is available. Clockwork and battery versions may be available for some instruments. However, these may operate for comparatively short periods without attention, so increasing the number of visits to the site to service them. The following can be measured automatically or manually: temperature, relative humidity, direction and speed of wind, sunshine, rainfall and radiation. All these measurements will not necessarily be required for tests on protective coatings.

Measurements can be made to determine sulphur dioxide and chlorides, the most important of the pollutants in the atmosphere. Deposits of solid matter and acidity can also be measured. Some of the methods for measuring these pollutants will be discussed below. However, there is no point in collecting an enormous amount of data unless it is to be related to the tests in hand. Such data can be processed fairly nowadays with the availability of computers, and are necessary for certain test programmes. However, some caution should be exercised by those responsible for test programmes because much of the information collected, at a cost, may not be used. Furthermore, the microclimate, i.e. the climate at the coating surface, may be more important than the general environmental climate, particularly on a large test site.

16.7.6   Methods of measuring atmospheric pollution

16.7.6.1   Sulphur dioxide

Volumetric methods can be used in which the air is bubbled through a suitable solution of hydrogen peroxide and the acidity is measured. The lead dioxide method is not automatic but is simple. A lead peroxide paste is applied to a gauze, which is wrapped round a porcelain cylinder. This is placed in a small louvred box. The sulphur dioxide in the air reacts to form lead sulphate, which is determined by suitable chemical methods. The gauzes are usually changed monthly. Other methods can also be used, e.g. LeClerc's method in which a strip of filter paper impregnated with a solution of sodium hydrogen carbonate in glycerol is exposed at the site. The amount of sulphate ions can be measured by standard analytical methods. Other ions such as chloride can also be determined. All of these methods give a reasonable estimate of sulphur dioxide but variations can occur depending on the exact siting of the detection devices. When practicable, they should be freely exposed, well above ground level.

16.7.6.2   Chlorides

Chlorides cannot easily be determined with any accuracy because (unlike sulphur dioxide, which is gaseous) they react as solid particles, e.g. sodium chloride (NaCl), and these have to be collected or made to react in some way so that chloride can be determined, for example, by the LeClerc method. However, more commonly a method devised by Ambler, which bears his name, is used. In the Ambler method, absorbent gauze or filter paper is used, partially immersed in distilled water containing glycerol. The concentration of chlorides in the solution is determined by standard methods. The Irsid method, named after a French research organisation, is also used. Chlorides and sulphates are collected in a small trough, connected to or close to a flask. Rain washes the deposits into the flask and the solution is analysed. Any particles remaining on the trough are washed into the flask before it is removed prior to analysis.

16.7.6.3   Solid matter in smoke

This can be determined by passing air through a filter paper, suitably clamped between two devices connected to the tubes carrying the air. The device can be fitted to the volumetric apparatus used to measure sulphur dioxide (noted above). The intensity of the stain on the filter is a measure of the solid particles collected and it can be compared with standards to obtain an estimate of the solid content of the air.

16.7.7   Conduct of field tests

The conduct of field tests is comparatively straightforward. Larger-scale tests are carried out by organisations with expertise in this area and they will not be considered here. Smaller test programmes may be carried out by those with less experience and a few points are worth noting.

(i) Field tests generally proceed for comparatively long periods, so the test site, even if it is small, should be secure.
(ii) Identification of the test specimens is essential. For site inspections the positions of the specimens on the racks will suffice but, when they are returned to the laboratory for examination, permanent identification marks are essential. For reasonably short-term tests, identification by painting small numbers on the face of the panel may be suitable. For longer-term tests, small identification holes should be drilled parallel to

the edge. The number then relates to the distance from the top edge or side (see Figure 16.7).

(iii) Before the tests are started, the frequency and type of inspection should be decided. It is not possible to offer specific advice because this will be determined by the purpose of the test. However, the criterion of performance should be clearly defined, otherwise time will be spent collecting a large amount of data which may prove difficult to process. Where the criterion of failure is protection by the coating until rusting of the underlying steel occurs, then this is reasonably straightforward. A degree of rusting, e.g. 0.1% or 0.5% of the steel surface, is predetermined and the time to reach this is measured. The comparison between the different systems is then based on the time taken for such breakdown to occur. It is worth pointing out that most observers tend to overestimate percentages of rust, particularly where it occurs at small discrete points. Some form of assessment chart should be prepared indicating the appearance of particular percentages. For example, on a specimen 25 cm × 25 cm, even 0.5% will cover 3 cm², which can seem fairly considerable when appearing as rust spots.

This straightforward approach will not always be acceptable, not least because of the prolonged periods that may be required before breakdown occurs. Other criteria may be adopted, e.g. blistering and flaking. However, some specialist knowledge may be required to determine the importance of such breakdown in relation to the overall protective perfor-

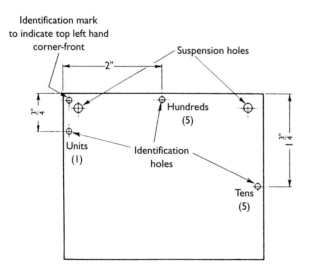

Figure 16.7   A method of identification of specimens for long-term test programmes. (Specimen No. 651, ignoring the first $\frac{1}{2}$ in from the edges, then using scale $\frac{1}{4}$ in = 1.)

mance. Furthermore, it must be appreciated that, in practice, the protective coatings will be repainted at a suitable point of breakdown and this, strictly, is the criterion of protection.

The properties of the film itself may be of interest, e.g. gloss and chalking, and there are standard methods for determining these properties.

(iv) Clearly, sound reporting procedures are essential, particularly for longer-term tests. Standard sheets with clearly defined requirements are preferable to descriptions, which are difficult to interpret. Furthermore, the various inspections may be carried out by different observers whose methods of expressing and describing breakdown may vary. British Standard 3900 provides standard methods of designating quantity and size of common types of paint defects, such as blistering, rusting and flaking. Photographs of the specimens at different stages provide a visual indication of the performance of coatings, which is useful over the test period. However, even the best photographs do not always provide a true picture of the situation so some notes should accompany them. All photographs should be clearly identified with date and specimen number, weather conditions and the face that has been photographed.

16.8   Service trials

Service trials are similar to field tests in that they are carried out in natural environments but, instead of exposing specimens on racks, actual structures are used. Examples of such tests are patch painting trials on bridges or gasholders. These may be carried out with a specific purpose in mind, such as the choice of suitable maintenance procedures and coatings. However, they may be carried out to obtain general information on coating performance under conditions that are more realistic than those for field tests.

There are advantages in carrying out tests on structures because the test areas can be quite large and the application of the coatings is more realistic. On the other hand, such tests are more difficult to control, particularly with regard to the comparisons of performance of different coating systems. The environment may vary markedly on different parts of a large structure because of prevailing winds and the effects of sheltering by the structure itself; for example, on large structures in the Northern Hemisphere, the north-facing areas and the undersides of beams will tend to remain damp for longer periods than areas that are south-facing or freely exposed.

Some points to be taken into account when considering service trials are listed below:

(i) Test areas should be chosen carefully, avoiding unusual features but taking into account representative ones such as welds and bolts.

(ii) Test areas should be grouped so that they are similarly orientated. A number of such groups may be used but the coatings to be tested should, so far as is practicable, be exposed to similar conditions.

(iii) Where tests are carried out on older painted structures, the condition of the paint substrate must be taken into account. It should be photographed before and after the test to check whether breakdown of the test systems is occurring on defects on the original paint system.

(iv) Test areas must be chosen to allow proper access for inspection and must be clearly marked for record purposes.

(v) The use of as many replicate test areas as can be conveniently accommodated is recommended. The position of replicates should be arranged randomly in relation to the structure.

(vi) All coating thicknesses must be measured. Generally, it is preferable to both weigh the paint before application and to measure the dry film thickness.

(vii) The trial should be adequately supervised and full information should be given to those involved with the structures. This should prevent accidental repainting of the test areas and unnecessary damage by operators who may be concerned with matters other than steel protection.

(viii) Although paint is usually applied to the structure, test panels may be used and fixed to the steelwork. This has the advantage that they can be removed for closer examination, and also eliminates problems arising from the presence of previously applied paint films on the structure. However, they suffer from the disadvantages of field test panels, being rather small and not taking into account the effects of steel mass. Such panels must be properly fixed to the structure, not least to avoid serious accidents.

16.9 Tests in water and soil

Tests in which specimens are buried in soils are generally undertaken by specialist organisations and it is recommended that advice should be sought before carrying out such work. Immersion tests in water follow the general lines of atmospheric tests so far as the preparation of the specimens is concerned. However, the exposure of such specimens requires more complex equipment such as rafts and is again usually carried out by specialists. Owing to the electrolyte present, usually seawater, it is more difficult to ensure that specimens are insulated from the test racks and from each other. Furthermore, fouling may occur on specimens, leading to problems with inspection. The siting of rafts or test rigs must take into account the tidal flow and localised pollution of the sea and it is often difficult to ensure that one set of specimens can be arranged so that they are not sheltered by others.

16.10   Formulating the test programme

The essential requirement when formulating a test programme is to be clear and precise about its objective. If this is not done, then there will be a tendency to collect data hoping that it will somehow be sorted out at the end of the programme. This will entail extra cost and time and may result in a mass of data which is of limited value.

If, for example, the decision is made to obtain data on six different primers, the objective of the test programme should be decided in advance. The following possibilities may be considered:

(i)   The effectiveness of the primers to protect steelwork before the final system is applied.
(ii)   The relative durability of the primers when included in a similar type of paint system.
(ii)   Relative durability under a range of different systems.
(iv)   The relative effectiveness in system(s) in a range of environments.
(v)   Their performance on steel cleaned to different levels and/or by different methods.
(vi)   Problems with overcoating after specific time delays.
(vii) Effectiveness of different methods of applying the primers.

There are clearly other possibilities but after due consideration the objectives of the particular test can be decided and the test programme can be formulated to achieve them.

Apart from defining the purpose of the programme, the programme must provide valid data.

The problem with field tests is the time taken to obtain results. To reduce this time it may, in some circumstances, be sufficient to apply only one coat of paint to a primer in comparative tests.

There is little point in carrying out field tests unless the results are of practical significance. If to achieve this the cost and time are considered to be too great, then it may be advisable to abandon this method of testing and to consider alternative ways of obtaining the basic data required.

16.11   Reporting the results of tests

Tests are carried out for a variety of purposes but, except where the tests have been conducted to provide copy for advertising a product, the method of reporting should follow certain well-defined lines. Whether the results are produced in an internal report or are published for information to a wider audience, sufficient detail must be provided to enable the reader to draw their own conclusions, based on a clear understanding of the purpose of the tests and the method of conducting them.

The report should generally contain the following:

(i) The purpose and objectives of the test programme.

(ii) Individuals and/or organisations responsible for the programme.

(iii) Procedures adopted to achieve the purpose.

(iv) Test site(s) used: an indication of why they were selected and some information about the location, climate, pollution, etc., at the site.

(v) Method of exposing test panels or selection of areas for service trials.

(vi) Size and composition of the test panels or areas for service trials.

(vii) Surface preparation: type, equipment used, and where carried out. Where appropriate, size and type of abrasive.

(viii) Method of coating application and where carried out.

(ix) Frequency and type of inspection methods used, e.g. BS 3900.

(x) Results preferably in tabular form with some indication of order of merit, etc. Although descriptions of breakdown are useful, it is easier to compare results if they are in some form of table.

(xi) A discussion of the results so that the reader can appreciate the reasons for the final conclusions that are to be drawn.

(xii) Conclusions should be reasonably short and to the point. They should be separate from the discussion.

(xiii) A summary or synopsis of the work should be included at the beginning of the report. This enables the reader to determine the type of work and his interest in it.

If information on periodic assessments is to be included then it may be advantageous to include this in an appendix or a separate part of the report. This allows the main points to be highlighted in the body of the report.

References

1. First Report of the Methods of Testing (Corrosion) Sub-Committee, *J. Iron Steel Inst.*, **158** (1948) 462–93.
2. Mills, D. J. and Mayne, J. E. O., *Corrosion Control by Organic Coatings*. NACE, Houston, p. 12.
3. Gardner, H. A. and Sward, G. G., *Paint Testing Manual*, 12th edn. Gardner Laboratory, Maryland, 1962.
4. Timmins, F. D., Avoiding paint failures by prohesion. *J. Oil and Colour Chemists Assoc.*, **62** (1979) 131–5.

5. Gardner, G. *Protective Coatings Europe* (Feb. 99) 86–90.
   ASTM D5894–96 'Standard Practice for Fog/uv Cyclic Test'.
6. Van Leeuven, B. *Protective Coatings Europe* (Nov. 96) 42–8.
7. Bayliss, D. A. and Bray, H., *Materials Performance*, **20**(11) (1981) 29.

Index